GOD'S PHYSICS
BEYOND THE EDGE OF QUANTUM-REALITY

Dipak Kumar Bhunia

Dedicated to

Dr. Cynthia Kolb Whitney
For her all assistances in publications of core ideas in this book as three articles in Galilean Electrodynamics

CONTENTS

Acknowledgement	7
List of Abbreviations	10
Prologue	11
Chapter-1: A Particle or System in its Inertia	31
Chapter-2: Universal Inverse Relations	89
Chapter-3: Quantization of Special Relativity	119
Chapter-4: Quantization of General Relativity	164
Chapter-5: Unification of All Basic Forces	217
Chapter-6: Duality in Every Quantum-Reality	246
Chapter-7: Causality in Every Quantum-Reality	269
Chapter-8: Objectivity in Every Quantum-Reality	285
Chapter-9: Beyond the Edge of Quantum-Reality	312
Chapter-10: Basic Form of Quantum-Reality	338
Conclusion	363
Inferences	368
Glossary	370
Bibliography	380
Index	384

ACKNOWLEDGEMENTS

My planning of writing this book was not so long compare to its developments of contents. That was extended over three decades starting from my undergraduate years. So many times, it was discontinued. Sometimes even more than one or more year. But finally, have emerged in a form of one unified mechanical theory and equated Special & General Relativity Theories to Standard Model of Particle Physics, and published as different articles in three issues of journal. But still I often surprised myself how a teen was found his pleasures in probing through the basic anomalies of physics than other worldly issues. All those boundless lash-greenly days under the deep blue skies without fearing anything about consequences.

In those days, I seem fortunate enough to become in contact with one exceptional personality as my teacher of physics. He often used to discuss the frontier topics beyond my undergraduate text book. From him, I listened about something calls 'energy of a vacuum'. An 'ocean' of untouchable kind or form of energy beyond our limits of all quantum-signal exchanges. I was also surprised by knowing, why only the 'light' has such an intrinsic constant (observer-independent and absolute) value in its inertial-motion unlike others in physical nature. But both 'light' and 'non-luminous' categories of inertial frames of reference are associated with matters, as universally common 'wave-corpuscular phenomena', are ultimately quantized in magnitudes. Then, I became inquisitive whether different wave-corpuscular phenomena would have different intrinsic values for inertial-motions. So, I am ever much indebted to Prof. Shisutosh Samanta being one of his students in those undergraduate years.

In proceeding years, another thought, that grab me, whether the unique property of constancy in magnitudes of motion for inertial-light, in basic postulates of Special Relativity Theory, would ultimately be one quantized magnitude for a photon-particle. Because, the same photon-particle, also as one wave-corpuscular phenomenon, having the wave-corpuscular duality similar to any fermionic scales of particles or systems-of-particles having with similar corresponding quantized mass-energies and wavelengths. Subsequently, whether all those fermionic wave-corpuscular particles or systems-of-particles would have also the similar scale-specific quantized magnitudes of inertial-motions as the photons have in physical nature. Moreover, could there be any inverse relationship in-between such quantized 'inertial motions' and 'inertial mass-energies' alike to one already known similar inverse relationship in-between quantized 'inertial mass-energies' and (de Broglie's) 'wavelengths' in all those same boson & fermion particles or systems-of-particles? Could there be any greater number of such inverse relationships still exiting in physical nature?

However, in later years, over a decade or so, a number of attempts were made to write these entire efforts in a form of journal articles. But were remained unsuccessful. Therefore, the developments were never been occurred in any prescheduled time plan to accomplish.

In next, sometimes during the year 1998, it was submitted to Galilean Electrodynamics (Massachusetts, USA). The submission scanned under process of reviews and correspondences over fourteen years up to the end of year 2012. That period also comprised many discontinuities and delays of correspondences from my side until I capable to reply back with thoughtful counter arguments. Finally, that was accepted for publication in a series of three separate independent papers [1]-[3] during years 2014-2018.

But personally, I never knew who were the reviewers of those papers. I am highly indebted to him/them for all those value-additions through such a long process of argumentations from different angles of the core issues, and then to defend, to re-organize, and to re-write before final publications.

I am also very much grateful to the Editor of said journal, for her great patience during those time-consuming reviewing cum editing processes, and also thankful for publishing the articles without any page charges.

Later, I was decided to write a monograph, to consolidate the whole ideas related to unification of physics in those published articles, within one cover during the years 2018-2020.

In doing that, it was my great opportunity to become in contact with Prof. S. N. Bhavsar (Space Scientist, Pune) through the process of review of my manuscript for such monograph. I am really indebted to him and learnt a lot through all his valued editorial advices on said manuscript, and those have precisely enhanced the level of fulfilment in manuscript to become a book.

I also like to thank my colleagues and friends to whom I was often shared my basic ideas and relentlessly debated, but never ever cared about whether they could have any other businesses in this planet than listening me. I am sincerely grateful to them for all their such gentle affections and patience on me. I also like to express my deep respects to everyone who had helped me in the course of all these developments over the decades to whom I forget to acknowledge here properly. And finally, to Gitanjali, my wife, without her sacrifice and unconditional tolerance on me it could never be accomplished in its present form.

Dipak Kumar Bhunia
4th July 2020
Email: *todipakforgodsphysics@gmail.com*

LIST OF ABBREVIATIONS

1. PSs = Particles or Systems of Particles
2. IFRs = Inertial Frames of Reference
3. SRT = Special Relativity Theory
4. GRT = General Relativity Theory
5. EMS = Electromagnetic Spectrum
6. SMPP = Standard Model of Particle Physics
7. CPI = Common-Properties-of-Inertia
8. CIPs = Common-Internal-Parameters
9. SSUCs = Scale-Specific-Universal-Constants
10. UCs = Universal-Constants
11. LAR = Left-Handedness-in-Rotation
12. LE-C = Left-Handed-Entropy of Cyclic-Rotation
13. RAR = Right-Handedness-in-Rotation
14. RE-C = Right-Handed-Entropy of Cyclic-Rotation
15. QM = Quantum Mechanics
16. QED = Quantum Electrodynamics
17. BBCOU = Bigbang-Bigcrunch Cyclic Oscillating Universe
18. RTs = Relativity Theories
19. $'\Delta'$ = Symbol for Quantized value

PROLOGUE

> "It has often said that mark of a good philosophy always begins with an observation which is so mundane is regarded as trivial and from it deduces a conclusion is so extraordinary that no one will believe it."
>
> \- George Gamow

Can 'constancy in magnitude' of free or *inertial* motion for 'light', that has postulated in foundation of Special Relativity Theory, be alternately assumed as an intrinsic 'quantized' [1] magnitude for motion of a 'free' moving photon-particle (similar to its quantized mass-energy)?

If so, that will not harm anything in the foundation of Special Relativity Theory. Because, an intrinsic quantized magnitude of inertial-motion for light or a photon-particle will remain be a universally constant and "independent" quantity respect to motions of its receiver and sender.

But, one such a tiny modification, in earlier postulate about inertial-motion of 'light', could make a huge difference in entire understandings of physics we known todays. That will make not only a quantized extension for Special Relativity Theory but also for the General Relativity Theory. Even, Standard Model of Particle Physics could be extended beyond the known fundamental forces in universe.

Moreover, there can be also a long list of predictions from this, and same can able to resolve many inconsistencies in current comprehensions of physics including long expected Grand   Unification of Everything (including gravitation)

---

1. Which is discrete and finite. An intrinsic quantized value which is also a universally constant, absolute and 'observer-independent' quantity.

as *quantum-real* [1] in the form of existences in this physical-nature [2].

Besides, such a new assertion for quantized inertial-motion of 'light' could even able to show us beyond the edge of such Grand Unification of Everything as quantum-real in same physical nature. That is a territory of all *Non-quantized* seems to be filled with all higher level of energies than quantum-reality.

That would also be a territory for all non-predictabilities within non-quantized continuum of such higher energies. The continuum of all non-quantized zero and infinity magnitudes of co-ordinates instead of non-zero and non-infinity magnitudes are involving with unified quantum-reality. But quantum-reality could create from or terminate into that non-quantized continuum through its Subjectivity.

Consequently, such a non-quantized subjective continuum seems to have all Non-causal 'Wills' instead of our conventional Causal Laws.

But, the Wills of what or whom? Is that merely a Will of a non-quantized 'continuum' with non-quantized higher state of physical energies beyond the effects of universal gravitation?

The 'continuum', that appears puzzlingly to spread up to infinity but within an absolute zero volume of space, out of the exchange limit of all our quantum-real cognition. A continuum of no space, no time and no matter.

Present monograph is an elaborative form of three earlier published articles [1]-[3] of author in different issues of journal during the years of 2014-2018. Those articles have revealed Unification of All Basic Forces in Nature (or a unification of entire quantum-reality) in 10-unfolded dimensions where each of the inertial-motions for particles or systems-of-particles are

---

1. The existence follows the sequence of cause and effect or *Causal Laws*, and also, basically having any intrinsic quantized-magnitude.
2. Conceptually, that could comprise everything including the whole finite universe itself being one macro-most quantum-real structure, and even the anything having infinity of magnitudes as well.

assumed to have quantized values. However, it has efforted to bridge the Unified Quantum-Reality with Non-quantum-Virtuality through all those unfolded 10-dimesions. Ultimately, the entire physical nature would have appeared in two folds – one is as quantum-real with *objectivity* following all Causal Laws of cause & effect sequences and another is as Non-quantum-virtual with *subjectivity* with Non-causal Wills having no cause & effect sequences.

However, prior to reach in such a Non-causal realm of Wills in physical nature there needs to have some obvious steps. So, the first eight Chapters will stepwise reveal a Grand Unification of entire quantum-reality of everything in physical nature along with so many new sets of predictions. Also, those have resolved many observational inconsistencies in current physics, starting from some common understandings of today.

Consequently, the Quantum Mechanics as well as the Theories of Relativity (Special & General), or particularly of the General Relativity Theory, though both are most successful theories of today in contexts of experimental supports, are now seem to comfortable in dealing with the respective arenas of micro and macro scales within such a quantum-real fold of physical nature. In that way, both Quantum Mechanics and General Relativity Theory can consider to have respective limitations in explaining the entire quantum-real-fold of physical nature in one unified manner. May be such limitations in both the theories are mutually reciprocal to each other if extended for one righteous unified theory of same quantum real-fold.

May be such limitation/s is/are due to missing of something remain common as in physical nature, and still need to be accounted in foundations of those two theories. Because now, the background of observational horizon in physics in last one hundred years or more, for such theoretical advancements, have a huge expansion after the outbreaks of Relativity Theories and Quantum Mechanics in first part of the previous century. Many of those post-Relativistic observational comprehensions in the new background can deliver now useful hints for finding any universally common missing link/s within same physical nature.

Almost same thing was happened in first and second decades of previous century when Relativity Theories and Quantum Mechanics were emerging. The ideas about inertial-light, as universally common invariant in all relative inertial frames of reference in Relativity Theories and one of key essentials in development for Quantum Mechanics.

That is, the assumptions, those were related to *finiteness, observer-independence, constancy, absoluteness, and universally invariance* in magnitude of inertial speed of light (as a wave), in foundation of the Special Relativity Theory in the year of 1905 had a long background of acceptance through a series of observational understandings. Those were developing over the past centuries. From the telescopic or other observations of Galileo Galilei in seventeenth century to calculation of Ole Rømer in 1676 (at least 11 years before publication of the Newtonian Mechanics) to conclude the speed of light as finite rather than any infinite. It was about $220000\ km/sec$ compare to current estimation as $2.99792458 \times 10^{10}\ cm/sec$. The observer-independence in constancy of magnitude for same speed of light realized in years of 1864 in Electromagnetic Equations of James Clerk Maxwell as well as 1887 in experiments while trying to measure the absolute speed of earth respect to 'aether' by Albert A. Michelson and Edward W. Morley. Where the aether was assumed in absolute 'rest' respect to all other relatively moving bodies or inertial frame of reference in Newtonian Mechanics. But Michelson and Morley every time got a 'null result' respect to both backward and forward of movements of earth along the rotation in orbit. That had shown, the speed of light is independent to its sender & receiver motions.

Also, the same inertial light, in development of Quantum Mechanics in almost same time, was found to have corpuscular property beside usual wave property. Although, corpuscular hypothesis about light had a long history dates back to Isaac Newton, but that was precisely revealed through the Black Body Radiation experiment of Max Planck in year 1900. That followed by Albert Einstein in the year 1905 by defining such corpuscles as photon-particles while explaining the Photoelectric Effects.

Similarly, those physical facts, including those have related to inertial-light, were seemed 'mundane' during emergence of Relativity Theories and Quantum Mechanics, may not be equally 'mundane' to us in todays context of observational understandings. Some of those physical facts may be similar now as it were earlier or have changed or even obsolete.

For example, the present comprehensions related to unique free or inertial-motion of light as well as other material-bodies. In today's contexts of observations in Particle Physics, it has now well perceived that, almost all classes or types of particles or systems-of-particles (as different scales of quantized mass-energies) having different fixed type intrinsic magnitudes in corresponding free motions. The free motions of specific molecules or atoms or the sub-atomic particles are correspondingly different. Moreover, all those different 'fixed' values [1] in corresponding scale-specific motions intrinsic to each particles or systems-of-particles could be similar kind of constants as one particle of inertial-light has in Special Relativity Theory. Therefore, any of those corresponding 'fixed' or 'constant' intrinsic values would also be the 'independent' kind of free or inertial-motions alike to the 'inertial speed of light' in Special Relativity Theory. But all such constants could have different intrinsic magnitudes.

Then, is the universal invariance in constancy for speed of 'light' as stated in Special Relativity Theory, intrinsic to the inertial speed of a photon-particle, merely one of all those

---

1. As it was only assumed with 'fixed' speed of light (for a photon-particle).

different kind of observer-independent intrinsic values for particles or systems-of-particles instead of any unique one?

Can such intrinsic values for quantized inertial mass-energies, inertial-motions and de Broglie wavelength be the three common internal parameters (CIPs) in every scale of particles or systems-of-particles in physical nature?

The same de Broglie's wavelengths have universally invariant inverse relationship [1] with all intrinsic values of quantized inertial mass-energies correspond to all scales of particles or systems-of-particles. Can there be any such universally invariant inverse relationship in between all same intrinsic quantized inertial mass-energies and intrinsic quantized inertial-motions correspond to all scales of particles or systems-of-particles in physical nature? Can such a new type of universally invariant inverse relationship unify the Relativity Theories and Quantum Mechanics?

Another was the non-précised ideas about the "material-bodies". That was very much mundane idea in foundations of Classical Physics or in Relativity Theories. But now, in last one hundred year, the mundane ideas of material-bodies have changed to the precisely intrinsic classes or 'scales' of any particles or systems-of-particles under broader divisions of bosons and fermions, micros and macros, sub-molecular and astronomical.

Current assessments regarding earlier material-bodies compare to hundred years before in universe is different. Where about 5% of total mass-energies in universe are visible type of matters which could have assumed as material-bodies. However, rest 27% and 68% are dark matters and dark energies respectively beyond the reach of 'light' (i.e. out of the scopes of Special Relativity Theory). Although, it is now suspected that both dark matters as well as dark energies may have precise

---

1. That is de Broglie's wave-corpuscular laws for all scales of particles or systems-of-particles, and an outcome inverse constant from it would be universally invariant inverse-constant.

scales of corresponding other particles or systems-of-particles.

Can intrinsic scale-specificness involving in every such earlier sense of 'material-bodies' could be universally common but any useful hint for unification in current physics?

Any of those scales actually comprises all identical type particles or system-of-particles are existing all over the universe. For example, all the normal hydrogen molecules, wherever those are existing in the physical nature, would have one universally identical intrinsic scale. Starting from one such a scale of all normal hydrogen molecules, if one proceeds gradually onward more and more micro scales, would pass subsequently through the scales of normal hydrogen atoms to neutrons to protons to quarks to electrons to neutrinos to photons to others (i.e. if there any further micro scales). Conversely, starting from same scale of normal hydrogen molecules, if one proceeds reverse gradually onward more and more macro scales, would reach to all increasing scales one after like heavier molecules to various scales of smaller mass-energy in todays List of Astronomical Objects tighten under electro-chemical bonding or surface tension force or viscosity up to the scale of Planetesimals [4] (having minimum mass $\geq 10^{12}$ kg and radius ≥ 0.5 km) to a scale of sub-planets to rocky planets to gaseous planets to brown dwarf planets to sub-solar stars to stars to cluster of few stars or nebula to galaxies to cluster of galaxies to universe (which can be considered as the macro-most system of all particles or systems-of-particles in physical nature).

Moreover, each of those scale-specific quantized magnitudes e.g. de Broglie's wavelength and mass-energy have the scale specific universal constancies. Hence, each of those 'scale-specific-universal-constancies' (SSUCs) as the quantized values are identical to everywhere in universe. But equally variable from one scale to another for particles or systems-of-particles. For example, the SSUC-magnitude in quantized mass-energy or wavelength of a normal hydrogen atom will be a universally constant but such constancies in magnitudes of quantized mass-energy and wavelength would be different if scale of that atom becomes fused to one helium-atom or even changed into an isotope of same hydrogen-atom. Because, the corresponding scales of all those SSUC-magnitudes would become different.

But alternately, that universally invariant inverse constant, that can emerge from the universal inverse relationship of Broglie's wave-corpuscular law in-between all SSUC-quantized magnitudes of mass-energies and wavelengths of all same particles or systems-of-particles, will be one Universal Constant (UC). A UC-magnitude will be in contrary to a SSUC-magnitude must be identical everywhere in same universe irrespective of the changes in scales of those particles or systems-of-particles.

Since the inertial light (as was assumed in Special Relativity Theory) as electromagnetic wave now merely perceives as any photon-particle with similar intrinsic property of observer-independent constant magnitude of c involves with its inertial motion. So, a photon would have one such SSUC-magnitude for such intrinsic constant magnitude of inertial-motion c.

In current observations of Particle Physics, although there are very little knowledges have gathered so far in relevance of intrinsic scale-specific quantized magnitudes of inertial-motions unlike corresponding scale-specific quantized magnitudes of inertial-mass-energies in same particles or systems-of-particles, but now it is one of growing conventions that various micro scale of particles or systems-of-particles would have their scale-specific intrinsic magnitudes for free or inertial motions. Now, it is evident that a specific scale of molecule or atom, a neutron or neutrino having their corresponding free or inertial motions with intrinsic scale-specific fixed kind of magnitudes. Even, in some experiments it has realized that one radio-wave (i.e. a radio-wave photon with lower range of mass-energies in electromagnetic spectrum) seems to travel with superluminal speeds [5]. Consequently, it can argue whether all other specific scales of particles or systems-of-particles (irrespective of all micro and macro scales) do have similar kind of quantization in corresponding magnitudes of intrinsic inertial-motions beside quantizations in corresponding magnitudes of inertial mass-energies and de Broglie's wavelengths with the same. If so, the inertial-motion, which is inseparable from the inertial mass-energies, could assume to involve simultaneously with all same specific scales of inertial particles or systems-of-particles in physical nature and consider as another CIP. Each of such scale-specific intrinsic quantized magnitudes of inertial-motion of particles or systems-of-particles then must have the corresponding SSUC-magnitudes. Since, onward micro scales of particles or systems-of-particles there seems to have higher intrinsic magnitudes of such free or inertial-motions with smaller mass-energies or vice versa.

If so, there could open further the new horizons in physics from the observational background in last one hundred years. That can provide the new hint for the unification of current physics.

Next is the growing observational knowledges about the existences of scale-specific intrinsic quantized magnitudes for radii in almost every scale of particles or systems-of-particles. Scale-specific intrinsic values of radii are not merely observing in micro scales but also seem to exist in macro scales of systems-of-particles. The radius of an atom or molecule or subatomic hadron as well as lepton are appeared to have their all scale-specific fixed magnitudes. The radius of an electron ($2.82 \times 10^{-15}\,m$), a normal hydrogen atom (Bohr radius $5.29177 \times 10^{-11}\,m$), and so on are appearing to have all scale-specific intrinsic quantized or fixed SSUC-magnitudes. Subsequently, in micro scales of physical universe, the volumes or three dimensions of 'space' would also to have their scale-specific intrinsic quantized SSUC-magnitudes. If those micro scales of particles or systems-of-particles are building blocks of the macro universe, then conceptually the macro universal space could also be quantized accordingly in scale-specific manners. Almost same thing is now also witnessing in astrophysical observations of various macro astronomical scales. A planetesimal with minimum mass with $10^{12}kg$, when gravitational force begins to dominate over other forces, it needs to acquire a radius at least 0.5 km or more. But to start thermonuclear reactions in it, while onward formation of any star, it further requires to accumulate a minimum concentration of scale-specific mass-energies within a corresponding minimum radius of space which can also assume to have intrinsic SSUC-magnitude compatible to the scale of a particular star. Actually, in every sequence all along the life span of one star, it seems now clear that there involves all intrinsic SSUC-magnitudes of radii with corresponding SSUC-magnitudes of mass-energies. Similarly, even in formations of further larger scale of structures, say in the range of galaxies or clusters of galaxies, the corresponding intrinsic minimum SSUC radii will require to attain corresponding intrinsic SSUC-magnitudes of mass-energies.

Hence, it may now also pertinent to assume through todays observational understandings that earlier concept of space in physical nature that was presumed universally to exist in omni-presence manner, with or without any physical mass-energies, is seemingly a sum of all SSUC-magnitudes of all discrete or quantized-spaces or volumes. Where one quantized volume or space correspond to a sum of all micro scales which are comprised it.

Alternately, no absolute void space, which was presumed earlier omni-present without involving any matter or particles or systems-of-particles, can be practically exchanged through any quantum-real signals (through which we are intrinsically limited to exchange/communicate only to anything beyond that limit in physical nature). Therefore, physical universal space may not be merely appeared to have quantized SSUC-magnitudes but same cannot practically measured without involving any specific scales of particles or systems-of-particles due such quantum-real limitations of signal exchanges with any observer like us.

Then, the present days might have reached to its efficient stature, after one hundred years from the emergence of Relativity Theories, to assume all such discrete or quantized SSUC-magnitudes of intrinsic space (in every scale of micro and macro particles or systems-of-particles) as another 'universally common thing' or CIP. That can ultimately substitute the earlier omni-present concepts of universal space in same physical nature.

Next is concept about the 'time'. The time was also one omni-present idea in Classical Mechanics. That flows from 'past' to 'future' in physical nature.

Since we cannot exchange with any absolute void or omni-present space without presence of any particles or systems-of-particles, therefore we cannot also exchange to omni-present but universally flowing time if any without such quantum-real signals of particles or systems-of-particles in same physical nature. Then time, that could be exchanged only through any such quantum-real signal, ultimately could perceive to us if and only if it is associated to any specific scale of particles or systems-of-particles. Because, any quantum-real signal exchange can only possible if there both of such clock as well as observer like us are fundamentally quantum-real in type.

Subsequently, any such particles or systems-of-particles having scale-specific intrinsic quantized inertial mass-energy, inertial-motion and inertial-space could have also the corresponding intrinsic quantized magnitude of inertial-time. In inertial state, the clock in a scale of normal hydrogen atom or a neutrino or a photon or in one earth-based clock or one clock with scale of a star or a galaxy would have corresponding scale-specific intrinsic values of 'durations' for such time, i.e. different intrinsic scale-specific magnitudes of time. Therefore, in such a fashion, time is also now appearing to have scale-specific intrinsic discrete or quantized magnitudes for all particles or systems-of-particles. Therefore, any earlier sense of universal omni-present time could not be practically communicable within our quantum-real limitations of signal exchanges. Even that may exist *there* anyway but beyond exchange limit of us in same physical nature. Then, such a scale-specific intrinsic quantized sense of time will also be another prospective CIP involving with all scales of particles or systems-of-particles in physical nature, irrespective of micro and macro scales.

Another new comprehension in current physics could be the observed intrinsic left-handed directions in axial rotations [6] of all same micro to macro scales of particles or systems-of-particles.

Consequently, each of those above-mentioned CIPs e.g. inertial quantized 'space', 'time' and 'mass-energy', which are linked to intrinsic 'left-handed axial rotation' of every scale of micro to macro particles or systems-of-particles, might have also one intrinsic left-handed directions.

Since, there could be such an 'intrinsic left-handedness' with every particle or system-of-particles, there might be a simultaneous intrinsic mirror-imaged 'intrinsic right-handedness' in the same particles or systems-of-particles. But conceptually, such an intrinsic right-handedness will be beyond all of our direct observational range of intrinsic left-handedness through exchange of similar intrinsic left-handed quantum-real signals. Similar to beyond tangibility of one mirror-image in the mirror.

Geometrically, in every co-ordinate system also, while to define every such left-handed particle or system-of-particles through one set of 3-positive (x, y, z) co-ordinates, there would be also a simultaneous mirror-imaged set of 3-negative $(-x, -y, -z)$ co-ordinates. That is total six numbers of co-ordinates involving with any such co-ordinate system. Where each of those co-ordinates in two sets are mutual mirror-images to each other in terms of their corresponding magnitudes. Consequently, the CIP like space is geometrically appearing in every particle or system-of-particles with 3+3 positive and negative mutual mirror-imaged dimensions instead our conventional three dimensions in Classical Mechanics.

But two such 3+3 positive and negative mutually mirror-imaged sets of dimensions of quantized space might have also an inverse relation. Then for conveniences, if the space with such three positive coordinates can imagine as 'space'. The corresponding mutual mirror-imaged magnitudes for negative coordinates of that 'space', could term as 'anti-space' simultaneously in same co-ordinate system or in every scale of particles or systems-of-particles in physical nature. So, that 'ant-space' could be also another CIP in same particles or systems-of-particles. But beyond direct exchange capability of any left-handed quantum-real signals.

Furthermore, since anti-space is a right-handed mirror-image of left-handed inertial space of all scales of particles or systems-of-particles, therefore would also be the left-handed quantized inertial 'time' involving with same particles or systems-of-particles and might have also simultaneous right-handed mirror-images of quantized 'anti-time'. Such concept of such anti-time might have also a simultaneous mirror-imaged or right-handed directional flow, in every same particle or system-of-particles, from future to past. Consequently, the anti-time could be another CIP in every same particle or system-of-particles.

Another new set of comprehensions those could derive in relevance of a conceptual model of Bigbang-Bigcrunch cyclic oscillating universe as the macro-most scale of all micro to macro scales of all particles or systems-of-particles in physical nature. That universe is now as if passing through its 'expansion phase' and such expansion once was initiated through a moment of Bigbang expectedly about 13-billion years ago. An expected opposite and alternate 'collapsing phase' will start through another moment of Bigcrunch in unknown future and that will again collapse-back into a moment after which another fresh Bigbang can occur. A cyclic oscillation from a Bigbang to Bigcrunch, then from a Bigcrunch to Bigbang, and again a Bigbang to Bigcrunch cyclic model of universe.

Such a cyclic oscillating model of universe can have also a newer outcome consequence further if it would have a similar left-handed direction in axial rotation as one macro-most scale for all particles or systems-of-particles. That is, it's all comprising scales of particles or systems-of-particles seems to have such left-handed & right-handed directional axial rotations as mentioned in above paragraphs.

Such a conceptual Bigbang-Bigcrunch Cyclic Oscillating Universe (BBCOU) is also ultimately a composition of all those scales of micro and macro particles or systems-of-particles including observers like us as the integrated parts.

But there would be another simultaneous direction for the same BBCOU including all its constituent micro to macro particles or systems-of-particles.

The same BBCOU would have an intrinsic direction of 'expansion' i.e. from the Bigbang to Bigcrunch in expansion phase. This could imagine to have the direction for its increments in entropy along with all those same BBCOU as well as all integrated scales of particles or systems-of-particles. That direction of incrementing *entropy* onward expansion of BBCOU can also be termed for convenience as a left-handed directional *expansion-entropy*.

Alternately, during collapsing phase of same BBCOU, conceptually from Bigcrunch to Bigbang, same left-handed direction expansion-entropy of BBCOU would decrement.

Consequently, one left-handed directional observer like us, who is cohesive to universe onward its direction of expansion-entropy, could merely able to observe the same BBCOU obviously in one cyclic oscillation. That will be a cycle of alternate increment (from Bigbang to Bigcrunch) and decrement (from Bigcrunch to Bigbang) phases of left-handed direction expansion-entropy for BBCOU.

Furthermore, every same scale of particles or systems-of-particles including that macro-most scale or BBCOU will have also the simultaneous right-handed or mirror-imaged axial rotational counterparts in above paragraphs. Therefore, an intrinsic right-handed or mirror-imaged decrement and increment cycle can also be perceived simultaneously within same macro-most scale of BBCOU along with all its scales of particles or systems-of-particles.

We can imagine one observer, who is say integrated with the same BBCOU but in opposite direction of us contrary to our intrinsic direction of expansion-entropy. Therefore, his intrinsic direction would be integrated with the right-handed opposite direction in same BBCOU. That is, onward the direction of observing all 'collapses' in BBCOU. Then, how the same BBCOU will appear to him from his such an opposite directional observation capability in comparison to our simultaneous direction of observation onward expansion-entropy?

That observer could see the same cyclic oscillation for same BBCOU from its Bigbang to Bigcrunch but only onward cyclic progresses in collapse. In the same Bigbang to Bigcrunch phase, the observer could see only decrementing values of collapses for that BBCOU. That is, everything what he could see in his such a collapsing surrounding would appear with any decrementing values of entropy or say the *collapse-entropy* or *anti-entropy* i.e. mirror-image of our expansion-entropy. Because, he never could directly see anything those are like our 'left-handed directional expansion-entropy' from that opposite 'right-handed directional collapses' of the same BCCOU. That is collapse-entropy would have decrements during Bigbang to Bigcrunch of BBCOU.

25

In alternate Bigcrunch to Bigbang phase, there would be an obvious increment in right-handed collapse-entropy of the BBCOU in contrary to the decrements of the expansion-entropy.

For convenience, if the left-handed directional cycle for expansion-entropy can term as *Lefthanded Entropy–Cycle* (LE-C) then it's simultaneous right-handed directional cycle for collapse-entropy might be termed as *Righthanded Entropy–Cycle* (RE-C) of same BBCOU.

Then, anything that is onward direction of LE-C in BBCOU cannot find anything that is onward direction of RE-C in same BBCOU. This can also otherwise describe anything which is integrated with RE-C that cannot be appeared onward LE-C with sustenance.

Conversely, onward direction of RE-C in BBCOU one cannot see anything onward LE-C in same BBCOU. This also depicts, anything that is integrated with LE-C could not withstand onward RE-C direction of BBCOU.

Then any left-handed directional particle or system-of-particles including observers like us, who are integrated with left-handed direction or LE-C of BBCOU, could not see simultaneously anything as right-handed directional or RE-C particle or system-of-particles or similar type of observers in steadily existing conditions. This may also be reasoned that why any anti-matter particle or system-of-particles do not appear to us in their stable existences in abundance onward current expansion-entropy phase of LE-C for same BBCOU. Because, all those presently observed anti-matters having intrinsic right-handed axial rotations. But whether due to the presence of such simultaneous intrinsic right-handed directional collapse-entropy in those anti-matters or anti-particles, along with similar right-handed direction of same BBCOU, are appearing to us as unstable in existences in our reverse way of left-handed sights can be questioned. This can also be one prospective consequence why within such intrinsic LE-C directional observations we ever observe an asymmetry in quantities of matters (as particles or systems-of-particles) over anti-matters (as antiparticles or systems-of-antiparticles) in BBCOU.

Then, due to gradual increase in observational horizon of the physics, compare to the same in about one hundred years ago, there could have now many new additions in earlier 'universally common invariants' irrespective of micro and macro scales of particles or systems-of-particles. Particularly, in relevance of those 10 number of CIPs (e.g. for inertial 3-space, 3-antispace, 1-time, 1-antitime, 1-mass-energy & 1-motion) and all prospective universally inverse invariant relationships irrespective scales of any particles or systems-of-particles or everything in physical nature.

As a consequence, there could have new scopes of unifications for entire quantum-reality in the realms of Relativity Theories and Quantum Mechanics. There could have many more new inferences including resolving the presently observed: asymmetries in quantities of total matter than anti-matter, E-P-R paradox, Objectivity vs. Subjectivity in creation / destruction of any event and so on. That will also trail to a Non-quantum-Virtuality fold of same physical nature, beyond that Unified Quantum-Reality of Causal Laws, which could have only the Non-causal Wills (with all zero & infinite values in same 10-CIPs) from where or into which quantum-reality can create or destroy.

In proceeding ten Chapters of this book, all those will be described in details. That will precisely sort out ten *common internal-parameters* or CIPs in all scales of micro and macro particles or systems-of-particles depend on some mundane phenomena of current understandings in current physics. That will extract seven effectively new universally common inverse invariants from those some mundane observational understandings of the current physics as key universal invariances.

Such universal invariances as new postulates will appear in the first two Chapters as foundation of this entire work. The next proceeding eight Chapters will be the subsequent eight sets of many inferences. Many of those can resolve the various observed inconsistencies in todays physics.

Most importantly, those inferences will derive new unified mechanical theory for physical nature. The gauge quantum-fields in Standard Model of Particle Physics (Quantum Mechanics) and the Einstein Field Equation (General Relativity Theory) could be extended as well as be equated, and finally be merged into a one single equation for everything we know as quantum-real in physical nature. Moreover, such a unification will not only deduce such a single equation for everything quantum-real but also express in unique manner the entire quantum-reality in total 5+5 inversely co-related 'unfolded' dimension of co-ordinates. Also, each of those unfolded dimensions will appear to oscillate in scale-specific magnitudes from 'zero' to 'infinity' where the particles or systems-of-particles would be the discrete units of quantum-reality with all scale-specific non-zero & finite magnitudes of 5+5 inversely co-related unfolded dimensions. Beyond of that non-zero & finite scale-specific discrete values those 5+5 inversely co-related unfolded dimensions of Causal Laws of Objectivity, there would be the zero & infinity values of Non-quantum-virtuality with all Non-causal Wills for Subjectivity.

As stated in above paragraphs, the Chapter-1 & Chapter-2 will be the foundation part. In first, total ten CIPs will excerpt from the current observations. In next, there will be a total seven universally invariant postulated inverse relationships with corresponding total seven inverse UCs from those 5+5 universally common internal-parameters or ten CIPs in all scales of micro to macro particles or systems-of-particles in inertial states.

Chapter-3 & Chapter-4 will derive quantized extensions of the SRT Equations and Einstein Field Equation in GRT in quantized manner for curved spacetime (gravitation) through those new inverse UCs.

Chapter-5 will deduce a unified quantum-real equation for everything. That equation will ultimately equate all three Supersymmetric basic forces or non-gravitational Gauge Fields of forces. All the smallest bound unescaped particles or systems-of-particles, those would trap due to quantized motions below the respective escape velocity for curved spacetime of gravitating-body, would be equal to scale-specific sum of same gravitating-body. Each of those smallest bound particles would also be any of quantum-fields in Standard Model of Particle Physics for quantized extension of Einstein Field Equation in gravitating-body.

Chapter-6 will reveal a mutual mirror-image duality with each of such unified quantum-reality of all particles or systems-of-particles and that will show a way to resolve the E-P-R Paradox as well as the asymmetries in total matter & antimatter in present universe.

Chapter-7 will infer *determinism* in every unified quantum-reality, in all diverse scales of particles or systems-of-particles (as integrated parts of the BBCOU), with duality in mutual left-handed and right-handed mirror-images.

Chapter-8 will depict further such determinism of unified quantum-reality as *objective* within the limits of all exchangeable quantum-real signals, observables irrespective of the scales of all micro to macro particles or systems-of-particles, and even in 'cognition' of observers like us which are discrete integrated parts of the BBCOU as macro-most scale of the physical nature.

Chapter-9 will trail to the *subjectivity* in the virtuality of Non-quantum-energy in physical nature, and then ultimately the subjectivity will appear as if a manifestation of any higher level of energy beyond the signal exchange limits of any corresponding observer who is belonging in one lower level of energy. This will find ultimately a subjective realm of the Wills in Non-quantum-virtuality with all zero & infinity magnitudes of 10 co-ordinates beyond the unified quantum-reality of objective Causal Laws with all non-zero & non-infinity magnitudes of same 10 co-ordinates.

Finally, Chapter-10 will derive one unified equation in most fundamental state of anything deterministic, objective and quantum-real in type but with all non-zero, finite and exchangeable discrete magnitudes as 5+5-dimensional *whirl* in an indeterministic, subjective and Non-quantized-virtual type continuum with all zero & infinite values. Such whirl only can subjectively create from/destroy into that indeterministic, subjective and Non-quantum-virtuality out of that higher energy level of all zero & infinity magnitudes of same 5+5 CIPs in the physical nature. Therefore, physical nature ultimately will appear in two folds – one as quantum-real inclusive of scale-specific quantized everything of BBCOU in lower energy level, and another as Non-quantum-virtual in higher energy level beyond quantum-real communication limits of signals including quantum-reality of gravitation.

Next will be the Conclusion of the whole text followed by the References.

Note-1: All the Equation numbers have started with respective Chapter Nos., then have appeared with respective number of Equation in sequence of that particular Chapter after dot.

Note-2: All the symbols are used in this book as either vectors or scalars. Scalar symbols are presented in non-bold but *italic* form while vector symbols are in **bold** but non-italic.

Note-3: The 'foot-notes' will appear within subscript brackets like $_{[..]}$ while 'references' within normal brackets [..].

CHAPTER-1:

A PARTICLE OR SYSTEM IN ITS INERTIA
[In search of common-internal-parameters]

"The task is not to see what has never been seen before, but to think what has never been thought before about you see every day."   - Erwin Schrodinger

1. Common Understandings in Universe:	32
1.1. Material-bodies as Scale-specific Particles or Systems	33
1.2. Common-Internal-Parameters in Particles or Systems	36
1.3. Common-Internal-Parameters of Scale-specific Values	37
1.4. Left-handed Entropy in Scales of Particles or Systems	41
1.5. Left & Right-Hand Duality in Particles or Systems	42
1.6. Particles or Systems have Left-handed Rotations	44
1.7. Quantum-Real Limit in Our Observations	47
2. Common-Internal-Parameters:	50
2.1. Inertial Mass-Energy	50
2.2. de Broglie Wavelength	51
2.3. Inertial Motion	52
2.4. Inertial Radius	65
2.5. Inertial Space	67
2.6. Inertial Anti-Space	70
2.7. Inertial Anti-Radius	72
2.8. Inertial Time	74
2.9. Inertial Anti-time	79
3. Common Definition of any Particle or System	81
Summary	87

In Prologue part, already we had a glimpse of some probable Common Understandings in physical universe. Those could emerge now through physical observations, almost in last one hundred years or so, after out breaks of Theories of Relativity and basic ideas of Quantum Mechanics. However, those Common Understandings are also appearing now to involve in each of the discrete units of matter or micro and macro scales of particles and systems-of-particles comprising that universe. Although, Theories of Relativity and Quantum Mechanics are most successful theories in todays' physics but are still mechanically well-suited to corresponding macro and micro scales of particles or systems-of-particles. A subsequent Grand Unified Theory for both micro and macro realms of scales of particles or systems-of-particles, by fusing both the theories, could develop on the basis of those Common Understandings in same universe.

The Chapter-1 will actually present a completely new comprehension for all same micro to macro scales of particles or systems-of-particles as discrete units of the universe in comparison of present views about the same. In the Chapter, a particle or system-of-particles would appear internally as a web of some Common-Internal-Parameters and each of those have all such Common Understandings in universe.

Sections 1 & 2 will reveal total seven Common Understandings in universe and twelve Common-Internal-Parameters respectively for all micro to macro scales of particles or systems-of-particles in universe. Section-3 will sketch a subsequent common ground for one Grand Unified Equation for same scales of particles or systems-of-particles in proceeding Chapters.

1. Common Understandings in Universe

The experimental or observational horizon of todays physics has not only now expanded but its rate of expansion seems much higher compared to previous century due to advances in technologies. Accordingly, many basic mechanical understandings in history, which were once assumed as universally true in the physical nature, are now appearing to have less importance in newer mechanical theory. The seven sub-sections below will sort some of those universally common understandings among diverse forms of 'matter' depending on various observational comprehensions of todays physics for any prospective foundations of any new unified mechanical theory:

1.1. Material-Bodies as Scale-Specific Particles or Systems

The basic perceptions about 'matter' or 'material-bodies', in mechanical foundations of Classical Physics, is now changed in previous one hundred years. The Classical meaning of material-bodies are now appearing with respect to any intrinsic *scales*. Because, those material-bodies are now showing to have the property of precise scale-specificness, i.e. with any intrinsic scale of quantized magnitude in its mass-energies. Each of those scale-specific quantized magnitudes of mass-energies also refers now to any corresponding scale of particles or systems-of-particles in terms of todays Particle Physics. One scale, for those particles or systems-of-particles, further means for all those identical particles or systems-of-particles are existing and spreading in the physical nature. So, there are all different groups or classes or scales of those corresponding particles or systems-of-particles in the same physical nature. Starting onward the 'direction' of all decrementing smaller and smaller mass-energies, there are all corresponding intrinsic scale-specific quantized magnitudes for those mass-energies. Those intrinsic micro scales are in practical sense are appearing from molecular to sub-molecular particles or systems-of-particles. However, onward opposite 'direction', heavier mass-energies than those molecular scales or say from the smaller astronomical objects there are also all precise macro scales of system-of-particles. Those systems are also now comprehending to have the property of scale-specificness in relevance of scale-specific range of internal mass-energies, spatial-extensions, and so on. Therefore, those astronomical objects, above the molecular scales being any systems-of-particles, are also emerging as any macro scales of 'material-bodies' unlike Classical Mechanics. In some smaller scales, those astronomical objects are observed in condensed forms under grip of electromagnetic force, but beyond of that there are all dominating gravitational forces. Therefore, it is now also observing in universe that all the micro to macro scales of particles or systems-of-particles are further divided in terms of internally dominating fundamental forces. In the scales of molecules or atoms those are internally dominated by the electromagnetic forces. Whereas, within the scales of atomic nucleus there are all dominating strong or weak forces. The domination of gravitational force in any system-of-particles has observed to begin from the scale of planetesimals up to the macro-most scale of whole universe.

Then, 'scale-specificness' or 'scale-specific quantization' in perceptions of all earlier 'material-bodies' as any precise form of particles or systems-of-particles will be one of the prominent "universally common physical comprehensions" in changing observational background of todays physics.

Then being an observer, if one is supposed to stand on a frame of molecular systems-of-particles, will 'see' in the physical nature, a long arrow of scales from micro to macro particles or systems-of-particles. That starts from one conceptual 'micro-most' scale to a 'macro-most' scale or vice versa. If that micro-most scale would conceptually be a 'quintessence' (in the range of dark energies) like particle/s in one side then obviously there would be the macro-most scale like Bigbang-Bigcrunch Cyclic Oscillating Universe (BBCOU) in opposite side. Where, our scale-specific position, being one observer within that long arrow, remains within molecular range of scales.

If such a molecular-observer starts from the scales of different molecules, towards gradual micro and more micro scales, will reach through all corresponding scales of atoms → a range of sub-atomic scales → a range of sub-nuclear scales → a range of scales for corresponding sub-nuclear particles → a range of quark scales → a range of lepton scales → a range of scales for bosons including photons (and obviously all these scales which are inclusively comprising the entire realm of visible-matter-scales merely 5% of the expected mass-energies of BBCOU), and next is → a conceptual realm of dark-matter-scales (seems to comprise 27% of entire mass-energies of same BBCOU), and finally there next is → a conceptual realm of dark-energy-scales (total 68% of entire mass-energies of the BBCOU) that includes the micro-most scale quintessence.

Conversely, on opposite direction, there are all corresponding macro scales where molecules are emerging to bond with other molecules to form other scales of larger molecules through any form of electromagnetic forces (like surface tensions, elasticity etc.) → then scales of planetesimals (when gravitational force becomes dominated over other natural forces within such larger scales of electromagnetic systems when achieves the minimum radius $\geq 0.5 \times 10^5 cm$ and mass $10^{15} gm$) → scales of sub-planets → scales of rocky planets → scales of gaseous planets → scales of red giant planets → scales of sub solar stars → scales of solar like stars → scales of super massive stars → scales of clusters-of-stars → scales of galaxies → scales of super galaxies → scales of clusters-of-galaxies → scales of super clusters-of-galaxies (and finally to) → BBCOU (macro-most scale of all particles or systems).

All the known natural forces like gravitation, electromagnetic, weak nuclear force and strong nuclear force are correspondingly appeared to dominate in those scales to configure those different particles or systems-of-particles. It is also not yet clear whether in respective realms of darks matter as well as dark energy scales there remains any other natural forces in universe excluding all those four known basic forces.

But in physical nature, the material-bodies, which are appearing to us as 'real' in existences, are also appearing to have all precise intrinsic quantized magnitudes of mass-energies by involving with any specific scale of particles or systems-of-particles.

Consequently, such intrinsic scale-specificness as well as quantization in magnitudes of any particles or systems-of-particles with real existences can be a *universally common understanding* in physical nature.

1.2. Common-Internal-Parameters in Particles or Systems

All scales of particles or systems-of-particles from micro to macro, with that universally common intrinsic quantization and scale-specificness in corresponding magnitudes, are further possessed *internally* some universally *common-internal-parameters* (CIPs).

For example, the intrinsic *inertial mass-energies, inertial-motions, inertial-wavelengths* [1], and so on are some of those universal CIPs which are involving in every micro to macro scale of particles or systems-of-particles according to todays observational comprehensions of physics.

Because, the CIP like inertial mass-energies or inertial motions or de Broglie wavelengths is universally integrated part not merely in any one particular scale but in every scale of particles of systems-of-particles irrespective of micro to macro in physical nature. Then, those CIPs will be another form of common universal parameters in all that same micro to macro scales of particles or systems-of-particles.

Therefore, each of those CIPs can be considered as another kind of *universally common understandings* irrespective of scales in same physical nature.

---

1. All the intrinsic scale-specific magnitudes of de Broglie's wavelengths.

1.3. Common-Internal-Parameters of Scale-Specific Values

In a conceptual inertial state, when no force is affecting the 'free' properties of a particle or a system-of-particles, the quantized magnitudes of same particle or system-of-particles must have any intrinsic value. Therefore, any such 'free' intrinsic values would have not only the universal quantization characteristic but same 'free' intrinsic values are actually any sum of quantized magnitudes or CIPs those are comprised each of those scale of particles or systems-of-particles in inertial state.

Then, each of those universally common CIPs would have all scale-specific intrinsic quantized magnitudes in all corresponding scales of particles or systems-of-particles.

Consequently, all those scale-specific intrinsic quantized magnitude for CIPs would have the universally 'common', 'invariant', 'constant' and 'observer-independent' characters in inertial state. For example, intrinsic inertial quantized mass-energy of one normal hydrogen atom (approximately $1.67 \times 10^{-24} gm$) would have similar universally common, invariant, constant and observer-independent value everywhere in the physical nature. This would also be true for the inertial-motion ($2.99792 \times 10^{10}\ cm.sec^{-1}$) of light, as a CIP in reference of one photon-particle in electromagnetic spectrum. The same thing can observe in inertial mass-energies and inertial-motions as CIPs in all other scales of particles or systems-of-particles (including those normal hydrogen atom and photon-particle).

Then, each of those scale-specific quantized magnitudes for all respective CIPs in all specific scales of particles or systems-of-particles will be another form of *universally common understandings* in physical nature.

Moreover, each of those scale-specific quantized magnitudes of all CIPs irrespective of the scales of particles or systems-of-particles must be also the scale-specific-universal-constants (SSUCs). Because, each of those values of SSUCs are universally invariant. Each of such SSUCs quantized magnitudes of the CIPs, for conveniences will be defined by a symbol in all proceeding text as 'Δ'. Therefore, to define scale-specific quantized inertial mass-energy (as one CIP) of any scale of particle or system-of-particles as Δm, or for inertial wavelength say $\Delta \lambda$, and so on. Then, inertial mass-energy of normal hydrogen atom will be like Δm_H and its inertial wavelength will be $\Delta \lambda_H$. But both of those are SSUCs respect to one specific scale of the normal hydrogen atom for corresponding CIPs like Δm & $\Delta \lambda$. Then, if any of those SSUCs would have all scale-specific universally constant or invariant magnitudes, and if the scales of particles or systems-of-particles become change automatically the corresponding magnitudes for those same SSUCs would be changed instantaneously. For instance, if the normal hydrogen atoms with same SSUC-magnitudes like Δm_H and $\Delta \lambda_H$ would fuse into a helium atom (i.e. of different scale), then all its corresponding SSUC-magnitudes would be automatically changed into say Δm_{He} & $\Delta \lambda_{He}$ with other two different universal constant & invariant magnitudes instantaneously.

Apart from the CIPs like Δm & $\Delta \lambda$ as SSUCs in quantized values in every scale of particles or systems-of-particles, it has now also comprehended that inertial-motions of some scales of particles or systems-of-particles having their SSUCs in discrete types of free or inertial magnitudes, say Δv. For example, the scale-specific non-relativistic or inertial-motions for a photon, a neutrino, a neutron, an electron, an atom, a molecule and so on having the corresponding SSUCs of discrete magnitudes or Δv. Although, all those possible SSUCs magnitudes for inertial-motions Δv for all corresponding scales of particles or systems-of-particles are not yet specifically measured. But, inertial mass-energies for all same scales of particles or systems-of-particles, which are inseparable from the corresponding inertial-motions involved in same particles or systems-of-particles, now established through observational comprehensions to have

SSUCs for all corresponding discrete magnitudes. Therefore, all the corresponding inertial-motions with same inertial mass-energies are also expected to have similar type of SSUCs in discrete magnitudes.

Next is about the intrinsic quantum radius for the different scales of same particles or systems-of-particles, say Δr. Particularly, it is now evident in the micro scales like the molecules or in further sub-molecular scales and alternately, also onward the macro astronomical objects like in a scale of planetesimals or in a scale of stars or in similar kind of other scales, there have the corresponding SSUCs in discrete magnitudes. Those SSUCs in discrete magnitudes like Δr, at least within our present range of observations, for any scale-specific non-relativistic or inertial electron, quark, neutron or proton, atom, molecule and planetesimal or specific type of star etc. to have SSUCs in corresponding discrete magnitudes.

Subsequently, for every such corresponding scale-specific SSUCs in discrete magnitudes for Δr, it can further assume that all the corresponding scale-specific SSUCs in discrete magnitudes would have three spatial coordinates say $(\Delta x, \Delta y, \Delta z)$ or volumes, say, (Δs) involves in each of those scales of particles or systems-of-particles. Since the entire quantum-real fold of physical nature i.e. BBCOU is now is conceptually appearing to comprise only by those scale-specific particles or systems-of-particles, therefore the entire existence of 'space' for that same BBCOU would be ultimately the sum of all those SSUCs for discrete magnitudes of scale-specific quantized spaces. The earlier concepts of infinite, omnipresent, 'flat', and continuous universal space in Classical Mechanics is now appearing as finite, occupied volume, 'curved' and discrete or quantized spaces involving with every individual scale of particles or systems-of-particles which is comprising the entire volume of the macro-most scale BBCOU of physical nature.

But, not only that Δs. In the same way, the duration of a 'tick' for the time also seems to be different in those different scales of inertial particles or systems-of-particles. The duration of such 'tick' in corresponding clocks in a free moving photon, neutrino, electron, hydrogen, helium molecules and so on will have different values [1]. Therefore, the magnitudes of time also seem as scale-specific, i.e. with all SSUCs in discrete magnitudes, say it as Δt. As a result, in todays comprehensions, the earlier infinitely flowing omnipresent but continuous concept of universal time in Classical Mechanics now can be assumed as a universal finite flows along the finite duration of creation-destruction for all those scale-specific intrinsic discrete or quantized SSUCs in magnitudes. That is, a free moving normal helium atom would have finite flow of its intrinsic SSUCs in discrete magnitudes ('tick') for its scale-specific Δt from its creation to destruction.

Therefore, from current observational understandings of physics it is now emerging that all such $\Delta v, \Delta r, \Delta s, \& \Delta t$ as new CIPs, beside two other conventional CIPs like $\Delta m \& \Delta \lambda$, and with all SSUCs in respective discrete values.

Furthermore, the de Broglie's wave-corpuscular law also suggested that both of the two conventional CIPs ($\Delta m \& \Delta \lambda$) with all SSUCs in quantized magnitudes are also inversely linked to each other through an inverse constant h/c (where h is Planck's constant and c is inertial constant motion for light or photon) which can term as an inverse universal constant (UC). The magnitude of such UC as h/c would be always unchanged irrespective of all scales from micro to macro particles or systems-of-particles in comparison of scale-specific changes in magnitudes of SSUCs. It is not yet known whether there is any other UCs in physical nature out of all those newly emerging other CIPs. The Chapter-2 would postulate such some possible newer UCs in background of todays understandings in physics.

---

1. Since all those different scales of particles have different magnitudes of Δv. Even the principle of time dilation in SRT has stated that amount of dilated time in inertial frames of reference is proportional to increasing (relative) inertial-motions. It also reveals that there would have such differences in duration of a 'tick' for time.

1.4. Left-Handed Entropy in Scales of Particles or Systems

Theory of the Bigbang is matching with current expansion of universe. The current astro-physical observational phenomena are also seemed supportive for this model of universe. But its presently observing expansions from Bigbang to Bigcrunch, to complete a course in any regular cyclic changes, there can conceptually be an alternate contraction for same universe next from Bigcrunch to Bigbang. Then, the cycle can repeat again and again. Then the universal time, that is related to that universe as the macro-most scale of BBCOU, would have a continuity rather than any scale-specific discrete values. Because that BBCOU cycle would appear as continuous. But it is still not observationally comprehended what was actually happened just before the moment of such Bigbang as well as Bigcrunch. Before the Bigbang, if the entire spacetime would collapse to zero then obviously the universal time of same BBCOU also could not be continuous cycle for that universal time anymore. That can be the discrete as well. In such a discrete reality of time for universe, the same BBCOU will have all separate cyclic cycles of Bigbang to Bigcrunch up to (a state before another possible) Bigbang, there will be all quantum-real causal steps in-between. But the same causality will break just before the next Bigbang, or say at the state of a *Big-collapse*, if the entire collapsed mass-energies along with spacetime of the universe become zeros. Then, the universal time would become discrete rather than any continuous cyclic entity.

Then, whatever would be the type of such universal time, continuous or discrete, during one conceptual cyclic oscillation of universe or BBCOU starting from the moment of a Bigbang to a Bigcrunch then again from the Bigcrunch to the Bigbang in cycle there will be always the quantum-real causal steps in that sequence of progression. The BBCOU, as macro-most scale through such a causal progression of oscillation, along with all its constituent scales of particles or systems-of-particles would have similar causal progressions. Once from such an incrementing 'expansion' (from Bigbang to Bigcrunch) and next from one decrementing 'expansion' (from Bigcrunch to Big-collapse). The same can also be described as the phase of incrementing 'entropy' onward from Bigbang to Bigcrunch and next as the phase of decrementing 'entropy' of same onward from Bigcrunch to Big-collapse.

Therefore, the entropy in BBCOU will have one intrinsic direction in that cycle, i.e. like the 'intrinsic direction of entropy' exclusively onward expansion in the cycle of BBCOU. To complete one such course within the expansion cycle, once that entropy will increment and alternately decrement in its values. If such direction of expansion is termed as left-handed, therefore intrinsic direction of the entropy is associated with it can also be termed for convenience as 'left-handed entropy'. If it is once incremented and then in next it will be onward expansion of cyclic oscillation can term as LE-C. Such LE-C will be another *universally common understanding* in current physics.

1.5. Left & Right-Hand Duality in Particles or Systems

In last paragraph, the observers like us would be also the LE-C type. Because those are also the integrated parts of the same BBCOU onward its progression of LE-C` direction for all alternate increments and decrements. Where all our sight of observations will have also an exclusive way of observations only onward the same direction of LE-C. Consequently, we never could see anything in different way, within the same BBCOU, if those have any different direction than LE-C.

Since that LE-C has an intrinsic left-handedness in direction for same BBCOU, that seems to have an intrinsic mirror-imaged direction. Consequently, such a mirror-imaged direction can have a simultaneous right-handedness and that can co-exist with the same BBCOU as macro-most quantum-real scale of all particles or systems-of-particles in the physical nature. But that right-handed mirror-imaged direction of entropy will be reversed in all aspects of ours. Subsequently, that would be always beyond all our conventional left-handed way of sights. If there any observer in that mirror-imaged way of sights, opposite of us, can also see all those same things or events simultaneously but in reverse direction. That is, those things would appear to him/her only onward the direction of causal progressions of simultaneous mirror-imaged cycle of 'collapse' for same BBCOU. If he or she goes to count their entropy of own surrounding, onward direction of same Bigbang to Bigcrunch, it would appear as a decrementing entropy in their surrounding onward progression of cycle of 'collapse'. But alternately, onward Bigcrunch to Big-collapse, the entropy of same would find incrementing just opposite to of ours onward LE-C direction.

For conveniences, such opposite entropy for progression of 'collapse' cycle can term as right-handed entropy or 'anti-entropy'; and that universal 'right-handed entropy' or 'anti-entropy' onward progression of collapse for same cyclic oscillation of BBCOU can also write as RE-C.

As the direction of RE-C is reverse on the line of sight of LE-C, then it is expected that anything that is RE-C directional cannot appear as steady or stay permanently on our line direction onward LE-C or vice versa. Through the observations of todays Particle Physics as well as in the present theoretical models as well, it is now expected that there would have simultaneous existences of antiparticles for every scale of particles or systems-of-particles including BBCOU itself. Although, antiparticles are observed so far only correspond to the micro scales of some particles or systems-of-particles but only in interventions of high energies and with existence for very infinitesimal moment after creations. After that, those have appeared to destroy spontaneously, i.e. never could have long sustenance onward our way of LE-C observations. It is now also evident in particle accelerators that, to create any such infinitesimal moment of existence for such heavier scales of antiparticles or systems-of-antiparticles there could require interventions of more amount of energies in such creations inside particle-accelerators.

Those antiparticles, which also have intrinsic right-handed axial rotations, can also be assumed to have RE-C direction. As a consequence, those antiparticles never could have sustainable existence onward our LE-C line of sight where we are involving being the observers.

Such RE-C will be another outcome *universally common understanding* in current physics beside LE-C.

However, both of those simultaneous LE-C & RE-C are assumed to involve not only with the BBCOU as macro-most scale in physical nature, but the same LE-C & RE-C must be also involved with all its comprising scales of particles or systems-of-particles. Then, in other way, every scale of those same particles or systems-of-particles (including BBCOU) could be assumed to have a simultaneous LE-C & RE-C intrinsic duality. Such a duality in simultaneous 'LE-C & RE-C' can be also another *universally common understanding* which is also involving with the same BBCOU.

1.6. Particles or Systems have Left-Handed Rotations

All the same scales of particles or systems-of-particles in above paragraphs are now observing to have intrinsic clockwise or left-handed axial rotation (LAR) in physical nature [6]. Therefore, such intrinsic LAR-direction for every particle or system-of-particles can also be considered as another *universally common understanding* in physical nature.

The fact of such intrinsic left-handedness in LAR-direction, which have in every scale of particles or systems-of-particles in BBCOU, seems incomplete. Because, it is not yet known whether same particles or systems-of-particles could have also simultaneous mirror-imaged intrinsic right-handedness in direction of axial-rotation for directional duality as like as the duality in simultaneous LE-C & RE-C within the same. If so, that can show one simultaneous left-handed and right-handed mutual images for particles or systems-of-particles or events in this physical nature, in terms of 3-positive and 3-negative spatial values for total 3+3 axes in a co-ordinate system, etc.

Hence, due to such universal intrinsic LAR-direction of axial-rotation for any particle or system-of-particles in the BBCOU, it is expected that there might have simultaneous mirror-imaged counterpart as intrinsic right-handed directional axial rotation (RAR). The antiparticles of all particles or systems of particles are now known [7] to have such intrinsic RAR-directional axial rotations.

Therefore, every particle or system-of-particles are now assumed to have antiparticle or system-of-antiparticles counterpart, and that intrinsic RAR-direction can also be another *universally common understanding* in the same physical nature.

Therefore, every same scale of particles or antiparticles as well as systems-of-particles or systems-of-antiparticles in BBCOU not only are existing in pairs but also having the simultaneous LAR and RAR intrinsic 'directional duality' in same pairs beside another simultaneous LE-C and RE-C intrinsic 'entropy-anti-entropy duality'. Such LAR & RAR 'directional duality' in every particle or system-of-particles can be also assumed as another *universally common understanding* in physical nature.

Furthermore, that intrinsic LAR-direction of any particle or system-of-particles can also be comprehended with some CIPs. Those are webbed the same particles or systems-of-particles, onward LE-C increments of BBCOU, with all SSUCs in respective scale-specific discrete magnitudes. For example, the three scale-specific discrete valued LAR-direction coordinates $(\Delta x, \Delta y, \Delta z)$ for space or volume or Δs as CIPs would have similar intrinsic LAR-direction. Also, all other CIPs with similar SSUCs in discrete magnitudes would have similar LAR-direction e.g. mass-energy Δm etc. those have occupied the same Δs. The same Δs can also be assumed have LAR-directional radius Δr ($\equiv \Delta x \equiv \Delta y \equiv \Delta z$) with scale-specific SSUCs in discrete magnitudes. Where all those Δs ($\Delta x, \Delta y, \Delta z$), Δr & m have increments for SSUCs in discrete magnitudes along the increments of LE-C.

Another CIP can be the time as Δt that also associates in all those same scales of particles or systems-of-particles. The time not only flows from the 'past' to 'future' onward clockwise direction but anything that is clockwise can assume as left-handed directional. So, a clockwise time as Δt would also be one LAR-directional in every same particle or system-of-particles. The same Δt might have also the increments as another SSUCs with discrete magnitudes along the increments of LE-C. Hence, the Δt, as one CIP, could assume to possess intrinsic LAR-direction.

Oppositely, the direction of other prospective CIPs like de Broglie Wavelength $\Delta \lambda$, inertial-motion Δv, and so on with all those same scales of particles or systems-of-particles with increments as SSUCs in discrete magnitudes onward RE-C of same BBCOU. Since there is simultaneous left-handed and right-handed intrinsic directional duality in each of those particles or systems-of-particles for simultaneous LE-C & RE-C intrinsic 'entropy – anti-entropy' duality in magnitudes. Therefore, all those LAR-directional and LE-C-directional incrementing magnitudes of CIPs seems to have also their corresponding intrinsic RAR-directional and RE-C-directional incrementing dual or mirror-imaged counterpart CIPs in all same particles or systems-of-particles. As a result, all those CIPs having increments as SSUCs in discrete magnitudes onward RE-C could be assumed as RAR-directional to web that intrinsic right-handed directional RE-C involving with every dual or mirror-imaged existence of particle or system-of-particles as integrated part of the same BBCOU in physical nature.

Hence, simultaneous LAR and RAR intrinsic directional duality in every scale of particles or systems-of-particles can also be assumed as another *universally common understanding* in physical nature.

1.7. Quantum-Real Limit in Our Observations

As mentioned earlier all the smaller scales of particles or systems-of-particles including the micro-most scale, as the constituent parts of macro-most scale BBCOU, are not only as integer but also as quantum-real [1]. Then, that macro-most scale can assume as the sum of all those integer quantum-realities for entire quantum-real fold of physical nature. Therefore, any quantum-real-observers like us, who are also integrated to such quantum-real fold, can exchange with any of those scales of particles or systems-of-particles through any quantum-real signals only.

Actually, such observers, along with all possible smart-enough instruments would have today, are intrinsically configured by any corresponding scales of such quantum-real particles or systems-of-particles. Therefore, all the quantum-real observers like us, along with all necessary instruments, as the integrated parts of that macro-most scale of BBCOU, would be also quantum-real in type.

In practical sense, such quantum-real observers along with their all corresponding instruments are now also witnessing to communicate with any of the observances only through exchange of quantum-real signals. That means, each of those 'observers-and-their-instruments', 'exchangeable-signals', and 'observances' as integer parts of the quantum-real fold of physical nature would have involvements with any intrinsic scale-specific quantum-real particles or systems-of-particles. If any one of those three, during such exchange, would have no

---

1. Because, each of those have progressions through causal steps starting from 'post-creation' to 'pre-destruction' existences.

such scale-specific quantized (non-zero, finite and discrete) magnitudes or have any non-quantized (infinite, zero and continuous) values, that signal communication never can be completed in quantum-real sense. Subsequently, an observer like us, who is basically such a quantum-real in type, never can observe any such part in the physical nature if that will possess only non-quantized, infinite and zero magnitudes.

But in context of todays observational understandings in physics, particularly in relevance of growing consensuses in particle physics and astro-physics, any quantum-real observers like us along with all our fine-tuned smart instruments, we cannot exchange directly through any quantum-real-signal in this physical nature to anything which have any non-quantized, infinite and/or zero magnitudes. Therefore, it is now appearing that every quantum-real observer as if possesses one intrinsic quantum-real limitation in their most basic level of observations that seems to imposed on us by the same physical nature. Within such a range of quantum-real limitation of us everything would be discrete, finite and/or non-zero in magnitudes. That is, observers like us are basically quantum-real limited and cannot interact with anything beyond of that limit. All our quantum-real observations/experiments in relevance of finding any event or observance beyond that limit remains a trail for reaching ultimately anywhere exceeding that quantum-real 'end'.

Because, our quantum-real limited exchangeable signals can only exchange between any observer and observances if and only if those two have any quantized, non-zero and finite magnitudes.

Then, conceptually, such quantum-real limitation in our observation can only have maximum reach within that entire quantum-real fold of the physical nature. Therefore, beyond that quantum-real fold of our observational limit, there might have another fold in same physical nature, which can be termed as *Non-quantized-virtual* in type. Conceptually, that can have all non-quantized, infinity and/or zero magnitudes. But that never can be exchanged directly through any quantum-real signals.

Accordingly, in same physical nature also perceives as if there two types of 'logical' sequences those are intrinsically involved to all its core dynamisms. Some logical sequences appear comprehensible to us as quantum-real observers if the same compatible with our intrinsic sequence of logic (i.e. *causal*). In contrary, if those are incompatible to ours sequence of logic obviously incomprehensible (i.e. *non-causal*) to our quantum-real limited sense of causality. Therefore, both of those two sequences of logic, no matter whether those would appear to us, in the form of causal or non-causal, are ultimately linked to same physical nature. Those two are also seemed to remain flow in deeper level of same physical nature as one unified sequence of logic that might be beyond our comprehensible limit. Because, all those existential everything, causal and non-causal as well as quantum-real-fold and non-quantum-virtual-fold, are ultimately integrated parts of that entire physical nature.

Hence, in deeper level, the physical nature seems to be unified in spite of its all apparent diversifications in such two parts or folds. One is compatible with us as quantum-real but with causal sequence of logic and another is incompatible to us as Non-quantized-virtual but with broken sequence of causality or non-causality. That is, one is with discrete, finite and non-zero values of causality and another with continuity, infinity and zero values of non-causality.

Therefore, the quantum-real fold, equivalent to BBCOU as macro-most system-of-particles, would be the maximum limit for all quantum-reality as well as quantum-causality for any quantum-real-observers like us. Such limitation can term for conveniences as 'quantum-real-limitation' that includes limitations related to both quantum-reality and quantum-causality.

Then, in all our concerned quantum-real observations, such quantum-real-limitation can be considered as another *universally common understanding* in physical nature.

2. Common-Internal-Parameters:

Earlier generalize concepts of all 'material-bodies' in Classical Mechanics, has now changed to any form of discrete particles or systems-of-particles with respective micro to macro scales in above Section-1. Therefore, whether all that micro to macro diverse scales of particles or systems-of-particles are internally uniform, through all those seven 'universally common understandings' of Section-1, would be the central issue of the Section-2. Because, all those scales of particles are also constituents of the macro-most scale of system like universe. Then, proceeding Sub-sections from 2.1 to 2.9) will find total ten or twelve numbers of common-internal-parameters or CIPs in all those diverse scale-specific particles or systems-of-particles in absence of forces by incorporating all those seven numbers of 'universally-common-understandings'. Those universal CIPs in every scale of particles or systems-of-particles would be:

2.1. Inertial Mass-Energy

Through Max Planck's wave-corpuscular inverse relationship (in year 1900), it was first time comprehended that the light could have not only the discrete or quantized magnitudes for its inertial mass-energies but such magnitudes have also an inverse relationship with corresponding wavelengths. Later, such inverse relationship as well as concept of quantization in magnitudes of inertial-mass-energies Δm was extended beyond the photons of electromagnetic spectrum or EMS up to the all scales of micro to macro PSs through the de Broglie wave-corpuscular relation

$$\Delta m \, \Delta \lambda = h/c . \qquad (1.1)$$

Where $\Delta\lambda$ as corresponding de Broglie wavelength, and h & c are the Planck constant and that constancy for inertial speed for light in SRT correspondingly. In Eq. (1.1), both of the Δm & $\Delta\lambda$, as two CIPs, are also possessed all scale-specific quantized magnitudes with SSUCs in discrete magnitudes. The Eq. (1.1) also describes all such scales of particles or systems-of-particles or PSs as individual wave-corpuscular-phenomena (WCP).

Then, the Δm in Eq. (1.1) is not only associated with every micro to macro scales of PSs irrespective of scales as SSUCs in discrete magnitudes but the same Δm with corresponding PSs would have intrinsic left-handed direction of axial rotations from above Sub-section-1.6. Similarly, from Sub-section-1.4 the Δm would possess also the increments for same SSUCs in discrete magnitudes onward micro to macro scales along the increments of left-handed entropy or LE-C of BBCOU. Therefore, in inertial states, that Δm, as one of CIPs in all PSs in Eq. (1.1), can be postulated to have:

（i) intrinsic SSUCs in quantized magnitude;

(ii) intrinsic increments as SSUCs in quantized magnitudes onward increments of the LE-C in expanding phase of BBCOU;

(iii) intrinsic left-handed directional axial rotation or LAR along the LAR of same PSs; and

(iv) the unit in gm.

2.2. de Broglie Wavelength

The Eq. (1.1) also reveals all scale-specific quantized magnitudes for the de Broglie wavelengths $\Delta\lambda$ involving in every scale of PSs in BBCOU. It could include scale-specific magnitudes for all wavelengths of light in EMS, the wavelengths for all other scales of bosons and fermions, and conceptually the wavelengths for all other scales of PSs even beyond the range of visible matters those are comprising the entire BBCOU. But, that Δλ also has appeared with an inverse co-relation with Δm in Eq. (1.1), where it shows a reverse direction of increments as SSUCs in all respective discrete magnitudes' onward macro to micro scales parallel to right-handed entropy or RE-C for same BBCOU. As a result, the Δλ as one CIP in all PSs in Eq. (1.1), and can be postulated to have:

(i) intrinsic SSUCs in quantized magnitudes;

(ii) intrinsic decrements in such quantized magnitudes as SSUCs but onward opposite to LE-C direction of BBCOU, i.e. will have the increments onward simultaneous mirror-imaged RE-C of the BBCOU;

(iii) intrinsic entanglement with a right-handed direction axial-rotation or RAR in all same PSs; and

(iv) the unit in *cm*.

2.3. Inertial Motion

As one of most basic assumptions of todays SRT as well as Quantum Mechanics, in absence of external forces or in an inertial state, the free moving 'mass-energy' and 'motion' are simultaneously involved with every PSs and conceptually two cannot be separated [1] from each other. That is, each of those relatively moving PSs irrespective of scales cannot be observed in absolute rest respect to all other PSs.

But, such free moving or inertial 'mass-energy' in any such PSs would have intrinsic scale-specific quantized magnitude in Eq. (1.1). Then it could be asked whether it's inseparable inertial or free moving 'motion' for same mass-energy' will have also

---

1. Means that, in inertial state, 'inertial-mass-energy' and 'inertial-motion' involving with any material-body cannot be separated from each other in basic assumptions of Relativity Theories as well as Quantum Mechanics. Conceptually, it seems impossible to imagine one inertial entity having only inertial-mass-energy without inertial-motion or vice versa in mechanical processes of both theories.

the similar scale-specific quantized magnitudes in all those same PSs.

Moreover, in basic assumptions of SRT, the physical-entities (which are ultimately nothing but the different PSs) in terms of the respective inertial or free 'motions' have also broad divisions in two broader categories: **(a)** one is with absolute or observer-independent constant magnitude c for inertial speed of light and **(b)** another is with all relative or observer-dependent non-constant type magnitudes for all non-luminous material-bodies.

As per present conventions of Particle Physics as well as from the Eq. (1.1), both of those two categories, 'light' and non-luminous 'material-bodies' in SRT, are now very precisely appearing as the specific scales of PSs as wave-corpuscles. As a result, each of those PSs, both light and material-bodies, are possessing the scale-specific intrinsic quantization in magnitudes of Δm (i.e. as the one of observer-independent CIPs). Therefore, the concept of 'light' in SRT, now as any wave-corpuscle or particle, not only possesses any such scale-specific intrinsic quantization in magnitudes of Δm but inseparable intrinsic magnitude of it's free-motion or inertial-motion $c = \Delta v_c$ (i.e. free-motion of a Photon) can also assume as quantized.

But now, such notions of intrinsic (or observer-independent) free-motion $c = \Delta v_c = (2.99792 \times 10^{10}\ cm.sec^{-1})$ can assume further correspond to one particular scale of free-photons in EMS. Consequently, there can be other different scales of intrinsic free-motions [5] for the respective other scales of free-photons in the same EMS and also even beyond of that EMS according to different direct / indirect estimations of experiments in todays Physics. Beyond the EMS, there are heavier bosons as well as fermions of varying scales of PSs and those could have different scale-specific intrinsic quantized magnitudes of free-motions say Δv with corresponding inseparable free mass-energies Δm. As like free-photons, the free-neutrinos, free-atoms, free-molecules and so on are also now unfolding to us in different ranges or spectra of scales with corresponding intrinsic mass-energies and de Broglie wavelength. Even among the astronomical objects, everything now is in similar way unfolding to us through such specific ranges of scales as well from planetesimals to galaxies or clusters-of-galaxies.

However, as per current comprehensions, the free-motions of such a varying range of scales for the neutrinos are now further appearing to have variable scale-specific quantized magnitudes of inertial-mass-energies. Such a range of varying magnitudes for inertial-motions of neutrinos are observed to have up to 99% or even 100% (or could be more than) the speed ($c = \Delta v_c = 2.99792 \times 10^{10} cm/sec$) of 'light' for one particular free-photon. This has not merely observed in the free-motions of neutrinos, but an analogous kind of scale-specific intrinsic magnitudes of free-motions are also perceived in gradually heavier inertial-mass-energies of other scale-specific fermions. For example, in all scales starting from leptons to hadrons, hadrons to atoms or molecules. Although in astronomical range of scales for PSs such scale-specific intrinsic magnitudes of free-motions are not yet evident.

But any PSs having all scale-specific existences, and particularly if would have intrinsic magnitudes of CIP like Δm, then such an intrinsic inertial magnitude of Δm must have co-existence of intrinsic inertial magnitude of another CIP i.e. Δv. In macro range of scales for PSs, due to the stronger presence of influencing non-inertial forces particularly universal gravitation, it is now almost impossible to measure directly any such scale-specific intrinsic quantized magnitudes of inertial-motions Δv as like as the difficulties in direct measurements of corresponding scale-specific intrinsic quantized values of $\Delta \lambda$. But, the magnitudes of all those corresponding Δm those are involved with the respective heavier (macro) scales of PSs in astronomical range also have definitely constituted by the corresponding sums of micro scales of PSs. Such constituent micro scales of PSs would have respective quantized mass-energies of lesser magnitudes and inseparable intrinsic quantized magnitudes of free-motions. Therefore, it is no matter, whether such different scale-specific quantized inertial-motions for those different scales of PSs are precisely measured yet or not, compare to all the corresponding assumptions of the quantized inertial mass-energies in same. Then, conceptually, there seems now all scale-specific intrinsic quantized magnitudes of inertial-motions involving with the astronomical range of those macro PSs.

Therefore, proceeding paragraphs of this Sub-section will describe in more detail about such different scale-specific intrinsic quantized magnitudes of inertial-motions in different SSUCs of magnitudes for wavelengths of 'light' (or photons) in EMS and bosons beside all fermions; and also designate same inertial-motion Δv as another universal CIP in all those same scales of PSs:

2.3.1. Inertial Motions of Photons

In three different occasions below, it seems that there would have three different scale-specific appearance of inertial-motions with intrinsic magnitudes correspond to three different wavelength-photons in EMS:

i) Inertial Motion of Gamma-Ray Photons

The intrinsic quantized inertial-motion Δv of 'light' or a photon, as any WCP in EMS from Eq. (1.1), has assumed in SRT as the 'observer independent' (i.e. 'intrinsic' to same photon) absolute sense of constant (i.e. a universally 'fixed' magnitude) c. Therefore, two gamma-ray-photons in same EMS might have magnitudes of inertial-motions c (= $2.99792 \times 10^{10}\ cm.sec^{-1}$) as assumed in SRT. Then in context of such two gamma-ray-photons it can be realized whether the same magnitude c might be remained higher than all evaluate fermions (as any non-luminous entities in SRT). This can judge through very well-known phenomenon like an electron and positron pair annihilation and a subsequent spontaneous formation of the pair of two gamma-ray photons

$$(e^- + e^+) \rightleftarrows (\Delta\gamma_1 + \Delta\gamma_2). \tag{1.2}$$

Where e^- & e^+ are two electron & positron in such a pair before annihilation, and $\Delta\gamma_1$ & $\Delta\gamma_2$ are wavelengths of two spontaneously created gamma-ray photons after annihilation respectively with all intrinsic quantized magnitudes. Due to conservation of mass-energies, there will be for such e^- & e^+, the corresponding quantized inertial mass-energies as say $\Delta m_{e^-} = \Delta m_{e^+}$ and wavelengths as say $\Delta\lambda_{e^-} = \Delta\lambda_{e^+}$ in Eq. (1.1). After annihilation, he corresponding inertial mass-energies for a newly evolved pair of gamma-ray-photons $\Delta\gamma_1$ & $\Delta\gamma_2$ are respectively ay $\Delta m_{\gamma_1} = \Delta m\lambda_2$. Then according to the universal laws of mass-energy and momentum conservations, we can re-write the Eq. (1.2) as

$$\Delta m_{e^-} \times \Delta v_{e^-} + \Delta m_{e^+} \times \Delta v_{e^+} = \Delta m_{\gamma_1} \times \Delta v_{\gamma_1} + \Delta m_{\gamma_2} \times \Delta v_{\gamma_2} \tag{1.3}$$

where $\Delta v_{e^-}, \Delta v_{e^+}, \Delta v_{\gamma_1}$ & Δv_{γ_2} are corresponding quantized inertial-motions of electron, positron, and two gamma-ray-photons. Then, in Eq. (1.3), there should be the corresponding scale-specific intrinsic quantized magnitudes for inertia-motions for both of those electron-positron in the pair $\Delta v_{e^-} = \Delta v_{e^+}$. Because, conceptually from Eq. (1.1) we already could have quantized values for both of those e^- & e^+ as $\Delta m_{e^-} = \Delta m_{e^+}$ and $\Delta\lambda_{e^-} = \Delta\lambda_{e^+}$. Since we could have such quantized values for $\Delta m_{\gamma_1} = \Delta m_{\gamma_2}$ in the pair of gamma-ray photons, there would be also the inertial (quantized) motions $\Delta v_{\gamma_2} = \Delta v_{\gamma_1}$ for same two photons. Accordingly, due to $\Delta m_{e^-} = \Delta m_{e^+}$, we also might have there for two photons $\Delta m_{\gamma_1} = \Delta m_{\gamma_2}$. Then, from Eq. (1.3)

$$\Delta m_{e^-} + \Delta m_{e^+} = \Delta m_{\gamma_1} + \Delta m_{\gamma_2}. \tag{1.4}$$

Since there $\Delta m_{\gamma_1} = \Delta m_{\gamma_2}$, we must have for wavelengths of those two gamma-ray photons in pair

$$\Delta \gamma_1 = \Delta \gamma_2. \tag{1.5}$$

Then ultimately, from Eq. (1.2) we can re-write further in the Eq. (1.4) as

$$\Delta m_{e^-} = \Delta m_{e^+} = \Delta m_{\gamma_1} = \Delta m_{\gamma_2} \tag{1.6}$$

and then, according to Eq. (1.6), we can also have from the Eq. (1.3)

$$\Delta m_{e^-} \times \Delta v_{e^-} = \Delta m_{e^+} \times \Delta v_{e^+} = \Delta m_{\gamma_1} \times \Delta v_{\gamma_1} = \Delta m_{\gamma_2} \times \Delta v_{\gamma_2} \tag{1.7}$$

where obviously we should have an equivalence among the values of inertial (quantized) motions

$$\Delta v_{e^-} = \Delta v_{e^+} = \Delta v_{\gamma_1} = \Delta v_{\gamma_2} \tag{1.8}$$

for each member of that electron-positron (before annihilation) as well as two gamma-ray-photons (after annihilation) in respective pairs.

But according to the basic assumptions in SRT, related to magnitude of c for light as any electromagnetic wave in EMS, the so called non-luminous inertial-motions of both electron and positron in Eq. (1.8) must be

$$(\Delta v_{e^-} = \Delta v_{e^+}) < c. \tag{1.9}$$

Then obviously in same Eq. (1.8) we should not have the magnitudes for inertial-motions of those two gamma-ray-photons, equivalent to those electron and positron as a form of the 'light' in EMS as per similar basic assumptions in SRT, which are appearing with magnitudes lesser than c

$$(\Delta v_{\gamma_1} = \Delta v_{\gamma_2}) < c. \qquad (1.10)$$

That is, in Eq. (1.8), actually it has appeared that magnitudes $(\Delta v_{\gamma_1} = \Delta v_{\gamma_2})$ of inertial-motions for those two bosons or particles of 'light' equal to the magnitude $(\Delta v_{e^-} = \Delta v_{e^+})$ of two 'non-luminous' fermions (having lower than speed of light) in contrary to basic assumptions of SRT. Consequently, for such an observer independent inertial-motion for two identical bosons or particles of 'light' must be $c = (\Delta v_{\gamma_1} = \Delta v_{\gamma_2}) < 2.99792 \times 10^{10} cm.sec^{-1}$ less than the usually assumed magnitude in SRT i.e. as $c = 2.9792 \times 10^{10} cm.sec^{-1}$. This essentially hints at, either inertial-motions of those two gamma-ray-photons in EMS are not in the sense of any such luminous-body (or boson) those need to have inertial-motion of $c = 2.9792 \times 10^{10} cm.sec^{-1}$ as per basic assumptions of SRT or there could have any other précised scale for different photons in same EMS those might have equivalent inertial-motion $c = 2.9792 \times 10^{10} cm.sec^{-1}$ in a sense of 'light' in SRT. If later is true, in EMS, there can be different constant magnitudes of inertial (quantized) motions for all corresponding scales of photons.

Therefore, in SRT, where c has assumed to have identical inertial-motions for all electromagnetic wavelengths in EMS. But from the Eq. (1.8) it has appeared that there can also be other electromagnetic wavelengths in EMS those may possess different though remain constant magnitudes like Δv_{γ_1} & Δv_{γ_2} less than $c = 2.99792 \times 10^{10} cm.sec^{-1}$. Then one such photons, to fit for inertial-motion of 'light' in SRT, with say $c = \Delta v_c = 2.99792 \times 10^{10} cm\ sec^{-1}$ must be any other but not gamma-ray-photon on EMS. So, if we can consider that $c = \Delta v_c$ for a different scale of photons other than the scale of those gamma-ray-photons in Eqs. (1.8) & (1.10), then obviously from Eq. (1.1), that scale of photons will have different scale-specific intrinsic quantized magnitude for its wavelength (say $\Delta \lambda_c$) as well as inertial mass-energy (say Δm_c), compare to the gamma-ray-photons respect to Eqs. (1.2) & (1.4).

Since, such photons with $c = \Delta v_c$ in SRT can consider as any non-gamma-ray photons and from the Eq. (1.1) all the gamma-ray photons have heaviest inertial mass-energies compare to all other scales of photons in EMS, the inertial mass-energies for the gamma-ray photons in Eq. (1.6) would have

$$(\Delta m_{\gamma_1} = \Delta m_{\gamma_2}) > \Delta m_c . \qquad (1.11)$$

Because, it is still unknown what is the exact scale of the quantized magnitude of inertial-mass-energy for Δm_c with corresponding photon in EMS. Accordingly, for corresponding wavelengths of the same pair of gamma-ray photons in Eq. (1.5) from the Eq. (1.1), there would have

$$(\Delta \lambda_{\gamma_1} = \Delta \lambda_{\gamma_2}) < \Delta \lambda_c . \qquad (1.12)$$

Ultimately, we can re-write Eq. (1.10) for convenience to define $c = \Delta v_c = 2.9792 \times 10^{10} cm\ sec^{-1}$ as

$$(\Delta v_{\gamma_1} = \Delta v_{\gamma_2}) < (\Delta v_c = c) . \qquad (1.13)$$

Now, it reveals in Eq. (1.13) that a specific scale of gamma-ray photon-PSs shows different intrinsic-quantized magnitude in its inertial-motion $(\Delta v_{\gamma_1} = \Delta v_{\gamma_2})$ compare to the 'speed of light' $(c = \Delta v_c)$ in SRT. Therefore, the $c = \Delta v_c$ will possess another scale-specific intrinsic-quantized magnitude of inertial-motion beside with corresponding magnitude of wavelength $\Delta \lambda_c$ in Eq. (1.12) on EMS along with specific magnitude of Δm_c in Eq. (1.11) instead of all scales of photon-particles in entire range of EMS.

However, due to the Eq. (1.13), it can be assumed that there are at least two kinds of different intrinsic-quantized magnitudes for inertial-motions correspond to two different scales of photon-particles with Δv_γ & Δv_c in EMS.

ii) Inertial Motion of Microwave Photons

Already, it has assumed in Eq. (1.13) that $c = \Delta v_c = 2.99792 \times 10^{10}\, cm\, sec^{-1}$ may not be a universally common inertial magnitude for the inertial speed of all scales of photons irrespective of wavelengths in EMS. Even in some current experiments [5] related to quantum tunneling experiments related to quantum entanglements, it has been claimed that some microwave photons, with respective magnitude of wavelength, have an intrinsic-quantized magnitude of inertial-motion much higher than the 'speed of light' $c = v_c$, say Δv_η. Where obviously we can write for that $\Delta v_\eta > \Delta v_c (= 2.99792 \times 10^{10}\, cm\, sec^{-1})$. Then, we can re-write Eq. (1.13) in EMS for all different quantized inertial-motions of photons with different scale-specific inertial mass-energies and wavelengths

$$(\Delta v_{\gamma_1} = \Delta v_{\gamma_2}) < \cdots < (\Delta v_c = c) < \cdots < \Delta v_\eta < \cdots \quad (1.14)$$

and there, obviously we can further re-write the corresponding wavelength (say $\Delta \lambda_\eta$) of such microwave-ray-photons from Eq. (1.12) as

$$(\Delta \lambda_{\gamma_1} = \Delta \lambda_{\gamma_2}) < \cdots < \Delta \lambda_c < \cdots < \Delta \lambda_\eta < \cdots \quad (1.15)$$

Then, due to an inverse relation in-between inertial mass-energies and wavelengths for all specific scales of PSs in Eq. (1.1), there will be also the similar corresponding intrinsic-quantized magnitude of inertial mass-energy (Δm_η) for that microwave-ray-photon in Eq. (1.14). Therefore, the same micro-wave-ray-photon would have from the Eq. (1.11) an inertial mass-energy say Δm_η which likely to have $(\Delta m_c > \Delta m_\eta)$ scale-specific magnitude. Since there already have for all corresponding inverse magnitudes of scale-specific magnitudes of wavelengths $(\Delta \lambda_{\gamma_1} < \Delta \lambda_c < \Delta \lambda_\eta)$ in Eq. (1.15). Then we can write from the above Eq. (1.11) as

$$(\Delta m_{\gamma_1} = \Delta m_{\gamma_2}) > \cdots > \Delta m_c > \cdots > \Delta m_\eta > \cdots \quad (1.16)$$

in the EMS. This kind of different scale-specific intrinsic quantizations in inertial-motions like $(\Delta v_{\gamma_1} = \Delta v_{\gamma_2})$, $(\Delta v_c = c)$, & Δv_η for different scales of photons in Eq. (1.14) are obviously creating somehow inconsistencies in primary assumptions of the SRT. However, such inconsistencies will be resolved in Chapter-3, and that will further reveal a possibility for existence of superluminal inertial-motions in some scales of PSs in EMS with non-negative values of time in the corresponding IFRs of SRT.

iii) Inertial Motions of Photons with c

The Eq. (1.14) has actually revealed the different scale-specific intrinsic inertial-motions in the EMS in corresponding scale-specific magnitudes of wavelengths for different scales of photon-corpuscles. The lower magnitudes of intrinsic inertial-motions are appearing to involve with photon-corpuscles with corresponding shorter wavelengths but heavier quantized mass-energies in EMS and higher intrinsic inertial-motions with longer wavelengths and lighter quantized mass-energies in same EMS.

Therefore, in similar way, that intrinsic or observer-independent constant inertial-motion in SRT i.e. $c = \Delta v_c = 2.99792 \times 10^{10}$ cm/sec would also be linked to any of those specific wavelength as well as specific mass-energy correspond to any of such photon-corpuscles in same EMS which must have any specific scale. Then, what will be the particular quantized mass-energy and wavelength for such corresponding inertia-motion $c = \Delta v_c = (2.99792 \times 10^{10}$ cm/sec) that can reveal the scale of those specific photon-corpuscles in EMS?

But from same Eq. (1.14), it is expected that that scale-specific intrinsic magnitude like $c = \Delta v_c = 2.99792 \times 10^{10}$ *cm/sec* for a particular scale of photon-corpuscles likely to be somewhere within the range of visible spectrum in EMS. Because, the convention of constancy involving with the 'speed of light' in SRT has mostly ever measured by using 'light' rays within that range of visible-spectrum in same EMS.

But such a narrow visible-spectrum also includes millions of different scale-specific values for individual wavelengths of visible-light-photons in that EMS. Hence, it is again hard to point out precisely any such specific wavelength within same range of visible-spectrum for such constant magnitude of quantized inertial-motion $c = \Delta v_c = 2.99792 \times 10^{10} cm/sec$ to explore the corresponding scale for visible-photon-corpuscles except through any direct measurements. Even, if that $c = \Delta v_c$ is grossly assumed to possess the its scale-specific wavelength somewhere middle of such visible spectrum of light, then that will be within the range of yellow color range of all visible range wavelengths. That yellow range within the visible-light spectrum has further comprised thousands of scale-specific yellow-light-wavelengths of photon-corpuscles in EMS. So, it needs ultimately any direct measurement to fix that particular wavelength respect to the inertial-motion of 'light' in SRT i.e. $c = \Delta v_c = 2.99792 \times 10^{10} \ cm/sec$.

Hence, in above paragraphs, in relevance of intrinsic inertial-motions of three different scales of photon-corpuscles for a gamma-ray, a visible light-ray and a micro-wave-ray in EMS, particularly due to the Eqs. (1.1) & (1.14), it has realized that there might have at least three different kinds of intrinsic or observer-independent magnitudes for the inertial-motions of three different scales of photon-corpuscles (i.e. of 'light'). But apart from those three scales of photon-corpuscles, there remains exist other bosons with heavier mass-energies with shorter wavelengths than the gamma-rays, and conversely with lighter mass-energies with longer wavelengths than the micro-waves in the same EMS. As a result, the corresponding magnitudes for all scale-specific intrinsic inertial-motions for bosons would have the different scale-specific quantized magnitudes of inertial-motions.

2.3.2. Inertial Motions of Fermions

Hence, from Eqs. (1.1), (1.4), (1.5) & (1.16), all the bosons (including photons in EMS) are not only possessing scale-specific intrinsic or observer independent magnitudes of inertial mass-energies and wavelengths but instead of only one single observer-independent or intrinsic constant magnitude like $c = \Delta v_c = 2.99792 \times 10^{10} cm.sec^{-1}$) there would be all higher and lower intrinsic or observer independent different magnitudes of corresponding scale-specific inertial-motions.

The same thing is now also observing in Particle Physics. Various fermions are, particularly in micro scales of those particles or systems-of-particles, also similarly possessing the corresponding intrinsic or observer-independent type scale-specific magnitudes of inertial-motions.

In different scales of PSs, from molecular to all sub-atomic scales, it is now evident that there are different magnitudes of fixed type inertial-motions. All the specific scales of neutrinos, electrons, neutrons, atoms, molecules etc. would have different intrinsic or observer-independent type fixed magnitudes of inertial-motions beside corresponding intrinsic quantized magnitudes of mass-energies and wavelengths.

Although, it is not precisely known whether the heavier scales of fermions up to the scales in astronomical range under influence of dominating gravitation could also have similar kind of scale-specific intrinsic magnitudes of inertial-motions. But conceptually, all those macro scales of fermion-PSs, including the macro astronomical scales, are fundamentally constituted by the corresponding micro scales of bosons & fermions those have distinctly so-called scale-specific intrinsic fixed type of inertial-motions.

Hence, conceptually, it seems that all those imprecise macro 'material-bodies' or 'astronomical-bodies' in sense of Classical Physics, as any macro PSs, could also have specific scales. As a consequence, from Eq. (1.1), those macro PSs would also have corresponding scale-specific intrinsic or observer-independent values in terms of mass-energies & wavelengths.

If so, those macro PSs also could have the respective intrinsic or observer-independent inertial-motions. No matter whether such scale-specific intrinsic inertial-motions of macro-PSs are yet measured or not.

2.3.3. Inertial Motions for All Scales of Particles or Systems

Finally, from the above paragraphs in Sub-section 2.3., particularly from the Eqs. (1.1), (1.14), (1.15) & (1.16), it could assume that there might have all scale-specific intrinsic quantized magnitudes for inertial-motions in every scale of bosons to fermions beside the corresponding scale-specific quantized mass-energies and wavelengths of same.

Therefore, any of such scale-specific intrinsic quantized magnitudes of inertial-motion say as Δv that is common to all scales of micro to macro PSs can be postulated as one of the CIPs and each of such fixed type intrinsic scale-specific magnitudes of Δv would be ultimately the SSUCs. The Eq. (1.14) through the Eq. (1.1), shows that there would be all micro to macro scales of decrements in scale-specific magnitudes of Δv, i.e. onward all the LE-C directional decrements of BBCOU in current expansion phase, and opposite to the direction of increments in all scale-specific magnitudes of Δm in Eq. (1.16). But parallel to the direction of decrements of $\Delta \lambda$ in Eq. (1.15). This seems not only appearing in the scale-specific magnitudes of bosons but also for all scales of fermions. Hence, the inertial motion Δv, as one of the CIPs in all scales of PSs in Eq. (1.1), can postulate to have:

> (i) intrinsic quantized magnitudes where each of those not only possesses the corresponding SSUCs in discrete magnitudes but decrements in it onward

opposite direction of Δ*m* but parallel to the direction of Δλ in current expanding phase of BBCOU;

(ii) in current expansion phase of BBCOU, such direction of intrinsic decrements in discrete magnitudes are opposite to increments of LE-C, i.e. parallel to the direction of the simultaneous RE-C of same BBCOU;

(iii) intrinsic entanglement with the right-handed directional axial-rotation or RAR or simply right-handed direction of same PSs (as the CIP that has increment as SSUCs in all discrete magnitudes onward RE-C); and

(iv) unit in cm/sec.

2.4. Inertial Radius

Every material-bodies onward LE-C in current phase of expansion for BBCOU are conceptually appearing as precise PSs having the specific scale. Particularly, all those PSs which are appearing to us as any bosons or fermions in the visible range. It can also assume, in the range of 'dark matters' as well as 'dark energies', there could have some different scales of PSs. But whatever may be those PSs, in between micro-most and macro-most scales, must have corresponding quantized magnitudes of mass-energies to constitute the total mass-energies within the 'shape' of BBCOU (as macro-most scale).

But within that range of micro to macro scales of fermion-PSs now observes to have scale-specific types of intrinsic valued all *inertial radii*. Such scale-specific different values of radii are also indicated that there could have all corresponding scale-specific inertial values of spatial extensions or volumes for all 'occupied' quantized inertial mass-energies of those fermion-PSs.

There are all such scale-specific type intrinsic values in fermion-PSs in scales of an electron ($2.82 \times 10^{-17} cm$), a neutron ($0.8 \times 10^{-17} cm$), a hydrogen atom ($5.29 \times 10^{-13} cm$), and even in astronomical scale of a planetesimal ($\geq 5 \times 10^4 cm$), a minimum radius to form a sub-solar star e.g. a red-dwarf ($\geq 1.67 \times 10^{10} cm$), and so on with corresponding scale-specific types of quantized inertial mass-energies, de Broglie wavelengths & inertial-motions.

Hence, such scale-specific intrinsic inertial radius (say Δr) for any fermion-PSs can consider also to have quantized magnitudes. The increments in such scale-specific quantized magnitudes of that Δr in fermion-PSs would have increments from micro-scales to macro-scales onward the direction of LE-C in current expansion (phase) of the BBCOU.

Since all other scales of PSs, within the dark matters and dark energies including bosons (beside those fermions) in the range of visible matters, are conceptually integrated constituents of same expanding BBCOU, it can be considered that in all those scales could have also scale-specific intrinsic values of radii.

Then theoretically, such Δr can be considered as another CIP in all same micro-most scale to macro-most scale of PSs irrespective of visible matters, darks matters & dark energies of the BBCOU, and the same Δr would have left-handed increments in scale-specific quantized magnitudes along the increments in scale-specific quantized magnitudes of Δm. That would be also an opposite directional increment in scale-specific quantized magnitudes compare to the Δv & $\Delta \lambda$.

Therefore, the same Δr, as one CIP in all those micro-most to macro-most scales of PSs in Eq. (1.1), can postulate to have:

(i) intrinsic scale-specific quantized magnitudes where each quantized magnitude as SSUCs;

(ii) intrinsic increments in those scale-specific quantized magnitudes along the direction of LE-C with the Δm in current expanding phase of BBCOU;

(iii) entangled with LAR direction of axial rotation of Δm involved with same PSs; and

(iv) unit in *cm*.

2.5. Inertial Space

Our existence in current expanding phase of BBCOU is seemingly onward all micro to macro structures formation. Such increments in structures is now observing parallel to the increments onward LE-C of same BBCOU. Then each of those individual structures is appearing quantized (due to Δr) in scale-specific manner.

Any one of such scale-specific macro-structures could be also an integrated sum of corresponding integer number of further scale-specific micro-structures or micro-PSs. Then, any of those spatial structures, micro and macro, would ultimately be the sum of all specific scale of PSs except micro-most.

Hence, such scale-specific spatial structures can consider its inertial space or volume. That could also imagine as if a sum of all individual scales of micro volumes or spaces for all its constituent-PSs which are also discrete ingredients of the BBCOU or macro-most scale.

Then, conceptually, any such inertial space, in current expanding phase of BBCOU, would be also quantized as like as quantizations in other CIPs like $\Delta r, \Delta m, \Delta \lambda$ & Δv along with corresponding scales of PSs.

As a result, such quantized inertial space or volume, say Δs, can also assume scale-specific. Therefore, exchange of any spatial co-ordinates in between two corresponding volumes or spaces of two PSs also need to refer like any other intrinsic internal quantized parameter like CIPs of those PSs.

Because, it is impossible to exchange with any space or volume without any such PSs or conceptually something like an 'absolute void' [1] that might configure by all zero and/or infinity magnitudes of co-ordinates. As a result, such an absolute-void never could be exchanged, in current expanding phase of

BBCOU, through any quantized-signals (also any PSs) with scale-specific quantized magnitudes of CIPs. Because, all such quantized-signals need to have non-zero & non-infinity quantized magnitudes of CIPs as those could ably exchange with any similar type of quantized magnitudes.

2.5.1. Common Realizations about Inertial-Space

Configuration of above inertial space, that seems to associate with every scale of quantized PSs in current expansions of BBCOU might have the following common properties:

i) inertial 'space' can become only exchangeable to quantum-real observers like us if and only if it is entangled to any Particle or System-of-particles as its curved, quantized and non-void-volume.

That scale-specific quantized inertial space is intrinsically associated with every corresponding scale of PSs only, and without any scale of PSs there no such space can be detected through exchange of any quantized-signal. Therefore, beside the CIPs like Δm, $\Delta \lambda$, Δv & Δr in all scales of PSs there would the inertial space or Δs that can be considered as another intrinsic CIP in all those same PSs. Any such Δs may define in three conventional, "spatial dimensions" or co-ordinates of space e.g. $\Delta x, \Delta y$ & Δz. Hence, the Δs as CIP for quantized volumes could also be imagined as a three quantized radii-CIPs for three Δr equivalent to $\Delta s = (\Delta x, \Delta y$ & $\Delta z)$ for convenience in every scale of PSs in expanding phase of BBCOU.

However, due to the scale-specific intrinsic magnitudes of Δr it appears that, there are all intrinsic discrete volumes for space $\Delta s(\Delta x, \Delta y, \Delta z)$, that envelopes corresponding PSs in every scale. Hence, such an 'envelope' of inertial space also as volume of every PSs will be obviously scale-specifically 'curved' in its geometric configurations.

---

1. with magnitudes $\Delta m = 0$, $\Delta s = 0$ and so on.

Since, values of Δr or all three axes $(\Delta x, \Delta y \,\&\, \Delta z)$ are quantized, such a curved space Δs or volume would also be naturally quantized.

In addition, since the magnitudes of $(\Delta x, \Delta y, \Delta z)$ as well as Δr are emerging as quantized, each of those quantized CIPs would have always any non-zero and non-infinity magnitudes. Then, there will be no meaning of space as Δs without referring any IFRs that does not involve with any PSs and also not having the non-zero and non-infinity magnitudes for all $(\Delta x, \Delta y \,\&\, \Delta z), \Delta r, \Delta m, \Delta v \,\&\, \Delta \lambda$ as specific CIPs. Hence, there would have no practically communicable absolute 'void' for any quantized observers like us anywhere within the expanding BBCOU through exchange of any such *real* sense of signals with same non-zero and non-infinity magnitudes.

ii) Inertial 'space' with of Left-Handed Rotations

It is now also realized through the current observations in both astronomy as well as particle physics that every micro to macro scales of PSs in expanding universe have intrinsic axial rotation or LAR [6] say in clockwise direction. Therefore, the respective scale-specific intrinsic magnitudes of every quantized and curved inertial-space Δs, which is intrinsically associated with all scales of PSs, would have also the left-handed rotation or in clockwise direction. Therefore, such quantized & non-void space Δs would have left-handed or clockwise intrinsic direction.

2.5.2. Defining Quantized Inertial Space

Then, for convenience, the scale-specific non-void intrinsic (i.e. also as an observer-independent) quantized magnitude of inertial volume of space for all PSs, irrespective of scales from micro to macro as $\Delta s(\Delta x, \Delta y, \Delta z)$ due to corresponding magnitudes of Δr, if consider as perfect spheres, would be

$$\Delta s = (3\pi \cdot \Delta r^3)/4. \qquad (1.17)$$

Not only that, the same Δs, also as an observer-independent CIP with scale-specific-intrinsic-quantization in magnitudes for all scales of PSs, would be another SSUC. The Eq. (1.17) has also defined, that Δs with all scale-specific intrinsic quantized magnitudes. The same Δs would be non-void, curved, left-handedly rotating non-zero & non-infinity values and pre-occupied by the PSs. Therefore, the same Δs, as one CIP of PSs in Eqs. (1.1) & (1.17), can postulate to have:

(i) scale-specific intrinsic quantized magnitudes where each of those quantized magnitudes as respective SSUCs;

(ii) intrinsic increments in those scale-specific quantized magnitudes are onwards the LE-C along with Δm of currently expanding phase of BBCOU;

(iii) possess an intrinsic LAR due to entanglement with the direction of axial rotation of Δm involves with every same PSs; and

(iv) unit in cm^3 due to unit of the Δr in cm.

2.6. Inertial Anti-Space

In Eq. (1.17), the Δs shows a two-fold of directions. The Δs is not only incrementing scale-specific quantized values onward the direction of LE-C but also having the intrinsic LAR around the axes of PSs or clockwise direction. One such intrinsic left-handedness of LAR needs to have a 'mirror-image' for symmetry in each of those PSs. Then the LAR or clockwise axial rotating space Δs as intrinsic volume of every PSs must have one simultaneous right-handed mirror-image axial rotation or RAR (i.e. with a conceptual anti-clockwise direction) can be termed as 'anti-space', say $Δs_u$. Conceptually, such a mirror-imaged $Δs_u$ can be also defined by the three mirror-imaged right-handed co-ordinates (similar to the negative axes) in a cartesian co-ordinate system for every such $Δs(Δx, Δy, Δz)$.

Hence, there will be the $\Delta s_u(-\Delta x, -\Delta y, -\Delta z)$ or for such anti-space as mirror-images of the Δs say $\Delta s_u(\Delta x_u, \Delta y_u, \Delta z_u)$ in same Eq. (1.17). Therefore, every scale of familiar clockwise rotating LAR for Δs (that has preoccupied by the PSs onward LE-C in current expansion of BBCOU) would possess the simultaneous mirror-imaged anti-clockwise or right-handed axial-rotation or RAR for Δs_u. Then, two of these LAR & RAR would conceptually have simultaneous co-existence in every scale of PSs within BBCOU as mutual mirror-images to each other. If the observers (like us) 'attached' with the LAR direction of the observations onward LE-C, then it would be practically impossible for such a left-handed observer to see that mirror-imaged-RAR in its anti-clockwise opposite side.

As in Eq. (1.17), the space $\Delta s(\Delta x, \Delta y, \Delta z)$ has considered as spherical volume. Therefore, its simultaneous mirror-imaged anti-space $\Delta s_u(\Delta x_u, \Delta y_u, \Delta z_u)$ can also be considered as spherical

$$\Delta s_u = (3\pi \cdot \Delta r_u^3)/4, \qquad (1.18)$$

where, Δr_u is the simultaneous mirror-imaged radius of the Δr in same PSs irrespective of scales. Since, the Δs in Eq. (1.17) has unit in cm^3, obviously the Δs_u will also have unit in cm^3. As a consequence, the Δs_u will also be another CIP and having the SSUCs in scale-specific discrete magnitudes due to the same for Δs in every scale of PSs. Then, the magnitudes of Δs and Δs_u, being mutual mirror-images to each other, would have inverse relationships.

The Δs has scale-specific increments onward LE-C in current expanding phase of BBCOU. Then its simultaneous mirror-imaged Δs_u would have the decrements in scale-specific quantized intrinsic (with similar kind of observer-independence) quantized magnitudes onward LE-C of current expanding phase of BBCOU alike other two CIPs like $\Delta \lambda$ & Δv.

Subsequently, both of those CIPs e.g. $\Delta \lambda$ & Δv, due to that simultaneous RAR or anti-clockwise axial rotation in Δs_u for every scale of PSs, seem to have links to the same RAR alike the links of the two CIPs e.g. Δr & Δm to the LAR of Δs in the same PSs.

The $\Delta s_u(\Delta x_u, \Delta y_u, \Delta z_u)$ in Eq. (1.18), being a mirror-image of $\Delta s(\Delta x, \Delta y, \Delta z)$ in Eq. (1.17), will have also the scale-specific intrinsic magnitudes of quantized, non-void, non-zero & non-infinity, curved, RAR (or anti-clockwise rotational) that also pre-occupied by any (anti-particles of) PSs in the BBCOU. Therefore, the same Δs_u, as one CIP of PSs in Eqs. (1.1) & (1.18), can postulate to have:

(i) scale-specific intrinsic quantized magnitudes where each of the quantized magnitudes are corresponding SSUCs as non-void, curved, non-zero & finite valued, and pre-occupied by any (anti-particle of) PSs;

(ii) intrinsic decrements from micro to macro scales or onward LE-E in current expanding phase of BBCOU i.e. alternately an intrinsic increment from macro to micro scales like CIPs Δv & Δλ of same PSs;

(iii) intrinsic entanglement with an intrinsic right-handed directional axial rotation or RAR or simply right-handed direction with the same PSs (as the CIP has increments in all SSUCs discrete magnitudes onward RE-C); and

(iv) unit in cm^3 due to the unit of Δs in cm^3.

2.7. Inertial Anti-Radius

As the Δr_u in Eq. (1.18) is appearing mirror-image of Δr in Eq. (1.17) since Δs_u is the mirror-image of Δs respectively in same equations irrespective of the scales of PSs in expanding phase of BBCOU. Consequently, in contrary to Δr, the Δr_u is also possessing the similar scale-specific intrinsic quantization in magnitudes. Since, Δr has intrinsic LAR directions, it would attribute similar LAR directions in all corresponding spatial axes of spherical space $\Delta s(\Delta x, \Delta y, \Delta z)$ which has described in Eq. (1.17). Its mirror-image Δr_u having intrinsic RAR directions would have corresponding anti-spatial axes of spherical anti-space $\Delta s_u(\Delta x_u, \Delta y_u, \Delta z_u)$ in Eq. (1.18). In Eq. (1.18), that mirror-imaged Δr_u would have decrements onward direction of LE-C in all scale-specific quantized magnitudes from macro to micro scales of PSs in same expanding phase of BBCOU as like as the $\Delta \lambda$ in Eq. (1.1), Δv, and Δs_u in Eq. (1.18). Therefore, the Δr_u, as mirror-image of the CIP Δr in every scale of PSs, would also be another CIP. Since, the Δr possesses unit in cm, its mirror-image Δr_u will also have unit in cm. Not only that, all the scale-specific quantized magnitudes for Δr_u as mirror-images of Δr will also be the SSUCs in all PSs.

The de Broglie's wavelength $\Delta \lambda$ in Eq. (1.1) possesses not only the unit in cm as like as Δr_u in Eq. (1.18) but also has similar LE-C decrements in scale-specific magnitudes (i.e. as observer independent magnitudes) as a CIP with SSUCs in current expansion phase of BBCOU. But it is not evident whether $\Delta \lambda$ possesses any intrinsic RAR direction in axial rotations in every scale of PSs. But the $\Delta \lambda$ in Eq. (1.1) has inverse relationship with the Δm, where that Δm has intrinsic LAR direction with left-handed axial rotation of PSs. Therefore, it seems that LE-C decrementing CIP like $\Delta \lambda$ in Eq. (1.1) would link to RAR direction with right-handed axial rotation of all corresponding simultaneous spherical anti-space $\Delta s_u(\Delta x_u, \Delta y_u, \Delta z_u)$ in Eq. (1.18). As a result, it can be postulated that in Eq. (1.1) for the CIP

$$\Delta \lambda \equiv \Delta r_u \qquad (1.19)$$

in Eq. (1.18) for all scales of PSs onward LE-C in currently expanding BBCOU. Consequently, the de Broglie's wavelength $\Delta \lambda$ can postulate equivalent to the conceptual anti-radius Δr_u in for every scale of PSs and the symbol $\Delta \lambda$ can use in places of Δr_u.

From the postulates in Eq. (1.19), de Broglie Wavelength $\Delta\lambda$ as inertial anti-radius Δr_u in above Sub-section 2.2 as another CIP of all PSs in Eq. (1.1), will postulate to have:

(i) intrinsic scale-specific or SSUCs in quantized magnitudes;

(ii) in expanding phase of BBCOU, it has all intrinsic decrements in all those same SSUCs in quantized magnitudes but opposite to LE-C of BBCOU, i.e. it would have increments onward simultaneous mirror-imaged RE-C of the same BBCOU in its reverse direction of 'collapsing phase';

(iii) intrinsic entanglement with an intrinsic right-handed direction of axial rotation or RAR or simple right-handed direction with same PSs (as the CIP has increments in all SSUCs discrete magnitudes onward RE-C); and

(iv) unit in *cm*.

2.8. Inertial Time

From the basic conventions of both SRT and QM, it is now also another well-accepted convention that, like the inseparable inertial mass-energies and inertial-motions, the inertial-space also cannot be separated from inertial-time or vice versa in any inertial frame of reference. However, we already have stated such inertial-space as $\Delta s(\Delta x, \Delta y, \Delta z)$ in Eq. (1.17) and mirror-image of same as the inertial-anti-space $\Delta s_u(\Delta x_u, \Delta y_u, \Delta z_u)$ in Eq. (1.18). Hence, those two have appeared as two intrinsic CIPs and both are associated with every scale of PSs in currently expanding phase of the BBCOU. Hence, in that context, the inertial-space as Δs, the perception of inertial-time in all those same scales of PSs could have unavoidable newer definition as well. Then, inertial-time can also be defined in every such scale of PSs in expanding universe below:

2.8.1. Realizations about Inertial Time

The magnitude of Δs in Eq. (1.17) primarily has postulated as any scale-specific intrinsic non-zero & non-infinity quantized magnitudes of inertial-volumes as inertial-space with all scale-specific inertial-PSs and the entire volume of the BBCOU has assumed as the integer sum of those all constituent inertial-volumes as inertial-space. Again, similar to the inertial-space Δs in Eq. (1.17), the concept of inertial-time would have also no practical meaning without referring any such inertial-space associating with any of those scales of inertial-PSs in BBCOU. Then, earlier concept of inertial-time, alike the inertial-space, referring any absolute 'void', would be practically impossible. It will carry no meaning due to lacking of scopes for any quantized-type signal exchanging with the same absolute void, even if there. Subsequently, any of our practical realization of inertial-time and inertial-space, must be always in reference to the scale-specific inertial-PSs in this expanding phase of BBCOU. Therefore, in all those same inertial-PSs, there would be a simultaneous realization of inertial-time as well. Then, the Δs in Eq. (1.17), which has scale-specific quantized values, would have also the simultaneous association with one inertial-time and which can measure in terms of such quantized inertial-space as the conventional time now measures through amount of spatial displacements of hands in conventional clocks. Consequently, the inertial-time, say Δt, alike inertial-space in all inertial-PSs would have also the scale-specific quantized magnitudes.

On other hand, as a general convention, the measurement of inertial-time in common sense is primarily nothing but a process of choosing any unit for the 'duration' which is ultimately any equivalent unit of 'distance' that travels 'anything' through the space. For example, it could be the travelling of 'specific spatial distance in space' by a hand in a clock equivalent to the 'specific duration in time' for same hand in that clock. It could be also the 'axial-distance' that covers through one 'axial-rotational distance in space' of earth equivalent to 'axial-rotational-duration in time' for a day. The same thing is conceptually happening due to axial-rotations of different scales of molecules or atoms or electrons or astronomical-objects. Therefore, the sense of 'durations' for the corresponding unit-time in every scale of PSs can conceptually be defined through all that corresponding sense of 'distances' for the corresponding unit-space through scale-specific movements in space of PSs. Subsequently, that unit-duration of time in all those specific-scales of quantized PSs can also be defined through the corresponding unit-distance of space in all those same specific-scales of PSs.

However, such a unit-distance of space in every specific-scales of PSs having the intrinsic quantized magnitude like $\Delta s(\Delta x, \Delta y, \Delta z)$ in Eq. (1.17). As a result, the subsequent unit-duration of time also requires to have in same specific-scales of PSs any similar intrinsic quantized magnitude like Δt. Moreover, such Δt also requires to define in the units of spatial 'distance'.

In the Eq. (1.17), it has described that the Δs has an intrinsic LAR direction of rotations around the respective axis of a PS. Then, we can define one such axial LAR of Δs in any PS as a unit 'duration' as equivalent as it's one 'equatorial distance', or, 'equatorial length' for every left-handed rotation of the same PS around its respective LAR axis. For conveniences, we can also imagine such 'equatorial distance', or 'equatorial length', as the spatial unit for time as like as it is in a clock (for twelve or twenty-four hours in seconds). However, as we have scale-specific-intrinsic-quantization in magnitude of the Δs in Eq. (1.17) in every scale of PSs, then obviously there will be also the scale-specific-intrinsic-quantization in magnitude of the equatorial distance or length for each of those same specific scale of PSs. Then, the Δt also has similar scale-specific intrinsic quantization in magnitude for equatorial distance or length as equivalent as its 'duration' of time. However, through our conventional measurement practice of time, we can define that Δt for convenience in every scale of PSs as an equatorial distance or length per LAR of its Δs in Eq. (1.17) as

$$\Delta t = 1\, unit = 360°, \qquad (1.20)$$

where we have the conventional unit of time in degrees or minutes or seconds. But it is also apparent in Eq. (1.20) that unit of time is in angular distances along the equator of the Δs in PSs in degrees. As a result, there will be no scale-specific sense of intrinsic quantization in magnitudes for the Δt in Eq. (1.20) in all scales of PSs where magnitudes of every unit of time ($\Delta t = 360°$) is appeared universally identical in every scale of PSs. As the angular distance of equator for Δs in every scale of PSs are universally identical, *i.e.* 360°. That is, the Δt in Eq. (1.20) becomes as one of the scale specific CIPs in all scales of PSs in universe with no scale-specific intrinsic quantization in magnitudes.

But it is also true that the Δs in Eq. (1.17) has scale-specific intrinsic quantization in magnitude. Then, obviously there would have the scale-specific intrinsic quantization in magnitudes for same 'equatorial distance' or 'length' for all PSs in Eq. (1.20). Then, the Δt as a scale-specific intrinsic quantization in magnitude in Eq. (1.20) as an 'equatorial distance' or 'length' per LAR of its Δs in Eq. (1.17) can be also defined as

$$\Delta t = 1\, unit = 2\pi \cdot \Delta r \qquad (1.21)$$

where Δt appears with a non-conventional unit in *cm* due to the unit of radius Δr. However, the Δt in Eqs. (1.20) & (1.21) ultimately would have both conventional and non-conventional units whenever they may be used for specific purposes.

2.8.2. Defining the Inertial Time

The Δt in Eqs. (1.20) & (1.21) also has a LAR or clockwise direction for axial rotation, due to the same for the Δs in Eq. (1.17). On the other hand, as a usual convention, the same Δt is 'moving', *i.e.* 'flowing' clockwise from 'past' → 'future'. We can consider such a convention of clockwise flow of time Δt is parallel to the convention of its intrinsic LAR or clockwise axial rotations. That is, finally, the Δt in Eqs. (1.20) & (1.21) has defined to possess the non-conventional spatial unit in *cm* with intrinsic scale-specific quantization in magnitude beside its conventional unit in *sec* with clockwise (*i.e.* left-handed) flow from past to future in all scales of PSs. Also, due to the scale-specific increment of *sec* in magnitudes of the CIP like Δr from micro to macro scales parallel to increments in LE-C in expansion phase of BBCOU, the Δt would be as one CIP in Eq. (1.21) will have also the LE-C increments in its scale-specific quantized magnitudes from micro to macro scales in same expansion phase of BBCOU. Due to the same scale-specific SSUCs in quantized magnitudes of same CIP like Δr in Eq. (1.21), the Δt as the CIP will also have the all scale-specific SSUCs in intrinsic (i.e. also observer-independent) quantized magnitudes in unit of *cm*. Therefore, the same Δt, as one CIP of PSs in Eqs. (1.1) & (1.21), can postulate to have:

(i) scale-specific intrinsic quantized magnitudes where each of quantized magnitudes are as corresponding SSUCs;

(ii) intrinsic increments in those scale-specific quantized magnitudes parallel to the increments with LE-C in current expansion phase of the BBCOU;

(iii) possess an intrinsic LAR direction with the intrinsic LAR direction of axial rotation of PSs; and

(iv) unit in *cm* due to the unit of Δr in *cm* in Eq. (1.21).

2.9. Inertial Anti-Time

Again, logically, since there is simultaneous mirror-imaged Δs_u in Eq. (1.18) for Δs in Eq. (1.17) as well as (Δr & Δr_u) in those same equations respectively for all scales of PSs, there must be a simultaneous mirror-imaged anti-time say Δt_u for the Δt in Eqs. (1.21) with all those same scales of PSs. That would be due to the mirror-imaged RAR direction of axial-rotations of such Δs_u in Eq. (1.18) along its respective equator. Then, for the mirror-image of Δt in Eq. (1.21), we can define further the conventional unit for an 'equatorial distance' or 'length' per RAR rotation of Δs_u in Eq. (1.18):

$$\Delta t_u = 1\ unit = 360°, \qquad (1.22)$$

where the conventional unit of Δt_u would be in degrees or minutes or seconds. But the same Δt_u in Eq. (1.22) has no scale-specific intrinsic quantization in magnitudes. But still there are Δs_u and Δr_u in Eq. (1.18) which have the intrinsic scale-specific quantization in magnitudes. Hence, there would be also a similar scale-specific intrinsic quantization in magnitude for such 'equatorial distance' or 'length' of the Δs_u. Then, for such intrinsic scale-specific quantization in magnitudes of 'equatorial distance' or 'length' in all scales of PSs, there will be an equatorial distances or lengths for the RAR direction of Δs_u in Eq. (1.18)

$$\Delta t_u = 1\ unit = 2\pi \cdot \Delta r_u \qquad (1.23)$$

and the Δt_u in Eq. (1.23) as like as Δt in Eq. (1.21) would have the non-conventional unit in *cm*, due to the unit of anti-radius Δr_u.

As we have the flow of Δt in Eq. (1.21) is intrinsically clockwise i.e. onward 'past' → 'future' along the intrinsic clockwise or LAR direction of axial rotations for Δs in Eq. (1.17); then obviously its simultaneous mirror-imaged Δt_u in Eq. (1.23) would have also the mirror-imaged or anti-clockwise directional flow onward 'past' ← 'future' along the intrinsic anti-clockwise or RAR direction of axial rotations for Δs_u in Eq. (1.18).

This simultaneous opposite directional flow of Δt_u can be experienced essentially in every scale of PSs in BBCOU. Practically, during transformation of any 'physical event' in between the frames of a sender and a receiver. For example, A and B are two frames for such sender and receiver respectively; and an event is say transforming from A to B through one quantized signal-particle, the Q, with another specific scale. Suppose the Q has started from A at a moment of time say Δt_A in the 'past' and has reached to B at another moment of time say Δt_B in the 'future'. Then B, by receiving the Q at Δt_B, became aware that the Q has traveled through space for the duration of time say $\Delta t = (t_B - t_A)$ from 'past' → 'future'. But, simultaneously, through receiving the same Q, that receiver B is also realizing the 'past' era (t_A) of time from his era of existing time in future (t_B), or as if he is feeling to reach simultaneously at that past era by moving from the future to past or ('past ← future').

Hence, in every transformation of physical events through any exchange of Q, there would always such type of a simultaneous flow of anti-time $\Delta t_u = (t_A - t_B)$ be perceived in reverse direction of above $\Delta t = (t_B - t_A)$. The flows of two such simultaneous Δt & Δt_u would be always realized during every astronomical observation through receiving the photon-signals (or Q) from distant objects. More and more one can receive that Q from the distant astronomical objects, *i.e.* also from the more 'distances' in space, one can simultaneously reach to the more and more to distant 'pasts' in time or anti-time from future. This can be imagined as if both (Δt & Δt_u) are flowing simultaneously in reverse directions of each other within same photon-signal or Q all along its history of travel in that specific scale of particle simultaneously. Therefore, any such history of travel for any scale of particle can also be the span of existence of the same. So, in every such span of any PSs there would be always that kind of simultaneous flow of Δt & Δt_u.

Therefore, finally the Δt_u in Eq. (1.23) not only flowing simultaneously in reverse direction from 'future' → 'past' in every scale of particles with Δt in Eq. (1.21) but is equally right-handedly incremented in its scale-specific quantized magnitudes in expanding phase of BBCOU as another CIP with all SSUCs in quantized magnitudes of all PSs. Therefore, the same Δt_u, as one CIP of PSs in Eqs. (1.1) & (1.17), can postulate to have:

(i) the scale-specific intrinsic (i.e. observer-independent) quantized magnitudes where each of those similar magnitudes are as corresponding SSUCs;

(ii) intrinsic decrement in those scale-specific quantized magnitudes parallel to the LE-C of BBCOU;

(iii) intrinsic entanglement with a right-handed directional axial rotation or RAR or simply right-handed directional with same PSs (as the CIP would have increments in all SSUCs discrete magnitudes onward RE-C); and

(iv) unit in cm due to the unit of Δr_u in cm.

3. Common Definition of any Particle or System

In current expanding phase of BBCOU, where an observer like us along with all its possible instrumentations (made out of any scales of PSs), onward LE-C direction along the micro to macro scales of PSs, are possessing CIPs $\Delta m, \Delta v, \Delta s(\Delta x, \Delta y, \Delta z), \Delta s_u(\Delta x_u, \Delta y_u, \Delta z_u), \Delta t, \Delta t_u, \Delta r$ and ($\Delta r_u \equiv \Delta \lambda$) those have postulated in corresponding Eqs. (1.1) – (1.23). Out of all these 12-CIPs, the Δr & ($\Delta r_u \equiv \Delta \lambda$) are as corresponding radius and anti-radius seem more elemental to define all other CIPs like inertial mass-energy in Eq. (1.1), inertial-motion in relevance of decrementing [1] values of Δr in SRT, space in Eq. (1.17), anti-space in Eq. (1.18), time in Eq. (1.21) & anti-time in Eq. (1.23). Hence, instead of total 12 CIPs there can be ultimately 10 CIPs like $\Delta m, \Delta v, \Delta s(\Delta x, \Delta y, \Delta z), \Delta s_u(\Delta x_u, \Delta y_u, \Delta z_u), \Delta t$ & Δt_u to describe every scale of PSs including BBCOU. Hence, there can be total 10 $[\Delta m, \Delta v, \Delta s(\Delta x, \Delta y, \Delta z), \Delta s_u(\Delta x_u, \Delta y_u, \Delta z_u), \Delta t$ & $\Delta t_u]$ or 6 $[\Delta m, \Delta v, \Delta s, \Delta s_u, \Delta t$ & $\Delta t_u]$ numbers of CIPs.

Moreover, out of all those total 10-CIPs in every scale of PSs there would be total intrinsic left-handed 5-CIPs $[\Delta m, \Delta s(\Delta x, \Delta y, \Delta z)$ & $\Delta t]$ in one set and also total intrinsic right-handed 5-CIPs $[\Delta v, \Delta s_u(\Delta x_u, \Delta y_u, \Delta z_u), \Delta t_u]$ in another set. Therefore, onward every increment and the alternate decrements in magnitudes of LE-C of BBCOU, those 5-CIPs $[\Delta m, \Delta s(\Delta x, \Delta y, \Delta z)$ & $\Delta t]$ would have similar increments and alternate decrements in magnitudes would be the intrinsic left-handed in terms of both magnitudes and directions. But conversely, rest of the 5-CIPs $[\Delta v, \Delta s_u(\Delta x_u, \Delta y_u, \Delta z_u)$ & $\Delta t_u]$, would have all simultaneous decrements and alternate increments in magnitudes as the right-handed in terms of both magnitudes and directions.

Conversely, along the simultaneous direction of RE-C observation for same BBCOU there would have also all those same 10-CIPs ($\Delta m, \Delta v, \Delta x, \Delta y, \Delta z, \Delta x_u, \Delta y_u, z_u, \Delta t, \Delta t_u$) or 6-CIPs ($\Delta m, \Delta v, \Delta s, \Delta s_u, \Delta t, \Delta t_u$) but in reverse manner.

Subsequently, onward simultaneous decrements and alternate increments in magnitudes onward RE-C of BBCOU, each of

---

1. In SRT, measuring rod is dilated or decremented onward its incrementing motions. If such rod is equivalent to scale-specific quantized magnitudes for Δr in PSs (that decrements onward RE-C), then there would be the increments in corresponding quantized magnitudes for Δv onward same RE-C. This can show inverse values of Δr respect to all such Δv in all PSs (it would describe in detail in Sub-Section-1.6 of Chapter-2).

those same PSs irrespective of scales would appear in reverse type. The intrinsic right-handed 5-CIPs (those are headed towards decrementing RE-C) would have now decrementing magnitudes and rest of intrinsic left-handed 5-CIPs (those are incremented towards LE-C) would have now all incrementing values onward same direction of RE-C.

Therefore, the whole BBCOU, inclusive of all those same scales of ingredient-PSs, if possible, to observe from its two simultaneous LE-C and RE-C directions by two conceptually opposite directional *observers* (who will also be the obvious ingredients of those two), would appear simultaneously opposite to each other. Towards the LE-C-observation, present expanding phase of BBCOU would appear to increment in its all its 5-LAR CIPs [$\Delta m, \Delta s(\Delta x, \Delta y, \Delta z)$ & Δt] but decrementing in all 5-RAR CIPs [$\Delta v, \Delta s_u(\Delta x_u, \Delta y_u, \Delta z_u)$ & Δt_u]. Simultaneously, toward the RE-C-observation, the same BBCOU in present expanding phase would appear with decrementing in all its 5-RAR CIPs [$\Delta v, \Delta s_u(\Delta x_u, \Delta y_u, \Delta z_u)$ & Δt_u] but incrementing in all 5-LAR CIPs [$\Delta m, \Delta s(\Delta x, \Delta y, \Delta z)$ & Δt].

Furthermore, during alternate collapsing phase, the same BBCOU will appear as decrementing in all 5-LAR CIPs [$\Delta m, \Delta s(\Delta x, \Delta y, \Delta z)$ & Δt] but incrementing in all 5-RAR CIPs [$\Delta v, \Delta s_u(\Delta x_u, \Delta y_u, \Delta z_u)$ & Δt_u]. Toward the LE-C-observation, but simultaneously incrementing in all 5-RAR CIPs [$\Delta v, \Delta s_u(\Delta x_u, \Delta y_u, \Delta z_u)$ & Δt_u] and decrementing in all 5-LAR CIPs [$\Delta m, \Delta s(\Delta x, \Delta y, \Delta z)$ & Δt].

That is, simultaneously, two opposite observers 'by standing alone' in two corresponding left-handed and right-handed sides of the same BBCOU along with all its constituent parts or PSs would appear LE-C and RE-C directional, and all those same scales of PSs including the macro-most scale BBCOU can describe correspondingly as

$$\xrightarrow{LE-C} \quad \& \quad \xleftarrow{RE-C} \tag{1.24}$$

and clearly each of those incrementing or decrementing PSs onward LE-C would be as

$$[\Delta v, \Delta s_u(\Delta x_u, \Delta y_u, \Delta z_u), \Delta t_u] \,\&\, [\Delta m, \Delta s(\Delta x, \Delta y, \Delta z), \Delta t] \approx \xrightarrow{LE-C} \tag{1.25}$$

and simultaneously, same decrementing or incrementing PSs onward RE-C would be as

$$\xleftarrow{RE-C} \approx [\Delta v, \Delta s_u(\Delta x_u, \Delta y_u, \Delta z_u), \Delta t] \,\&\, [\Delta m, \Delta s(\Delta x, \Delta y, \Delta z), \Delta t_u] \tag{1.26}$$

where symbols "$\rightarrow$" & "$\leftarrow$" are showing for the corresponding intrinsic left-handed and right-handed simultaneous directions of same PSs. The Eqs. (1.25) & (1.26), for convenience, can further rewrite jointly within a single bracket to define simultaneous LE-C & RE-C in every ingredient-PS including the BBCOU as

$$\left\{ \begin{array}{l} [\Delta v, \Delta s_u(\Delta x_u, \Delta y_u, \Delta z_u), \Delta t_u] \,\&\, [\Delta m, \Delta s(\Delta x, \Delta y, \Delta z), \Delta t] \approx \xrightarrow{LE-C} \\ \xleftarrow{RE-C} \approx \Delta v, \Delta s_u(\Delta x_u, \Delta y_u, \Delta z_u), \Delta t] \,\&\, [\Delta m, \Delta s(\Delta x, \Delta y, \Delta z), \Delta t_u] \end{array} \right\}. \tag{1.27}$$

The Eq. (1.27) will be ultimately an observer-independent common definition for every scale of ingredient-PSs of BBCOU including itself as macro-most scale due to all intrinsic SSUCs in discrete or quantized magnitudes involving with all associated 5+5 CIPs those have webbed each of those PSs.

The Eq. (1.1) suggests that, perceivable everything in current expanding phase of universe or BBCOU are intrinsically quantized in all scale-specific manner. It also assumes that, without quantization as well as any scales of PSs those cannot exist within that BBCOU unlike earlier conventions of any 'physical-bodies' in Classical Mechanics. All those same PSs, conceptually, are also the unit discrete constituent parts or ingredient-PSs of that whole BBCOU. Moreover, such fundamental intrinsic property of quantization in each ingredient-PSs are due to the presence of SSUCs in discrete magnitudes of all those 10-CIPs in Eqs. (1.25) & (1.26).

However, all those scales of ingredient-PSs (as any quantum-real-observables) in Eq. (1.24) by following Eqs. (1.25) & (1.26) are only communicable or exchangeable to observers (equally quantized) like us through any interactions of the force carrying *real*-particles or quantized *real*-signals. This seems a natural quantum-real-limitations for any quantum-real-observers like us (and also that would be same if the observers stand in right-handed 'side' of same BBCOU) with all non-zero, non-infinity and non-overlapping magnitudes of all those 10-CIPs. If there anything or any scales of PSs, which are beyond the exchange or communicable range of our quantum real-signals say including gravitation, that never can be perceived by us through any of our quantum-real capacity of observations. Then, all our perception about BBCOU, defined by Eqs. (1.24) - (1.26), would be based on any exchange of such quantum-real-signal particles and would ultimately create one *quantum-real-horizon* for such observations in physical nature. We can perceive anything within such quantum-real-horizon would be any entity of *quantum-reality* involving with same BBCOU in physical nature. As a result, any such quantum-real signal must have any non-zero, non-infinity and non-overlapping quantized or discrete magnitudes as SSUCs for all associating 10-CIPs. Hence, such a non-zero, non-infinity and non-overlapping quantum-real-signal cannot be exchanged or communicated with anything which is non-quantized in type i.e. having zero & infinity overlapping magnitudes. Then, all *observers* like us (and one in that opposite RE-C side of BBCOU), *observances* and communicable *signals*, which are possessing any quantized magnitudes for those 5+5 total 10-CIPs in Eqs. (1.24) - (1.26), or more precisely are associated with any non-zero and non-infinite magnitudes for all those 5+5 total 10-CIPs, can only be perceivable to us, i.e. if and only if those are only quantum-real.

Furthermore, as per present understanding in physics, there can also be possible another form or state in physical nature, which can exist beyond our all such exchangeable or communicable reach of the quantum-reality of signals (even beyond the effects of gravitation) in physical nature. That would be the *virtual* or *virtuality* involving with some form of energies in *Non-quantized state* with the zero and/or infinity values for all those same 10-CIPs. If so, such Non-quantized state or beyond exchangeable zero valued space cannot be exchanged through any quantum-real signals with all non-zero, non-infinite and non-overlapping magnitudes of conventional signals. Consequently, such virtual energies in Non-quantum state must be ever Non-quantized virtual to us even beyond the interaction/exchange limits of quantum-real gravitation. Therefore, such Non-quantum-virtuality of energies can describe only with all same 5+5 total 10-CIPs but with all non-quantized zero, infinite and overlapping magnitudes in same physical nature, and that would not be any quantum-real ingredient parts of the similar quantum-real BBCOU in physical nature.

Therefore, all our perceivable signal exchange range in physical nature is limited and seems to be limited up to the direct and indirect effects of gravitation. Those quantum-real everything, in any form of PSs, must have the responses to the gravitational effects if there only any non-zero, non-infinite and non-overlapping quantized scale-specific magnitudes for all 10-CIPs and also those are definable through the Eqs. (1.24) - (1.26). Subsequently, all our finest perceptions could be limited only up to that quantum-real-horizon that includes all PSs from micro-most to macro-most scales.

Therefore, any such quantum reality from micro-most to macro-most scales in physical nature must be ultimately a mesh of $5[\Delta m, \Delta s(\Delta x, \Delta y, \Delta z), \Delta t] + 5[\Delta v, \Delta s_u(\Delta x_u, \Delta y_u, \Delta z_u), \Delta t_u]$

two sets of 10 number CIPs; and each of those 10-CIPs are "unfolded" or tangible; and entangled to all other ingredient-PSs of BBCOU.

Summary

The earlier universally common or general conventions of 'material-body' in Classical Mechanics have now changed into all precise and diverse scales of quantized particles or systems-of-particles with appearance of Quantum Mechanics.

The Chapter-1, on the basis of above three Sections, has depicted again some specific common or general universal conventions among all those diverse scales of particles or systems-of-particles. As a consequence, all those common or general conventions can be considered as the new foundation, on the basis of new observations in last one hundred years of physics, for any new unified theory of everything in universe.

In above paragraphs, many newer universally common phenomena have revealed out of various mundane observational comprehensions in current physics. Such observational comprehensions have sorted out in Section-1 as total seven 'Common Understandings in Universe' for the everything. That term everything means for the all diverse scales of particles or systems-of-particles but assumed to have all those seven common properties universally. So, finding the universally common properties among all diverse forms of particles or systems-of-particles from micro to macro scales was basic objective of the Section-1 on the basis of current observations.

Another common or general thing that has envisaged in internal configuration of all same micro to macro diverse scales of particles or systems-of-particles is a set of 'common-internal-parameters' (CIPs). Each of those CIPs are as if individual fabric in common internal texture of those particles or systems-of-particles and having all the seven 'Common Understandings'. That comprises the Section-2, and in Section-

3 it creates a new foundation for any new unified definition of all scales of particles or systems-of-particles.

The next Chapter-2 would deduce some universally common inter-relationships among all those CIPs in every scale of particles or systems-of-particles for unification of quantum-real everything.

CHAPTER-2:
UNIVERSAL INVERSE RELATIONS
[Emergence of Seven New Universal Constants]

"To raise new questions, new possibilities, to regard old problems from new angle, requires creative imagination and marks real advance in science." — Albert Einstein

1. Inertial Inverse Relationships:	91
1.1. Mass-Energy and Wavelength	93
1.2. Mass-Energy and Motion	94
1.3. Radius and Anti-Radius	99
1.4. Space and Anti-Space	101
1.5. Time and Anti-Time	103
1.6. Radius and Motion	105
2. A Unified Equation for Inertial Everything	106
3. Unfolded 10-Dimensions of Quantum-Reality	110
4. Magnitudes of Inverse Universal Constants	114
Summary	116

Universal common ideas about 'matter' or imprecise scales of 'material-bodies' in Classical Mechanics have changed to all diverse precise scales of micro to macro particles or systems-of-particles in todays comprehensions of Particle Physics. Moreover, in current astro-physical considerations, such 'matters' in universe have further stretched into the forms of 'visible-matters', 'dark-matters' & 'dark-energies'. Visible-matters are comprising different scales of 'fermions' and 'bosons'. Fermions and bosons have also numerous types of scales, and each of those scales are different from other in terms of physical properties.

In last Chapter, at least in inertial state, all those diverse scales of particles or systems-of-particles in universe are assumed to have some universally common internal properties in terms of ten or twelve common-internal-parameters or CIPs. That list includes all scales including one conceptual 'micro-most' scale of particles and one 'macro-most' scale for all scales of systems-of-particles as conceptual Bigbang-Bigcrunch Cyclic Oscillating Universe or BBCOU. That can also be assumed as a conceptual list for all quantum-real entities in physical nature.

Hence, whether there would have any basic connections among all those ten or twelve CIPs in all such micro to macro diverse scales of particles or systems-of-particles could be also another prospect for any emerging unified mechanics for everything.

The Chapter-2 would hypothesize total seven universally invariant inverse-relations among all those ten or twelve CIPs in all micro to macro diverse scales of particles or systems-of-particles. That would include one similar universally invariant inverse-relationship in existing (de Broglie's) wave-corpuscular duality law in-between two CIPs e.g. de Broglie's wavelengths ($\Delta \lambda$) and inertial mass-energies (Δm) in all same diverse scales of particles or systems-of-particles. All those seven universally invariant inverse-relationships will present a single unified equation for all those diverse scales of particles or systems-of-particles in physical nature.

Section-1 would postulate total seven of those different universally invariant inverse-relationships starting from de Broglie's universal wave-corpuscular inverse-relationship among all those 10 or 12 CIPs as assumed in the last Chapter. There would be also the outcome of corresponding seven new universal constants (UCs) out of all those universal invariant inverse-relationships. Consequently, the Section-2 would deduce one complete new unified equation for all scales of particles or systems-of-particles including BBCOU in inertial sate state with an intrinsic mirror-image duality. The Section-3 would describe all those inversely inter-linked CIPs in very scale of particles or systems-of-particles as corresponding numbers of unfolded or tangible dimensions. Finally, Section-4 would reveal approximate magnitudes for all those seven corresponding UCs.

1. Inertial Inverse Relationships:

In Chapter-1, there has assumed five [$\Delta m(1), \Delta s(3)$ & $\Delta t(1)$] left-handed CIPs and five [$\Delta v(1), \Delta s_u(3), \Delta t_u(1)$] right-handed CIPs have webbed each of particles or systems-of-particles or PSs within the range of all scales from micro-most to macro-most scales. The two other CIPs like left-handed radius $\Delta r(1)$ and right-handed radius as de Broglie's wavelength $\Delta r_u(1) = \Delta \lambda$ (1) have appeared more elementary compare to all those total 5+5 inversely related 10-CIPs. So, each of those diverse scales of particles or systems-of-particles or PSs are considered with 5+5 ten-CIPs as unfolded dimensions. Then, each of those PSs within it has appeared as if a mesh of two such simultaneous left-handed & right-handed sets of mutually mirror-imaged 5+5 total ten CIPs.

Therefore, within all those PSs, all left-handed 5-CIPs would have intrinsic **(i)** left-handed directions due to direction of the LAR and simultaneously all right-handed 5-CIPs would oppositely have intrinsic **(ii)** right-handed directions due to direction of the RAR.

Beside this, all those same quantized PSs, as ingredient of the macro-most scale BBCOU, also have simultaneous **(i)** left-handed directional entropy or LE-C and **(ii)** right-handed directional anti-entropy or RE-C for scale-specific magnitudes.

Moreover, in Eqs. (1.25) & (1.26) the same PSs, within itself there would have those two sets of left-handed and right-handed (5+5) entangled CIPs, and simultaneously in outside those are also the integrated units of macro-most scale for the system-of-particles like BBCOU. As a result, all those same PSs are appearing as two simultaneous left-handed and right-handed directional sets of CIPs through the same Eqs. (1.25) & (1.26).

Furthermore, in expanding phase of BBCOU:

(a) an observer like us who is seeing all its ingredient-PSs onward increments of LE-C direction would observe the progress in scales from micro to macro along with increments in scale-specific quantized values for the 5 left-handed CIPs [$\Delta m(1), \Delta s(3)$ & $\Delta t(1)$] and simultaneous decrements in all scale-specific quantized values for 5 right-handed CIPs [$\Delta v(1), \Delta s_u(3), \Delta t_u(1)$] in Eq. (1.25); and

(b) another observer who is opposite to us is simultaneously observing the same ingredient-PSs from RE-C direction would see the same progress in scales from micro to macro along the decrements in scale-specific values of 5 right-handed CIPs [$\Delta v(1), \Delta s_u(3)$ & $\Delta t_u(1)$] and increments in scale-specific quantized values for 5 left-handed CIPs [$\Delta m(1), \Delta s(3)$ & $\Delta t(1)$] in Eq. (1.26).

In alternate collapsing phase of same BBCOU everything might be occurred in the mirror-images of above expanding phase. In such a mirror-imaged like collapsing phase of the same BBCOU:

(a) observer who is seeing all its ingredient-PSs onward RE-C direction would find a progression in scales from macro to micro and subsequently all those ingredient-PSs in Eq. (1.26) would have increments in scale-specific quantized values of 5 right-handed CIPs [$\Delta v(1), \Delta s_u(3)$ & $\Delta t_u(1)$] but decrements in scale-specific quantized magnitudes of 5 left-handed CIPs [$\Delta m(1), \Delta s(3)$ & $\Delta t(1)$]; and

(b) conversely, another observer (like us from opposite direction of same collapsing BBCOU) would simultaneously observe the same ingredient-PSs onward LE-C direction would see the similar progression in scales also from macro to micro but all those same ingredient-PSs in Eq. (1.25) would have decrements in scale-specific quantized values of 5 left-handed CIPs $[\Delta m(1), \Delta s(3)\ \&\ \Delta t(1)]$ and increments in scale-specific quantized values of 5 right-handed CIPs $[\Delta v(1), \Delta s_u(3)\ \&\ \Delta t_u(1)]$.

However, in both of these LE-C & RE-C directions of observations reveals that there seems to subsist one universally mutual mirror-imaged relationships in between those two sets of 5-left-handed & 5-right-handed sets of 10-CIPs in every scale of ingredient-PSs as integer parts of same BBCOU. An inverse relationship between any two CIPs would also like be one mutual mirror-image relationship. Consequently, both of those two mirror-imaged sets 5-left-handed and 5-right-handed 10-CIPs also have a similar inverse relationship.

Then total seven universally invariant inverse relationships in all those scales of ingredient-PSs, including the macro-most scale BBCOU, can be postulated from the corresponding universal mirror-imaged relationships of all 5+5 CIPs. The sub-sections (1.1) - (1.6) would postulate total six inverse-relationships [1] (in inertial state) out of those two sets of left-handed & right-handed (5+5) CIPs in every scale of ingredient-PSs:

1.1. Mass-Energy and Wavelength

The wave-corpuscular relation of de Broglie in Eq. (1.1), not merely has defined every scale of quantum-real ingredient-PSs as the wave-corpuscular phenomena or WCP but also has revealed a unique universally invariant feature for all same PSs.

1.The seventh inverse-relationship as inertial unified definition would be deduced in the next Section-2.

That is, one universally invariant *inverse relationship* first time ever comprehended of in physics.

The universal inverse relationship between two CIPs having all SSUCs in quantized magnitudes of Δm & $\Delta \lambda$ irrespective of micro to macro scales for WCP or PSs. Hence, Eq. (1.1) can re-write as

$$\Delta m\, \Delta \lambda = h/c = k_1 \qquad (2.1)$$

where k_1 is one universal constant or UC with a universally invariant magnitude for all scales of PSs or WCP in contrary to Δm and $\Delta \lambda$ as two CIPs have all SSUCs in quantized magnitudes.

In Eq. (2.1), the unit for Planck's constant h, if, considers in $gm \cdot cm^2/sec$ instead of erg/sec, and the unit for that constant in SRT for inertial-motion of light as c in cm/sec, then obviously unit for that k_1 as UC will be in $gm \cdot cm$.

As a result, from de Broglie's wave-corpuscular Eq. (1.1), we have a universal inverse relationship in-between two those CIPs Δm & $\Delta \lambda$ in Eq. (2.1) in all scales of PSs or WCP including cyclic oscillating universe as

$$\Delta m \propto 1/\Delta \lambda \qquad (2.2)$$

The Eq. (2.1), excluding the UC like k_1, is a very well accepted basic inverse relationship in current physics. Not only that, it is a pillar of modern Quantum Mechanics (QM) as well as Standard Model of Particle Physics (SMPP). But the same Eqs. (2.1) & (2.2), those have related merely two CIPs Δm & $\Delta \lambda$ out of the total 5+5 ten CIPs within every scale of ingredient-PSs including BBCOU yet to be revealed and explored the subsequent impacts of the same in todays physics.

1.2. Mass-Energy and Motion

Beside the scale-specific intrinsic quantized magnitudes of those two CIPs like Δm & $\Delta \lambda$ in Eqs. (1.1) & (2.1), the intrinsic inertial-motions, in all same micro to macro scales of particles or systems-of-particles (PSs), is another CIP. That inertial-motions would have all SSUCs in corresponding quantized magnitudes as Δv, including all those scales of photons in EMS in Eq. (1.14) and in paragraph 2.3.3 of Chapter-1. But Δv would decrement onward SSUCs in quantized magnitudes of LE-C in expanding phase of BBCOU which seems identical to the $\Delta \lambda$ in Eq. (1.15) for the scales of bosons or photons on EMS. The alike decrements of SSUCs in quantized magnitudes for Δv also now realizing not only in such bosons or photons but also in various scales of fermions (onward LE-C). That is, the Δv quantized magnitudes in all scales of PSs.

On the other hand, the CIP like Δm in Eq. (1.16) has an inverse relationship with the $\Delta \lambda$ in Eq. (1.15) in same scales of photon-PSs as well as other fermion-PSs in Eqs. (1.1) & (2.1).

The Δv in Eq. (1.14) shows an identical intrinsic RE-C increments or LE-C decrements in corresponding quantized magnitudes along with the $\Delta \lambda$ in Eq. (1.15) in every same scales of photons as PSs on EMS. Then both of those CIPs with all discrete SSUCs in quantized magnitudes having increments onward RE-C or decrements onward LE-C, would have one direct proportional relationship

$$\Delta \lambda \propto \Delta v \qquad (2.3)$$

not only in scales of photons on the EMS but also in every scale of PSs universally due to universal invariant Eqs. (2.1) & (2.2).

In Eq. (2.3), since the $\Delta \lambda$ as one CIP has an inverse relationship with Δm in Eqs. (2.1) & (2.2) for all scales of same PSs, therefore the Δv in Eq. (2.3) that has direct relationship will have an inverse relation with same Δm.

Such an inverse relationship, in-between Δm & Δv, would be limited not merely within the all scales of photons in EMS but conceptually within all micro to macro scales of PSs which are definable through the Eqs. (1.1) & (2.1). Consequently, the inertial mass-energies of those fermion-PSs in such an inverse relation in-between Δm & Δv, which were once considered as non-luminous and non-scale-specific 'material bodies' in Classical Mechanics, are now not only emerged as Δm having all SSUCs in quantized magnitudes but its simultaneous (inseparable and inverse related) inertial-motions Δv would also have similar SSUCs in respective discrete magnitudes. In the range of micro-scales for fermions (or the so-called 'non-luminous' micro scales) like a neutrino (say with scale-specific lower quantized magnitude of inertial mass-energy Δm_φ) shows its higher magnitude of inertial-motion say Δv_φ compare to the inertial-motion say Δv_ϕ for a free moving neutron (with scale-specific quantized magnitude of inertial mass-energy say $\Delta m_\phi > m_\varphi$). A specific scale of hydrogen atom with inertial-motion say $\Delta v_H < \Delta v_\phi < \Delta v_\varphi$ (where inertial mass-energy say Δm_H for same hydrogen atom appears to have comparative magnitudes $\Delta m_H > \Delta m_\phi > \Delta m_\varphi$).

All these are showing the way that there might have all similar kind inverse quantized magnitudes with SSUCs in inertial mass-energies Δm as a CIP correspond to all discrete SSUCs in magnitudes of inertial-motions Δv in all scales of fermion-PSs.

Since, all those Eqs. (1.1), (2.1), (2.2) & (2.3) have universal invariances irrespective of scales for all micro to macro PSs (including all those fermions and bosons), there could be postulated another new inverse relationship for the two CIPs Δm & Δv. That could have universally invariant applications irrespective of the realms of "Light" & "Material Bodies" (i.e. luminous & non-luminous) in Classical Physics; bosons and fermions in QM; and visible matters, dark matters and dark energies in Astrophysics; and ultimately to the all scales of ingredient-PSs including the macro-most scale BBCOU as

$$\Delta m \propto 1/\Delta v . \qquad (2.4)$$

But, in Eqs. (1.14)-(1.16) respect to the different scales of photon-PSs in EMS, say with corresponding CIPs like ($\Delta v_c = c$), Δm_c & $\Delta \lambda_c$ when the $c = 2.99792 \times 10^{10} cm$ for all scale-specific quantized magnitudes, we can again re-write the Eq. (1.1) as

$$\Delta m_c \times \Delta \lambda_c = h/\Delta v_c. \tag{2.5}$$

Similarly, for a scale of gamma-ray-photons with $\Delta v_\gamma < \Delta v_c$ in Eqs. (1.14) - (1.16), the Eq. (1.1) can further re-write in scale-specific manner as

$$\Delta m_\gamma \times \Delta \lambda_\gamma = h/\Delta v_\gamma, \tag{2.6}$$

and also, for a scale of radio-wave photons in the EMS with $\Delta v_c < \Delta v_\eta$ in same Eqs. (1.14) - (1.16) we can re-write further the Eq. (1.1) as

$$\Delta m_\eta \times \Delta \lambda_\eta = h/\Delta v_\eta \tag{2.7}$$

Then in Eqs. (2.5), (2.6) & (2.7), the corresponding magnitudes of c in Eq. (1.1) are appearing different due to differences in all SSUCs in quantized magnitudes of $\Delta v_\gamma < (\Delta v_c = c) < \Delta v_\eta$ in Eq. (1.14) while the other two CIPs like $\Delta \lambda_\gamma < \Delta \lambda_c < \Delta \lambda_\eta$ in Eq. (1.15) and $\Delta m_\gamma > \Delta m_c > \Delta m_\eta$ in Eq. (1.16) respect to same different scales of photons in EMS. As a result, the magnitudes of h in all those three Eqs. (2.5), (2.6) & (2.7) cannot have any UC like constancy in magnitude as well due to all SSUCs in quantized magnitudes for those three CIPs like quantized inertial mass-energies Δm, discrete de Broglie's wavelengths $\Delta \lambda$ and quantized inertial motions Δv in same equations. Therefore, in each of those three Eqs. (2.5), (2.6) & (2.7), the constant h must have one such SSUC in magnitudes. If there all such differences of SSUCs magnitudes for h in Eq. (1.1), then, due to Eq. (2.5) it would be say h_c, due to Eq. (2.6) it would be say h_γ, and due to Eq. (2.7) it would be say h_η and so on. Then obviously there will be the $h_c \neq h_\gamma \neq h_\eta$. Since the corresponding magnitudes of all $\Delta v_\gamma < (\Delta v_c = c) < \Delta v_\eta$ in Eq.

(1.14) in respective Eqs. (2.6), (2.5) & (2.7), there will be in same Eqs. (2.6), (2.5) & (2.7)

$$h_\gamma < h_c < h_\eta \tag{2.8}$$

respectively. Consequently, the conventional h in Eq. (1.1) as well as in Eqs. (2.6), (2.5) & (2.7) are appearing with all SSUCs in magnitudes as like as $c, \Delta m$ & $\Delta\lambda$ in all scales of photons in the EMS. Moreover, since the Eq. (1.1) is universal irrespective of scales for PSs, that includes all scales of bosons as well as photons in EMS, the magnitudes of h would also having the all corresponding SSUCs in magnitudes in all other micro-most to macro-most (BBCOU) scales of PSs (e.g. fermions, and even beyond the realms of visible-matters that includes dark-matters & dark-energies) in physical nature. That is, the h with all SSUCs in magnitudes would have all the scale-specific magnitudes say as Δh as like as all the $\Delta m, \Delta v$ & $\Delta\lambda$ have in Eq. (1.1). Then, the same Eqs. (1.1) and (2.1) will ultimately appear as

$$\Delta m \cdot \Delta\lambda = \Delta h/\Delta v, \tag{2.9}$$

and from same Eq. (2.9), the Eq. (2.1) can re-write for all micro to macro scales of ingredient-PSs of BBCOU as

$$\Delta m \cdot \Delta\lambda = \Delta h/\Delta v = k_1 \cdot \tag{2.10}$$

So, from this Eq. (2.10), all the Eqs. (2.6) (2.5) & (2.7) can also re-write in all corresponding scales of photons as

$$\Delta m_\gamma \cdot \Delta v_\gamma = \Delta h_\gamma/\Delta\lambda_\gamma = k_2, \tag{2.11}$$

$$\Delta m_c \cdot \Delta v_c = \Delta h_c/\Delta\lambda_c = k_2, \tag{2.12}$$

$$\Delta m_\eta \cdot \Delta v_\eta = \Delta h_\eta/\Delta\lambda_\eta = k_2, \tag{2.13}$$

and k_2 must be another inverse UC irrespective of all three scales of photons in EMS. Since, the Eq. (2.10) is applicable to all scales of bosons, fermions, and beyond only those three scales of photons in EMS from Eqs. (2.11) to (2.13), there can be a universal inverse equation for all scales of PSs

$$\Delta m \cdot \Delta v = \Delta h / \Delta \lambda = k_2, \quad (2.14)$$

where the k_2 being one inverse UC which would have the unit in $gm \cdot cm/\sec$ due to corresponding units of $\Delta m, \Delta v, \Delta h$ & $\Delta \lambda$. Then, the Eq. (2.14) would be a new universal inverse relationship that co-relates all SSUCs or scale-specific quantized inertial mass-energies to all SSUCs or scale-specific quantized inertial-motions of every scale of PSs in BBCOU. The mechanical foundations of Special Relativity Theory that primarily based on constant (quantized) inertial-motion of light $c(= \Delta v_c)$ without such co-relationship with quantized inertial-mass-energy. In contrary, the mechanical basis of Quantum Mechanics that based on primarily co-related quantized magnitudes of inertial-mass-energy and inertial-wave-length but without co-relationship with quantized inertial-motion. Therefore, the Eq. (2.14), would have unique consequences in both mechanical foundations of Relativity Theories as well as Quantum Mechanics.

1.3. Radius and Anti-Radius

The inertial radius Δr as another CIP in all PSs, as stated in Eq. (1.17), has an intrinsic association to LAR or left-handed axial-rotation or left-handed-spin for the 'volume' of PSs. That also possesses a scale-specific increments of SSUCs in quantized magnitudes onward left-handed entropy or LE-C in current expansion phase of the BBCOU. The Δm in Eq. (1.1) is similarly associated to that intrinsic LAR or left-handed axial rotation or spin of a 'volume' of PSs and is also incremented in SSUCs of quantized magnitudes onward LE-C in same current expansion phase of the BBCOU. The Δr in Eq. (1.17) has also the similar LAR direction and increments in SSUCs quantized magnitudes onward LE-C in all scales of PSs as like as Δm.

Therefore, the CIPs like Δm in Eq. (2.1) and Δr in Eq. (1.17) in every same scale of PSs are appearing to have a direct proportional relationship onward LE-C increments in SSUCs quantized values irrespective of scales

$$\Delta m \propto \Delta r \qquad (2.15)$$

On other hand, the anti-radius $\Delta r_u \equiv \Delta \lambda$ in Eq. (1.19) is a mirror-imaged CIP of the same Δr in Eq. (2.15) in all same PSs, and also has described in Eq. (1.18) as an integrated part of the mirror-imaged intrinsic RAR or right-hand axial-rotation or right-handed-spin direction of 'volume' with increments of SSUCs in quantized magnitudes onward RE-C of ever PSs. Then oppositely, same Δr_u would appear left-handedly decrementing onward all SSUCs in quantized magnitudes from LE-C (as the mirror-image of RE-C) in Eqs. (1.1) & (2.1) in same current expanding phase of the BBCOU.

Since, the CIPs like Δm & Δr have direct proportional relationship in Eq. (2.15) and Δm & $\Delta \lambda$ have inverse proportional relationships in Eq. (2.2) in every scale of PSs, therefore both of those mutually mirror-imaged Δr and $\Delta r_u \equiv \Delta \lambda$ would have an inverse proportional relationship in each of those same scales of PSs

$$\Delta r \propto 1/(\Delta r_u \equiv \Delta \lambda) \qquad (2.16)$$

onward LE-C or RE-C in BBCOU in both of its expanding and collapsing phases. Therefore, from the Eq. (2.16) there would be

$$\Delta r \cdot \Delta \lambda = k_3 \qquad (2.17)$$

for every scale of PSs in BBCOU and the k_3 would be another new UC (beside above two k_1 & k_2) in magnitudes irrespective of micro to macro scales. The unit of k_3 would be obviously in cm^2 due to the corresponding units of the Δr & $\Delta \lambda$. The Eq. (2.17) also depicts that each of the micro to macro scales of PSs, as ingredient parts of the current expanding phase of the BBCOU are now approaching onward observations of LE-C, would have 5-left-handed-CIPs those have increments for SSUCs in quantized magnitudes and 5-right-handed-CIPs having decrements for SSUCs in quantized magnitudes. Onward such a LE-C observation, the Δr as similar left-handed-CIP is incrementing in its all SSUCs in quantized magnitudes but $\Delta\lambda$ as right-handed-CIP is decrementing in its all SSUCs in quantized magnitudes in all scales of PSs.

1.4. Space and Anti-Space

The inertial three-dimensional space $\Delta s(\Delta x, \Delta y, \Delta z)$ in Eq. (1.17) also has assumed as one CIP (or three CIPs) with scale-specific quantized magnitudes with 3-spatial dimensions. That $\Delta s(\Delta x, \Delta y, \Delta z)$ can also be imagined to have spherical shapes in its configuration for conveniences and also equivalent to the resultant volume acquired by SSUCs in quantized magnitudes of inertial-mass-energies increments (for PSs) onward LE-C. The $\Delta s(\Delta x, \Delta y, \Delta z)$ has also intrinsic LAR directional axial-rotation in all PSs.

The Δr in Eqs. (1.17), (2.15) & (2.17) is also possessed similar characteristics alike to $\Delta s(\Delta x, \Delta y, \Delta z)$ beside Δm in Eqs. (1.1), (2.14) & (2.15).

In Eq. (1.18), there is also an inertial anti-space $\Delta s_u(\Delta x_u, \Delta y_u, \Delta z_u)$ CIP which has also emerged as a mirror-image of the $\Delta s(\Delta x, \Delta y, \Delta z)$ in Eq. (1.17) in all same scales of ingredient-PSs of BBCOU. Then the $\Delta s_u(\Delta x_u, \Delta y_u, \Delta z_u)$, as mirror-imaged counter part of that $\Delta s(\Delta x, \Delta y, \Delta z)$, would have an intrinsic right-handed direction for the axial-rotation or RAR as anti-space. That would conversely be imagined to have spherical shapes in its convenient configurations. But that could also be imagined as if equivalent to the resultant volume for the respective curved paths along the conceptual geodesic circle or circumference of the corresponding inertial-motions of Δm within BBCOU with SSUCs in quantized magnitudes. But that must have the corresponding SSUCs in quantized magnitude with such anti-radius Δr_u for geodesic circumference equivalent to the de Broglie wavelength $\Delta \lambda$.

There would be also all other right-handed decrementing CIPs onward LE-C of all same PSs in current phase of expansion of BBCOU. The $\Delta \lambda$ in Eqs. (1.18), (2.16) & (2.17) also possesses similar RAR direction and right-handed decrements in magnitudes identical to the $\Delta s_u(\Delta x_u, \Delta y_u, \Delta z_u)$ along with Δv.

However, the Δr & $\Delta \lambda$, which are not only inversely related in Eq. (2.17) but are also defining both of those corresponding CIPs like $\Delta s(\Delta x, \Delta y, \Delta z)$ and $\Delta s_u(\Delta x_u, \Delta y_u, \Delta z_u)$ as respective volumes in Eqs. (1.17) & (1.18) in an inverse manner. Consequently, from Eq. (2.17), both $\Delta s(\Delta x, \Delta y, \Delta z)$ and $\Delta s_u(\Delta x_u, \Delta y_u, \Delta z_u)$ would have also a universally invariant inverse proportional relationship in every scale of PSs in BBCOU. Consequently, from the inverse relation in-between both of those Δr and $(\Delta r_u \equiv \Delta \lambda)$ in Eq. (2.16), the corresponding two mutual mirror-imaged CIPs like Δs and Δs_u in Eqs. (1.17) and (1.18) in every scale of PSs, there would have

$$\Delta s \propto 1/\Delta s_u \qquad (2.18)$$

and from the same Eq. (2.18) another new UC might be emerged as

$$\Delta s \cdot \Delta s_u = k_4 \qquad (2.19)$$

where k_4 would have UC in magnitude irrespective of all scales of PSs. The same UC would have the unit in cm^6 due to corresponding units of Δs and Δs_u in Eqs. (1.17) & (1.18).

In Eq. (2.19) all those same scales of PSs, as part of the current expansion of BBCOU, are now incrementing SSUCs in quantized magnitudes of Δs onward LE-C and simultaneously decrementing SSUCs in quantized magnitudes of Δs_u onward same direction of LE-C.

1.5. Time and Anti-Time

In Eq. (1.21), the time Δt as another discrete CIP with all SSUCs in quantized magnitudes is also entwined with every scale of same ingredient-PSs of BBCOU that has an obvious non-void origin. That time Δt in Eq. (1.21) also has scale-specific quantized magnitudes because there all scale-specific intrinsic quantized magnitudes of Δr. Then Eq. (1.21) actually shows that that Δt has **(i)** a left-handed or clockwise direction along the left-handed axial rotation or LAR in all its entwined PSs, **(ii)** a flow from the 'past' to 'future' as the left-handed directional flow, **(iii)** a spatial unit in *cm* beside its conventional angular unit [1] in *sec*, and **(iv)** most remarkably the left-handed increments of SSUCs in quantized magnitudes (i.e. the Δt could get faster and faster in all scale specific clocks in PSs) onward LE-C from one micro to macro scales in current expanding phase of the BBCOU. Then conversely, the same Δt would get slower and slower in all scale-specific (left-handed) clocks are imagined to attach in PSs in reverse direction of LE-C, i.e. from one macro to micro scales in current expanding phase of the same BBCOU.

The SRT also has stated about clocks those could be dilated in corresponding 'time' with incrementing relative motions. If those PSs are attached with all corresponding left-handed clocks, and if gradually have the discrete changes in the scales' onward macro to micro, all corresponding SSUCs in quantized magnitudes of Δv (in all those clocks as well as PSs) would also become changed accordingly. The Eq. (2.14), that describes the SSUCs in quantized magnitudes of Δv would have also the increments onward such changing in scales from macro to micro of PSs. That would also make the changes of SSUCs in quantized magnitudes of Δt. As a result, that Δt would become

---

1. That is also an angular distance of any spatial unit.

gradually slower and slower or become dilated in respective time-scales through all corresponding clocks in all those scales of same PSs.

As a result, the duration (or distance) of a 'tick' defines as quantized time in a scale-specific "clock" within one macro scale of PSs will be faster (i.e. longer) compare to a 'tick' in another scale-specific clock within a micro scale of PSs. That would be due to the respective differences in scale-specific quantized values of Δr in those two scales of PSs. If one person could move in a spacecraft, where everything including himself is supposedly made homogeneously by all carbon atoms, would have a longer 'ticks' or slower magnitude of quantized left-handed time say $\Delta t_\otimes$ compare to the shorter 'ticks' for magnitude of quantized left-handed time say $\Delta t_⊞$ in another spacecraft which has made homogeneously by any heavier element than carbon say silicon atoms from the Eq. (2.14). Because, such a heavier scale-specific inertial-mass-energy atom say $\Delta m_⊞$ correspond to that quantized time scale $\Delta t_⊞$ would have scale-specific lower quantized magnitude of the inertial-motion say $\Delta v_⊞$ of aluminum-atoms compare to the inertial-motion say $\Delta v_\otimes$ of the lighter carbon-atoms.

In contrary, there seems to have the mutual for corresponding mirror-imaged anti-time as CIP which also associated with every same scale of PSs in Eq. (1.23) which would possess **(i)** also a non-void origin, **(ii)** a right-handed or anti-clockwise direction i.e. simultaneous flow from 'future' to 'past', **(iii)** a spatial unit, and **(iv)** a left-handed decrement in scale-specific magnitudes (i.e. such anti-time would get slower and slower) onward LE-C from micro to macro scales in current expanding phase of the BBCOU or universe.

Consequently, the duration (or distance) of any simultaneous 'tick' of such quantized anti-time in a scale-specific clock that is counting anti-time will be slower i.e. shorter in all gradual macro scales of PSs compare to a 'tick' in a micro scale of PSs. It is due to the corresponding scale-specific different quantized magnitudes of the $\Delta r_u \equiv \Delta \lambda$ in Eq. (1.23). As a result, the Δt_u in said carbon-atom would be faster compare to Δt_u in a said aluminum-atom.

This shows that there will be another new inverse relationship in-between those two mutually mirror-imaged CIPs like Δt and Δt_u in all corresponding Eqs. (1.21) & (1.23) due to the inverse relationship in-between Δr and $\Delta r_u \equiv \Delta \lambda$ in Eq. (2.16)

$$\Delta t \propto 1/\Delta t_u \qquad (2.20)$$

and also, from the Eq. (2.20) there will be

$$\Delta t \cdot \Delta t_u = k_5 \qquad (2.21)$$

where the k_5 is another new inverse UC with universally invariant magnitude irrespective of micro and macro scales for PSs, and its unit will be also in cm^2 due to the units of Δt and Δt_u in respective Eqs. (1.21) and (1.23).

The Eq. (2.21) has also defined that every scale of PSs, as integrated part of current expanding phase of BBCOU onward LE-C, has increments for Δt but decrements for Δt_u in SSUCs in quantized magnitudes onward LE-C.

1.6. Radius and Motion

Next, the Δr in Eq. (2.16) which has appeared inversely related to $\Delta \lambda$ in every scale of PSs. The same $\Delta \lambda$ in Eq. (2.14) also shows an inverse relation with Δm in Eq. (2.2) but a direct relation with Δv in Eq. (2.3) in all same scales of PSs.

The same Δr and Δm would have all scale-specific increments in quantized magnitudes onward LE-C in current expansion of the BBCOU. Therefore, the Δr appears to have a direct relationship with the Δm in every scale of PSs in Eq. (2.15).

Hence, there would be another inverse relationship in-between two CIPs like Δr and Δv in all same scales of PSs

$$\Delta r \propto 1/\Delta v \qquad (2.22)$$

due to the inverse relationships between (Δr & $\Delta\lambda$) in Eq. (2.16) and also between (Δm & Δv) in Eq. (2.14) but a direct relationships for (Δr & Δm) in Eq. (2.15) and ($\Delta\lambda$ & Δv) in same Eq. (2.4). However, from Eq. (2.22) there would be

$$\Delta v . \Delta r = k_6 \qquad (2.23)$$

where, the k_6 being another new UC and would possess a universally invariant constancy in magnitude irrespective of micro and macro scales for PSs. The unit of that UC will be in cm^2/sec, or if the unit of time is considered in unit of cm then the same k_6 would also have the unit in cm, due to corresponding units of Δv in Eq. (2.9) and Δr in Eq. (2.17).

The same k_6 in Eq. (2.23) can also be derived alternately from the UCs like k_1 in Eq. (2.10), k_2 in Eq. (2.12) and k_3 in Eq. (2.17) as

$$(k_2 \times k_3)/k_1 = (\Delta m. \Delta v).(\Delta r. \Delta\lambda)/(\Delta m. \Delta\lambda) = \Delta r. \Delta v = k_6. \qquad (2.24)$$

The Eq. (2.23) also states that every same scale of ingredient-PSs as integrated part of current expansion of BBCOU onward LE-C increments for SSUCs in discrete magnitudes of Δr and decrements for SSUCs in discrete magnitudes of Δv for those two CIPs.

2. A Unified Equation for Inertial Everything

All those 5+5 CIPs, those are appeared inversely related in above Section-1 within every scale of ingredient-PSs of BBCOU, had no such inverse link to each other in the Eqs. (1.25), (1.26) & (1.27). But those are now assumed to have inverse relations in all corresponding Eqs. (2.1), (2.14), (2.17), (2.19), (2.21) & (2.23). As a result, in inertial state, every inertial-PSs irrespective of scales seems to have one common unified inverse definition or unified inverse equation in all corresponding Eqs. (1.25), (1.26) & (1.27) from all those same universal inverse relationships in Eqs. (2.1), (2.14), (2.17), (2.19), (2.21) & (2.23) in Section-2.

Since, the Δr in Eqs. (1.17) & (1.21) is an inertial radius with all SSUCs in quantized magnitudes and the $\Delta r_u \equiv \Delta \lambda$ in Eqs. (1.18) & (1.23) is mirror-imaged anti-radius also with all SSUCs in quantized magnitudes where both are entangled with all same scales of ingredient-PSs of BBCOU. Therefore, these two mutually mirror-imaged CIPs would obviously be more elementary compare to the corresponding CIPs like (Δs & Δt) in Eqs. (1.17) & (1.21) as well as (Δs_u & Δt_u) in Eqs. (1.18) & (2.23) respectively. Then any PSs irrespective of scales can assume in two inverse mirror-imaged sets of CIPs with 5-CIPs in each. The left- handed set would comprise 5-CIPs from Eqs. (2.14), (2.19) & (2.21) as

$$[\Delta m \cdot \Delta s(\Delta x \cdot \Delta y \cdot \Delta z) \cdot \Delta t] \qquad (2.25)$$

and the right-handed set would have rest of 5-CIPs from same Eqs. (2.14), (2.19) & (2.21) as

$$[\Delta v \cdot \Delta s_u(\Delta x_u \cdot \Delta y_u \cdot \Delta z_u) \cdot \Delta t_u]; \qquad (2.26)$$

and from those Eqs. (2.4) (2.18) & (2.20) all scales of PSs can be unifiedly defined in such inverse way

$$(\Delta m \cdot \Delta s \cdot \Delta t) \propto 1/(\Delta v \cdot \Delta s_u \cdot \Delta t_u) \qquad (2.27)$$

where $\Delta s = (\Delta x . \Delta y . \Delta z)$ and $\Delta s_u = (\Delta x_u . \Delta y_u . \Delta z_u)$. As a result, from the Eq. (2.27) there would be an ultimate common inverse definition for all scales of PSs in BBCOU as macro-most scale for all PSs in physical nature

$$(\Delta m \cdot \Delta s \cdot \Delta t) \cdot (\Delta v \cdot \Delta s_u \cdot \Delta t_u) = (k_2 \cdot k_4 \cdot k_5) = k \qquad (2.28)$$

and the k is the ultimate UC for every scale of all PSs, where the unit of k would be in $gm^1cm^9sec^{-1}$ or gm^1cm^8. The Eq. (2.28) also defines every PSs irrespective of all micro and macro scales in a unified way in inertial conditions [1] or in absence of any kind external forces. Hence, the Eq. (2.28) can also be considered as the inertial unified inverse equation for all scales of PSs irrespective to the domains of Particle Physics in micro scales and also of the Astronomy in macro scales.

However, all those PSs, as defined in above Eq. (2.28), are also as ingredient-PSs in currently expanding phase of BBCOU onward LE-C having an increment in all SSUCs in quantized magnitudes of 5 left-handed CIPs i.e. ($\Delta m, \Delta s$ & Δt) in Eq. (2.25) and a simultaneous decrement in all SSUCs in corresponding quantized magnitudes of 5 right-handed CIPs i.e. ($\Delta v, \Delta s_u$ & Δt_u) in Eq. (1.26).

Then Eq. (2.28) can also write in inverse manner from the Eq. (1.25) as

$$k/(\Delta v. \Delta s_u. \Delta t_u) \cdot (\Delta m. \Delta s. \Delta t) \approx \xrightarrow{LE-C} \qquad (2.29)$$

for every scale of ingredient-PSs onward LE-C from the incrementing 'side' of the BBCOU. Therefore, simultaneously same ingredient-PSs in Eq. (2.29) onward RE-C 'side' of observation of same BBCOU would appear

$$\xleftarrow{RE-C} \approx (\Delta v. \Delta s_u. \Delta t_u) = k/(\Delta m. \Delta s. \Delta t); \qquad (2.30)$$

then from Eq. (1.27). Then, the simultaneous Eqs. (2.29) & (2.30) can finally re-write through all the inverse relationships of all those 5+5 CIPs in one single bracket for all unified ingredient-PSs and also by describing both LE-C and RE-C simultaneously as

$$\begin{cases} k = (\Delta v \cdot \Delta s_u \cdot \Delta t_u) \cdot (\Delta m \cdot \Delta s \cdot \Delta t) \approx \xrightarrow{LE-C} \\ \xleftarrow{RE-C} \approx k = (\Delta v \cdot \Delta s_u \cdot \Delta t_u) \cdot (\Delta m \cdot \Delta s \cdot \Delta t) \end{cases}. \qquad (2.31)$$

---

1. In Chapter-4, the same would define unifiedly in non-inertial way.

That is, if it is considered that all scales of ingredient-PSs in BBCOU are in their inertial states in absence of all forces or in free motions, the Eq. (2.28) would be a universally common or unified inverse definition for all those same PSs irrespective of scales including the whole universe or BBCOU as macro-most scale of PSs. Since all those same PSs along with macro-most scale or BBCOU are parts of one universal cyclic oscillation, each of those same PSs should have a simultaneous two universally common definitions correspond to the mutually mirror-imaged 'entropy' or LE-C and 'anti-entropy' or RE-C in above Eqs. (2.29) & (2.30).

Therefore, the Eq. (2.28) that defines every PSs in inertial states would be universally unified. But as if integrated units of that universal cyclic oscillations of BBCOU, the same PSs cannot exist in isolation but might be always in the form of simultaneous left & right-handed pairs due to the two different ways of observations of LE-C or RE-C observers as part of the same BBCOU.

As a result, the Eqs. (2.29) & (2.30) are actually existing in a pair of unified equations for every scale of PSs in Eq. (2.28) in inertial state. But the Eq. (2.29) defines PSs can only be appeared in 'steady' existence i.e. sustained as left-handed when it would come along the way of LE-C observation of any similar LE-C observer. Conversely, the Eq. (2.30) that defines PSs can only appeared in 'steady' existence when it would come along the way of RE-C observation of any similar kind of RE-C observer.

Subsequently, on the way of LE-C observation & observer the PSs as described in Eq. (2.30) would never appear as steady in existence and conversely, on the way of RE-C observation & observer the PSs as defined in Eq. (2.29) cannot be stable in existences.

3. Unfolded 10-Dimensions of Quantum-Reality

The Eq. (2.28) has appeared as a unified or common inertial equation for every scale of ingredient-PSs of BBCOU. It is also as an inverse product of total five left-handed incremented CIPs in Eq. (2.25) and other five left-handed decremented (i.e. right-handedly incremented) CIPs in Eq. (2.26) in current expansion phase of same BBCOU onward LE-C observation.

Conventional transformation of any 'event' in between two inertial frames of reference now needs 3+1 four 'unfolded' dimensions of space and time.

But in Eq. (2.28), any of such 'event' should have to be linked with any non-void entity which is associated with any scale of PSs anywhere in the current expanding phase of BBCOU. Actually, the 3+1 space and time are appearing as non-void and non-omnipresent entity with any of those scales of PSs unlike earlier conventions of transformations in Classical Mechanics.

Actually, such 3+1 space and time are associated with any those scales of PSs. If there is no such exchangeable space and time without any of those scales of discrete PSs, the same PSs cannot be communicated or exchanged in any of our conventional Mechanical Transformation processes.

That is, any such 'event' equivalent to any specific scale of PSs, or say any PS-event, definable by the Eq. (2.28), are emerged as a mesh of all scale-specific observer-independent quantized (i.e. finite and non-zero) magnitudes of total (5+5) 10-CIPs where all such 10-CIPs are inversely co-related through seven UCs ($k_1, k_2, k_3, k_4, k_5, k_6$ & k).

Therefore, any such PS-linked 'events' would be of total 10 numbered co-ordinates and for any transformation of such PS-events will actually need total 5+5 co-ordinates inclusive of 3+1 co-ordinates of space in Eq. (1.17) and time in Eq. (1.21). But both of such 3+1 space & time having scale-specific quantized magnitudes unlike in Lorentz Transformations.

However, not only those 3-space + 1-time dimensions are physically tangible or 'unfolded' but rest of other 6-dimensions are also similar physically tangible or 'unfolded' through Eq. (2.28). Consequently, any such PS-events could be exchanged as 'observance' to the observers like us through any quantum-real signals if and only if those are any quantum-real observances and can define by the Eq. (2.28). But, any of those PS-event if non-PSs in type, i.e. in such a void (or a Non-quantized-virtual in type), would be always beyond our all quantum-real exchange limits of signals, where all those same 5+5 CIPs have zero and infinity magnitudes as described (in Section-3 of Chapter-1). Consequently, all our exchangeable PS-events must be a quantum-real in type within non-zero and non-infinity limits of magnitudes within the similar limits of quantum-realities of all ours like observers as well as exchangeable signals.

For all such quantum-real PS-linked events with unfolded 5+5 CIPs, those may define through the Eq. (2.28), irrespective of micro to macro scales in the current expanding phase of BBCOU, can also assume instead to have 5+5 ten co-ordinates and each co-ordinate are inversely linked to each other through all those total 7 numbers of UCs. As a result, all those (5+5) CIPs, as 10-co-ordinates of any such quantum-real PS-events, can further consider as 10-dimensional. This seems completely a new outcome in current physics. For instances, when one such quantum-real PS-event occurs in our conventional unfolded (3+1)-dimensional space and time, which are actually the two CIPs with scale-specific quantized magnitudes (i.e. with observer-independent SSUCs magnitudes as well). Both of those $\Delta s(3)$ & $\Delta t(1)$ can define further by an elementary CIP like Δr in corresponding Eqs. (1.17) and (1.21) which is also inversely linked to the inertial antiradius $\Delta r_u \equiv \Delta \lambda$ in Eq. (2.17). Such an anti-radius CIP can further define other inverse CIPs in three ($\Delta x_u \cdot \Delta y_u \cdot \Delta z_u$) in anti-space Δs_u in Eq. (1.18) and one in anti-time Δt_u in Eq. (1.23).

Then, any such unfolded (3+1)-dimensional PS-linked events would appear to us as if an unfolded (6+2) or (4+4) spatial dimensional events through any similar quantum-real signal. That signal also exchanges with any finite quantity of all scale-specific 5+5 CIPs in-between any such PS-linked event and quantum-real observer like us.

Moreover, inertial anti-radius $\Delta r_u \equiv \Delta \lambda$ is also linked inversely to inertial-mass-energy Δm as one CIP in Eq. (2.2), and inertial radius Δr is inversely connected to the CIP like inertial-motion Δv in Eq. (2.23) within any of those same quantum-real PS-linked events. Consequently, in any one of such signal exchanges, (where any signal-PS would have also such 5+5 inverse 10-dimensions), in-between a PS-linked event and an observer like us. Therefore, every PS-linked event would ultimately be exchanged on total (6+2+2) or (5+5) or 10 unfolded dimensions including those 3+1 conventional dimensions of space & time. Finally, all those unfolded (5+5) inverse 10-dimensions for any quantum-real PS-linked event, signal and observer (like us) would have 3-space, 3-anti-space, 1-time, 1-ant-itime, 1-inertial-mass-energy & 1-inertial-motion with all non-zero and non-infinity quantized magnitudes as SSUCs.

In an inertial condition, when all the external forces are considered absent, the magnitudes of all those (5+5) inverse related dimensions are scale-specifically quantized as SSUCs. So, each of those non-zero & non-infinity quantized magnitudes are scale-specific observer independent values i.e. SSUCs or absolute magnitudes. If there are two relatively moving inertial frames of reference or IFRs say F_1 & F_2 in a straight line with corresponding scale-specific two such PS-linked events say Q_1 & Q_2 at the respective centers say p_1 & p_2 with all (5+5) 10-dimensions, and if a signal exchanges from F_1 to F_2, that signal would have also the similar (5+5) inverse 10-dimensions. If that signal is $\Delta \beta$, then, after transformation of that signal the scale-specific IFR ΔF_1 will appear in the scale-specific IFR ΔF_2 as

$$\Delta F_1' = \Delta F_1 - \Delta \beta \tag{2.32}$$

and the ΔF_2 will become transform after receiving that same signal as

$$\Delta F_2' = \Delta F_2 + \Delta \beta \tag{2.33}$$

in all corresponding (5+5) inversely related CIP-elements for each $(\Delta F_1, \Delta F_2, \Delta \beta, \Delta F_1' \,\&\, \Delta F_2')$ in matrices of such PS-linked events. Then, in Eq. (2.32) the $\Delta F_1'$ will be with scale-specific quantized magnitude in metrices as

$$\Delta F_1' = \begin{bmatrix} \Delta m_1 & \Delta v_1 \\ \Delta s_1 & \Delta s_{u1} \\ \Delta t_1 & \Delta t_{u1} \end{bmatrix} - \begin{bmatrix} \Delta m_\beta & \Delta v_\beta \\ \Delta s_\beta & \Delta s_{u\beta} \\ \Delta t_\beta & \Delta t_{u\beta} \end{bmatrix} = \begin{bmatrix} \Delta m_1' & \Delta v_1' \\ \Delta s_1' & \Delta s_{u1}' \\ \Delta t_1' & \Delta t_{u1}' \end{bmatrix} \tag{2.34}$$

with all corresponding (5+5) scale-specific quantized magnitudes for the CIP-elements of $\Delta F_1, \Delta \beta \,\&\, \Delta F_1'$ in matrices respectively. Similarly, in Eq. (2.33), after receiving the scale-specific quantized magnitude of the signal $\Delta \beta$ with all corresponding (5+5) CIP-elements the ΔF_2 will transform into another scale-specific quantized PS-event as

$$\Delta F_2' = \begin{bmatrix} \Delta m_2 & \Delta v_2 \\ \Delta s_2 & \Delta s_{u2} \\ \Delta t_2 & \Delta t_{u2} \end{bmatrix} + \begin{bmatrix} \Delta m_\beta & \Delta v_\beta \\ \Delta s_\beta & \Delta s_{u\beta} \\ \Delta t_\beta & \Delta t_{u\beta} \end{bmatrix} = \begin{bmatrix} \Delta m_2' & \Delta v_2' \\ \Delta s_2' & \Delta s_{u2}' \\ \Delta t_2' & \Delta t_{u2}' \end{bmatrix} \tag{2.35}$$

where all those physically tangible or unfolded CIP-elements in corresponding 10-dimensional PS-linked events, as individual $\Delta F_1, \Delta F_2, \Delta \beta, \Delta F_1' \,\&\, \Delta F_2'$ matrices, must be scale-specific observer-independent values as SSUCs.

Then, in Eq. (2.28), every quantized PS-linked event with scale-specific quantization in magnitudes is basically appearing as a product of two inverse 5-left-handed and 5-right-handed simultaneous sets of dimensions or CIP-elements as corresponding $(\Delta m \,.\, \Delta s \,.\, \Delta t)$ in Eq. (2.25) and $(\Delta v \,.\, \Delta s_u \,.\, \Delta t_u)$ in

Eq. (2.26). Consequently, since those 5-dimensions in left-handed set: [$\Delta m = 1$, $\Delta s = 3$, $\Delta t = 1$] and 5-dimensions in right-handed set: [$\Delta v = 1$, $\Delta s_u = 3$, $\Delta t_u = 1$] are inversely interlinked to each other in respective Eqs. (2.10), (2.14), (2.17), (2.19), (2.21), (2.23) & (2.28) in any quantized PS-linked event, that would be exchangeable only through any similar kind of quantum-real PS-linked signal in Eqs. (2.34) & (2.35).

As a result, there would be all instantaneous changes in all interlinked CIPs in scale-specific quantized magnitudes if in any one of those CIP-elements or dimensions becomes changed in its scale-specific quantized magnitude. That is, rest of the 9-CIP-elements or dimensions would be changed instantaneously and subsequently, through that process of PS-linked signal exchange there would be also an automatic change in specific scale of that PS-linked event.

Therefore, any such quantum-real PS-linked event would always be a 5+5 unfolded dimensional in place of earlier only 3+1 space and time dimensions, and also might be definable through Eqs. (2.28), (2.34) & (2.35) in current expanding phase of BBCOU onward LE-C or conversely onward RE-C.

4. Magnitudes of Inverse Universal Constants

The inverse constants k_1 in Eq. (2.10), k_2 in Eq. (2.14), k_3 in Eq. (2.17), k_4 in Eq. (2.19), k_5 in Eq. (2.21), k_6 in Eq. (2.23) and finally k in Eq. (2.28) are deduced as UCs i.e. with constant magnitudes irrespective all scales of PSs where all 10-CIPs in those same equations are possessing SSUCs in discrete magnitudes in all PSs. Approximate magnitudes for each of those UCs can be calculated, from some of currently available relevant data, in next two pages:

4.1. The Magnitude of k_1

In Eq. (1.1), it will be approximately equal to $2.2009 \times 10^{-37} gm \cdot cm$, if we consider the magnitude of $h = 6.6252 \times 10^{-27} erg/sec$ or $gm \cdot cm^2/sec$, and $c = 2.99792 \times 10^{10} cm/sec$.

4.2 The Magnitude of k_2

If the magnitude of $\Delta v = \Delta v_c = c = 2.99792 \times 10^{10} cm/sec$ in Eq. (1.14) and $\Delta \lambda_c$ is a mean wavelength of the white-light-ray-photon-WCP on the VIBGYOR range of EMS for visible-light equals to $\Delta \lambda_c = 5.85 \times 10^{-5} cm$ for the ($c_1 > c > c_2$), where c_1 and c_2 are considered for two most red and violet ends of visible wavelengths in same EMS respectively, then in Eq. (2.14) there will be $k_2 = \Delta m . \Delta v = \Delta m_c . (\Delta v_c = c) = (k_1/\Delta \lambda_c) . c = 1.1254246 \times 10^{-22} \; gm. cm/sec$.

4.3. The Magnitude of k_3

If the scale specific intrinsic-quantized magnitude for radius of a normal hydrogen atom in its ground state of energy of the orbiting electron is approximately considered as $\Delta r_H = \Delta r = 5.2917720859 \times 10^{-13} cm$, and the magnitude of its de Broglie's wavelength $\Delta \lambda_H = 1.3232934 \times 10^{-13} cm$, considering from Eq. (1.1) if the mass-energy of that hydrogen atom is $\Delta m_H = 1.67 \times 10^{-24} gm$; then in Eq. (2.17) the $k_3 = \Delta r_H . \Delta \lambda_H = 7.0025671 \times 10^{-26} cm^2$.

4.4. The Magnitude of k_4

From Eq. (1.17) we can write Δs_H = spherical volume for a normal hydrogen atom in ground state of mass-energy $= (3\pi \times \Delta r_H^3)/4 = 3.49292513 \times 10^{-37} cm^3$, and similarly in Eq. (1.18) Δs_{u_H} = anti-volume for the same hydrogen atom in same ground state of mass-energy $= (3\pi \times \Delta \lambda_{u_H}^3)/4 = 7.22787262 \times 10^{-39} cm^3$. Then in Eq. (2.19) we could have $k_4 = \Delta s_H \times \Delta s_{u_H} = 1.31892695 \times 10^{-79} cm^6$.

4.5. The Magnitude of k_5

In Eq. (1.21), if the radius Δr for a normal hydrogen atom in ground state of mass-energy is $\Delta r_H = \Delta r = 5.2917720859 \times 10^{-13} cm$, then its corresponding time scale Δt for the same will be, say, $\Delta t_H = 2\pi \times \Delta r_H = 3.32625668 \times 10^{-14} cm$; and in Eq. (1.23), if the respective anti-radius = de Broglie wavelength for that normal hydrogen atom, say $\Delta \lambda_H = 1.3232934 \times 10^{-13} cm$ for the same hydrogen atom in ground state of mass-energy, then its anti-time scale would be $\Delta t_{u_H} = 2\pi . \Delta \lambda_H = 4.15873543 \times 10^{-13} cm$. Hence, in Eq. (2.21), $k_5 = (\Delta t_H \times \Delta t_{u_H} = 1.38330215 \times 10^{-28} cm^2$.

4.6. The Magnitude of k_6

Since in Eq. (2.24) we have $k_2 . k_3/k_1 = k_6$, then from the above known magnitudes of k_1, k_2, k_3 we can calculate

$$k_6 = 3.56616194 \times 10^{-33} \, cm^2/sec.$$

4.7. The Magnitude of k

Now from all the above approximate magnitudes for $k_2 = 1.1254246 \times 10^{-22} \, gm. cm. sec^{-1}$, $k_4 = 1.31892695 \times 10^{-79} cm^6$ and $k_5 = 1.38330215 \times 10^{-28} cm^2$ there will be finally in the Eq. (2.28) an approximate magnitude for

$$k = 2.05331 \times 10^{-129} \, gm \cdot cm^9/sec.$$

Summary

The Chapter-2 reveals one 'unified inverse equation' in Eq. (2.31) through total 7 numbers of UCs for all micro to macro scales of particles or systems-of-particles (including the BBCOU as macro-most scale) or for all quantum-real entities in conceptual inertial state (when forces are assumed not affecting to those). There is already one such UC which has been deduced from the conventional universally invariant inverse-relationship within universal wave-corpuscular duality law of de Broglie between two CIPs Δm & $\Delta \lambda$ for all those same particles or systems-of-particles in Eqs. (2.1) & (22). Subsequently, there are next 6 UCs for all other CIPs in same particles or systems-of-particles $\Delta s(\Delta x, \Delta y, \Delta z)$ & $\Delta s_u(\Delta x_u, \Delta y_u, \Delta z_u)$, Δt & Δt_u, Δr & $\Delta r_u (\equiv \Delta \lambda)$, and Δv & Δr in corresponding Eqs. (2.14), (2.17), (2.19), (2.21), (2.23) & (2.28).

That Eq. (2.31) in inertial state of quantum-real everything is also a final form of its earlier one gross definition in Eq. (1.27) of previous Chapter.

Therefore, in inertial state, such a unified equation also has ultimately emerged out in mutual mirror-imaged form of two equations for simultaneous opposite directions of LE-C and RE-C pair existence with all scales of PSs, and those LE-C & RE-C are appearing to have an 'entanglement' in-between those two through such mutual mirror-images.

However, the Chapter-2, as a continuity of the Chapter-1 which has assumed all ten or twelve CIPs in all scales of particles or systems-of-particles, is actually appearing as one new unified foundation for the micro and macro states of physics through all those 7-UCs when everything is conceptually considered in absence of forces.

As a consequence, next Chapters from 3 to 10 would be the corresponding eight sets of new inferences or predictions from this two initial Chapters 1 & 2. Most importantly, there would be an obvious inference (in Chapter-6) of a non-inertial unified equation for all those same scales of particles or systems-of-particles as well as quantum-realities (in presence of forces) from the inertial unified equation in Eq. (2.31). That would actually have derived a grand unification through quantized extensions of the Einstein Field Equations in General Relativity Theory and Gauge Fields of Forces in Standard Model of Particle Physics. This same would also proceed, through same unified equation in Eq. (2.31), to beyond the limit of quantum-reality where everything could occur through Non-causality or subjective Wills of virtuality (in Chapters - 9 & 10).

However, in next two Chapters, both of Special Relativity Theory and General Relativity Theory would be extended in quantized manner. The Chapter-3 would be the first set of new inferences or predictions in relevance of Special Relativity Theory from this two initial Chapters 1 & 2.

CHAPTER-3:
QUANTIZATION OF SPECIAL RELATIVITY
[Considers Inertial Speed of Light a Quantized value]

"The Theory of Relativity confers an absolute meaning on a magnitude which in classical theory has only a relative significance: the velocity of light. The velocity of light is to the Theory of Relativity as the elementary quantum of action is to the Quantum Theory: it is its absolute core."

— Max Planck

1. Progresses in Concept of Invariance	122
2. Quantized Speed of Light and Special Relativity:	132
2.1. Relativistic Equivalence of Mass & Energy	134
2.2. Relativistic Increment of Mass	136
2.3. Relativistic Contraction of Space	138
2.4. Relativistic Dilation of Time	139
3. Quantized Speeds and Localized Special Relativity:	141
3.1. Respect to Inertial Speed of a Visible-Light photon	146
3.2. Respect to Inertial Speed of a Radio-Wave Photon	150
3.3. Respect to Inertial Speed of a Neutrino	150
3.4. Respect to Inertial Speed of a Hydrogen Molecule	151
3.5. Respect to Inertial Speed of any Scales of Particles	153
4. Quantized Speeds and Invariant Special-Relativity	154
5. Consequences:	157
5.1. Energy as Sum of Discrete Smaller Masses	158
5.2. Equal Mass Can Release Different Energies	159
5.3. Superluminal Quantized Motions	160
Summary	162

Magnitude of the inertial speed of light, as an observer-independent absolute constant, has postulated as universal invariant to all other relatively moving inertial frames of reference in Special Relativity Theory. The same observer-independent absolute constancy in such speed of (wave-corpuscular) light in Chapter-1 has emerged as an intrinsic quantized magnitude for one of scales for photon-particles in electromagnetic spectrum. Not only the photon-particles having intrinsic quantized magnitudes in inertial motions, but in same Chapter-1 it has also postulated that all the (micro to macro) scales of particles or systems-of-particles as any wave-corpuscular phenomena would have similar scale-specific intrinsic quantized magnitudes. As a consequence, each of those scale-specific intrinsic quantized inertial-motions would have similar observer-independent absolute constancies in corresponding magnitudes (alike that inertial-speed of light in Special Relativity Theory). That is, all of those intrinsic inertial-motions, as individual observer-independent absolute constants, are ultimately appearing as individual Scale-Specific Universal Constants (SSUCs) and would also be universally invariant to all corresponding relatively moving inertial frames of reference. Therefore, the structure of Special Relativity Theory, that has constructed respect to one of those SSUCs magnitude for the inertial speed of light or a photon-particle $c \equiv \Delta v_c = 2.99792 \times 10^{10} cm.\,sec^{-1}$, further could be constructed respect to all those SSUCs magnitudes of inertial-motions of particles or systems-of-particles. As a result, there would also be the all scale-specific structures for same Special Relativity Theory correspond to all such SSUCs of intrinsic quantized inertial-motions. Therefore, the Special Relativity Theory respect to one such SSUC in magnitude $c \equiv \Delta v_c = 2.99792 \times 10^{10} cm.\,sec^{-1}$ would be one local instead of universal as it is assumed now. There could be possible all similar localized structures of such Special Relativity Theories respect to all individual SSUCs in magnitudes of quantized inertial-motions for all micro to macro scales of particles or systems-of-particles.

But the Chapter-3 would ultimately universalize all those scale-specific (or localized) structures of Special Relativity Theory, due to all such SSUCs in magnitudes for quantized inertial-motions of particles or systems-particles, with the help of a new universally invariant inverse-relationship in Eq. (2.14) and corresponding universally invariant inverse constant k_2.

The Section-1 has a historical trail through major developments in *ideas* of 'universal invariances'[1]. Those universal invariances were postulated in corresponding foundations of separate mechanical models based on respective observational understandings for same physical nature in different ages of history. The Section-2 would imply the concept of quantization in present universal invariance of Special Relativity Theory, i.e. in the absolute constancy for inertial of speed of light as c. Consequently, the Section-3 would find a series of scale-specific Special Relativity Theories respect to all scale-specific quantized magnitudes for inertial motions of every particle or system-of-particles including the conventional Special Relativity Theory respect to c (for the quantized inertial speed of light or photon). In such a manner, ultimately the Special Relativity Theory respect to the quantized inertial speed of light would appear as one of scale-specific or 'local' theory rather than any universal invariant. Subsequently, the Section-4 would deduce one universally invariant Special Relativity Theory irrespective of all scales of quantized inertial-motions. Finally, the Section-5 would make some additional predictions without violating the basic rules of Special Relativity Theory, e.g. scopes for super luminal quantized inertial-motions of some scales of particles in universe if those particles have quantized inertial mass-energies smaller than a photon with inertial speed of light c, and scope of liberating more energies from one same quantity of mass.

---

1. The result of any physical experiments or consequences of any physical law in one Inertial Frame of Reference should be identical or invariant universally in all other relatively moving Inertial Frames of Reference.

1. Progresses in Concept of Invariance

What would be the exact 'behaviour' of any luminous or non-luminous material-body or physical-body (or particle or system-of-particles) irrespective of the scales, in inertial state or in isolation from the effects of any interacting forces [1], still one fundamental mystery in any Mechanical Theories of Physics. Because, the behaviour of any such conceptually isolated material-body can still be imagined as if a 'pure' state of same. Hence, in absence of any direct quantum-real-signal exchange with such a conceptual 'pure state' is difficult. So, to draw any direct comprehension about the same is more difficult.

But any such 'pure state' of behaviour can be assumed as the intrinsic (as universally invariant) behaviour or property of such inertially isolated material-body.

Hence, any indirect proposition or postulate, based on indirect evidences from other observational sources of understandings or findings about such 'pure' material-body in inertial state, could be the only options.

So, it seems ultimately as one metamorphosis in gradual development of postulates to comprehend indirectly such conceptual pure-state of behaviour for any inertially isolated material-bodies in different periods of history for mechanics. If any such conceptual inertial-behaviour seems to be universally invariant property in any material-body can be assumed through such indirect postulates, the inertial-frames-of-reference or IFRs with same material-body would be also appeared as any universally invariant IFRs respect to all other non-isolated IFRs.

So, the history of progress in Mechanical Theories over the previous centuries is eventually a parallel history of progresses in all such indirect speculations (in the forms of emerging postulates) in search for any such prospective 'invariant'

---

1. Alternately, when any quantum-real-signals cannot be exchanged with it in such inertial isolation.

features in all IFRs. Furthermore, metamorphoses in such prospective 'invariant' features in all IFRs had the direct links with all corresponding background of observational understandings those were available in all particular ages.

Actually, the basic mechanical assumptions through basic postulates of any prospective 'invariant' features for inertial material-body, especially related to the inertial-motions of material-bodies with IFRs have modified number of times in course of history. As a result, we have all corresponding transformations in Mechanical Theories from the Galilean Theory of Relativity to the Einsteinian Special Theory of Relativity.

Alternately, when such mechanical assumptions have emerged in relevance of intrinsic quantization in magnitudes of inertial-matters and inertial-wavelengths (with inverse relationship) with same inertial material-bodies with same IFRs, the result was basically the emergence of Quantum Mechanics.

But Special Relativity Theory (respect to mechanical assumptions related to inertial-motions) and Quantum Mechanics (respect to mechanical assumptions related to inverse co-relationship in-between quantized inertial-matters & wavelengths) within same inertial material-bodies in same IFRs seems now to have any link to unify the both mechanical theories in current physics.

The Eq. (2.14) have presented some newer mechanical assumption or postulate as newer invariant universal constant k_2 that links inversely inertial-motions (in mechanical foundation of Special Relativity Theory) to the same quantized inertial-matters & wavelengths (in foundations of Quantum Mechanics) within every inertial material-body of IFRs. That could have enormous prospect not merely in developments of newer mechanics but such one new mechanical theory can unify the Relativity Theories with Quantum Mechanics by ultimately unifying the physics.

However, in this Section, through following paragraphs, the major historical shifts in basic mechanical postulates (on the basis of corresponding changes in observational backgrounds over the ages) for relevant universal invariances within IFRs have trailed to reach in one quantized extension for Special Relativity Theory in proceeding Sections (as first step for unification of physics in Chapters 5 & 6).

Galilean Invariance: The concept of such universal invariance, to define any physical 'event' invariantly respect to all relatively moving physical bodies or IFRs along a straight line, was first ever precisely known in Galilean Theory of Relativity. Although, before Galileo Galilei, the searching of such universal invariances though was obscured but observed. Starting from the eras of Geo-centric concepts for universe of Ptolemy to the Solar-centric universe of Copernicus, all those had various indirect and imprecise attempts [1] to define the physical nature in one universally invariant manner.

In his book *Dialogue Concerning the Two Chief World System* in 1632, Galileo described the physical laws or events as universally invariant if two relatively moving IFRs would follow a straight line. He had conceptually defined such a universal invariance by comparing the physical laws or event linked (to an observer) within a locked cabin of a smoothly sailing ship in straight line by assuming no 'jerks' for no deviation in path equivalent to another (observer) in a similar locked room of a stationary land on the shore. The results of same physical experiment or occurrence of event in both locked cabin of a moving ship and a stationary room on the shore as universally invariant or identical. Although each of the observers are in mutual relative motions respect to each other. Because they could not recognize (without such 'jerk') whether they are in relatively stationary state on shore or in relatively moving with

---

1. Conventions for universal invariances in physical laws in reference of a conceptual Geocentric (by Ptolemy) geometric-point or Solar-centric (by Copernicus) geometric-point in universe.

a sailing ship. That is, two or more IFRs along the rectilinear relative motions will have such universally invariant physical laws or events until there is any jerk i.e. diversion from the (inertial) straight line through intervention of any force.

Then, such universal invariance of physical laws does not depend on any relative motion (or intrinsic-motion of the IFRs) but on the rectilinear relative motions of the IFRs in the Galilean Theory of Relativity (Mechanics).

Therefore, in principles of the Galilean invariance, any physical law remains would be a universally invariant in all relatively moving physical bodies or IFRs till those are moving in the straight line. Any deflection of such relative motions from the straight lines (due to intervention of forces) causes jerks (i.e. accelerations or decelerations). Due to those jerks or diversions from the straight lines, a physical law or an event would be any more universally invariant to all other relatively moving physical bodies or IFRs in rectilinear motions.

Moreover, in the Galilean Theory of Relativity, it also seems that all those individual relatively moving rectilinear inertial-motions, along the straight line and conceptually intrinsic to all relatively moving physical bodies or IFRs, would have also the absolute sense of motions. That was not accounted in such principles of universal invariance for the physical laws or events in corresponding material bodies or IFRs.

Then Galilean Theory of Relativity having the *rectilinear* universal invariant physical laws when all IFRs in relative motions of inertial-motions intrinsic (absolute) to same IFRs.

Newtonian Invariance: The Galilean principle for such invariance of physical laws in relatively moving IFRs in straight line was next modified by introduction of 'absolute sense magnitudes' in same inertial-motions but equally relative respect to other material-bodies or IFRs. Such absoluteness in inertial-motions of inertial material-bodies or IFRs, in Newtonian Invariance, were assumed in three broader groups of magnitudes: *zero* (those could be in absolute rest), *infinity* (as an highest possible absolute inertial-motion), and *finite* (those have all non-zero & non-infinite absolute inertial-motions) through Newtonian Laws of Motions. Then, in basic conventions of Newtonian Theory of Relativity, the principle of invariance for physical laws or events could be achieved through exchange of signals with motions up to the magnitude of infinity in between two or more relatively moving material bodies or IFRs. Through exchange of any such infinite-motion signals, there would be no practical differences in time, space & motion for an occurred 'event' in universe to all relatively moving IFRs.

In 1687, Isaac Newton was also assumed, in his *Philosophiæ Naturalis Principia Mathematica*, the space & time as universally omni present type of absolute parameters. The universal omni present space as if 'fills' by another omni present entity 'aether'. The universal omni-present time, as separate thing, 'flows' through such universal space. Such universal space and time are associated to all inertial material-bodies centred in each IFRs. Any physical law or 'event', that occurs in any of such material-bodies centred with IFRs at anywhere of that universal space and time, could be exchanged universally through signals of infinite motions to all other relatively moving IFRs with no differences in time and space. Consequently, the physical laws or event in one IFR or results of one physical experiment in a material-body centred with IFR would be universally invariant (through such infinity motion of signals) to all other relatively moving inertial material-bodies or IFRs.

Einsteinian Invariance: In 1676, it was confirmed through calculations of Ole Romer that light has finite speed through the space instead of infinite. Therefore, it was known during the Newtonian era that any fastest possible real signal (like light) in universe through space could move with finite speeds, and could not be exchanged in infinite speeds in-between the relatively moving inertial material-bodies or IFRs. Later, that

126

inertial finite speed of light, with great technical precision, has measured today approximately $2.99792 \times 10^{10} cm.sec^{-1}$.

However, after 218 years of Romer, two very important observational comprehensions have emerged in physics with enough potentials in creation of doubt in basic ideas of such Newtonian Invariances. James Clerk Maxwell, in year 1865, had revealed in his Electromagnetic Equations an observer-independent type constancy in the speed c for propagation of an electromagnetic wave in the spectrum or EMS though the space. The light wave is also one of such electromagnetic waves on EMS. Next was a 'null' result that was found in experiments, during the year 1887, while trying to measure the absolute motion of earth respect to that conceptual aether (assumed to fill the entire universal space) in absolute state of rest, by Albert A. Michelson and Edward W. Morley. Such a "null" result shown in their experiments could be inferred as if either there no existence of aether in absolute state of rest or the used light for signal exchange in-between two relatively moving concerned senders and receivers in experiments having an independent magnitude of speed.

That null result was not only raised questions for existence of conceptual universal aether in absolute state of rest but equally for existence of the absolute universal space in universe. Alternately, the same null result had also hinted about whether the magnitude of such inertial speed of light in free space (under negligible effect of very low gravitational field strength of earth on it) would be invariant or 'observer independent' constancy respect to all other relatively moving IFRs (also stated in Maxwell's electromagnetic law). Later, this hint was very precisely taken into considerations by Albert Einstein in year 1905 through his paradigm shifted article *Electrodynamics of Moving Bodies*.

Therefore, conventions about such observer-independent or absolute or intrinsic magnitude of the inertial speed of light, based on the background of all above theoretical as well as observational comprehensions in physics up to end of the nineteenth century (starting from astronomical calculations of Romer as finite value to the experiments of Michelson and Morley), have finally turned in one concrete postulate in foundation of the Special Relativity Theory or SRT. That has not only now assumed as unique value for inertial-motion of light with finite, constant and observer-independent magnitude $c = 2.99792 \times 10^{10} cm.sec^{-1}$ but also has considered as only universal invariant value irrespective any other (relative) inertial-motions for corresponding non-luminous inertial material-bodies in IFRs.

Then, in comparison to the Newtonian Theory of Relativity, the inertial-motions for all material bodies or IFRs in SRT can further categorize in two broad categories, e.g. *'luminous'* and *'non-luminous'*. This would exclude the ideas of earlier universal aether (due to experimental inability to detect it). The luminous-IFRs move in constant observer-independent inertial-motion c respect to all other relative inertial-motions of material-bodies or IFRs as its senders & receivers. The inertial-motions of non-luminous-IFRs, in contrary, are not only have dependences on respective sender & receiver inertial-motions' but also having the relative inertial-motions respect to each other.

As a result, in SRT conventions, there would be no IFRs to have any 'inertia of absolute rest' instead of the 'inertia of relative rest' (for two equal inertial-motions). Because, at least through any practical experimentation, it would be impossible to detect any such absolute rest in universe compare to the finite exchange speed of light in assumptions of Laws of Motions in Newtonian Theory of Relativity. Similarly, there would be also no infinite inertial-motions for any IFRs. Because anything with infinite inertial-motions never can be exchanged through any such finite inertial-motion signals or speed of light. Then, in SRT, all the possible inertial-motions would be non-zero and finite within the range of finite exchange limits of motions for any of those finite speed signals. Therefore, all those non-zero and finite inertial-motions within the realms of SRT are ultimately appeared as either non-luminous observer-dependent relative-motions below that magnitude of c or luminous observer-independent absolute-motion equals to magnitude of c.

Consequently, due to absence of any communicable signals with infinite inertial-motions, as were presumed in Newtonian Relativity, any 'event' that occurs in an IFR and involves with the other physical parameters like space, time, mass, energy etc. would appear in other relatively moving IFRs must be exchanged with all variable magnitudes. That is, in absence of any infinite motion signal in physical nature, all events in physical nature would appear in all other corresponding relatively moving IFRs must be 'local' or variable rather than any universal invariant.

But the same physical laws or events would be further universally invariant in all those relatively moving IFRs, if compare to that finite, universally invariant, observer-independent, absolute-inertial-motion of c for light through the principles of SRT.

Such principle of invariance for physical laws or events, in-between two non-luminous relatively moving IFRs, can exchange universally invariant respect to that c through the Lorentz Transformations.

Quantum Invariance: However, not only the inertial-motion, but there is also the inertial-matter (or inertial mass-energy) which is also intrinsically involving with each of those same corresponding material-bodies in IFRs those have mentioned in above paragraphs. Such inertial mass-energies have later emerged in the forms of any particles or systems-of-particles. That is, the same IFRs have started to appear with 'intrinsic quantized properties in magnitudes' in observational background from the beginning of twentieth century.

It was initiated through the Black Body Radiation experiments of Max Planck in 1900. But overwhelmed through the descriptions of Photo Electric Effect by Albert Einstein in 1905, Wave-corpuscular Duality hypothesis by Louis De Broglie 1924, replacement of earlier Electromagnetic Theory of Maxwell by the Quantum Electrodynamic (QED) Theory and so on. The emergence of Quantum Mechanics, as a new Mechanical Theory for same physical nature by incorporating the principle of quantizations in magnitudes, particularly for magnitudes of any inertial matter. The consequence is Quantum Mechanics.

But Quantum Mechanics is only Mechanical Theory in physics that did not fundamentally accounted any inertial-motions in its basic principles of universal invariance. Instead the intrinsic quantized-magnitudes of inertial-mass-energies (Δm) and (de Broglie) inertial-wavelengths ($\Delta \lambda$) for every same inertial material-body in the centre of those IFRs. It has also assumed that those two parameters, intrinsic inertial-mass-energies and intrinsic inertial wavelengths in each of those inertial-matters, have a universally invariant inverse relationship (de Broglie wave-corpuscular relationship) as new universal invariance for all IFRs in universe in Eqs (1.1) & (2.1).

However, each of those intrinsic scale-specific quantized magnitudes like Δm & $\Delta \lambda$ in Eqs. (1.1) & (2.1) for the IFRs are also the individual SSUCs in the respective discrete magnitudes. That is, each of those SSUC quantized magnitudes would be also always any universally invariant magnitude correspond all IFRs in universe. But, if the specific scale of particles or systems-of-particles are changed then automatically the magnitude of the same universal invariant would be changed to another. That is why each of those quantized magnitudes as universal invariant has stated as SSUCs in previous Chapters 1 & 2. quantized magnitude in contrary to other types of universally invariant inverse relationships like k_1 in Eq. (2.1) as UC and such UCs would remain be invariant irrespective of changes in scales of SSUCs.

The Quantum Mechanics constructs corresponding Wave Equation for every such change in scales (SSUCs) of quantized 'events' involving with inverse inertial mass-energies or wave-corpuscular phenomena. But Each of those SSUCs wave equations would be universally invariant respect to the UC k_1 in Eq. (2.1) from the de Broglie wave-corpuscular inverse relationship.

The Quantum Mechanics requires no differentiations in categories of IFRs through same de Broglie wave-corpuscular laws in terms of luminous and non-luminous as in basic principles of SRT. But instead of that, the same Quantum Mechanics considers each of those scale-specific quantized magnitudes of Δm & $\Delta \lambda$ as SSUCs and the outcome k_1 as UC two different types of universal invariants in universe.

Inverse Invariance. The same k_1 in Eq. (2.1), as universal invariance, has assumed as one of UCs among total seven of such UCs in Chapter-1. Then, that k_1 as UC could be termed as *Inverse Invariance* for convenience.

But after emergence of the SRT in the beginning of twentieth century, the background observational understandings about same physical nature have many progresses through observations in particle physics and astronomy. The inertial light or luminous IFRs that was once considered as wave in SRT is now assumed simultaneously as wave and corpuscles in Quantum Mechanics.

All other (non-luminous type) different IFRs once in SRT are also now emerged as simultaneous waves and particles (corpuscles) through de Broglie's wave-corpuscular law. Moreover, in various observations of the Particle Physics, it has now also perceived that different scales of particles or systems-of-particles (or IFRs) having the different intrinsic magnitudes of inertial-motions similar to one photon can have for its intrinsic inertial-motion c which has assumed in SRT as universally invariant. Therefore, the same SRT equations, respect to that intrinsic inertial-motion c of a photon, could be also deduced respect to all such different SSUCs intrinsic magnitudes of inertial-motions of those particles or systems-of-particles. Consequently, the SRT correspond to all those different magnitudes of SSUCs intrinsic inertial-motions would become variable or as SSUCs universal invariants.

Then inertial-motions of all scales of particles or systems-of-particles, including $c = 2.99792 \times 10^{10} cm.\,sec^{-1}$ of a photon, could consider as intrinsic SSUCs in quantized magnitudes (Δv). Most importantly, the same intrinsic SSUCs in quantized magnitudes of Δv for every scales of PSs are postulated to have one inverse relationship with the Δm in Eq. (2.14) irrespective of IFRs. Then that emerging constant k_2 in same Eq. (2.14) as one completely new UC in physics that includes the CIPs like inertial-motions Δv in the mechanical foundations of SRT with the Δm in similar mechanical foundation of Quantum Mechanics. Therefore, the UC like k_2 in Eq. (2.14) would be another Inverse Invariance alike k_1 in Eq. (2.1) for any unified mechanical foundations of another new Mechanics.

Beside these two, there are also other UCs ($k_3, k_4, k_5, k_6\, \&\, k$) those are in identical manner could have Inverse Invariant properties in all those same IFRs in corresponding Eqs. (2.17), (2.19), (2.21), (2.23) & (2.28). Although, all of those UCs could have no immediate consequences right now in this Chapter.

2. Quantized Speed of Light and Special Relativity:

The inertial-motion $c\,(= 2.99792 \times 10^{10} cm.\,sec^{-1})$ has postulated in SRT as an intrinsic magnitude of speed that involves with one inertial light-wave is finite, universally invariant, observer-independent and absolute constant unlike the inertial-motions of all relatively moving IFRs. Alternately, the same c as an inertial-motion as one CIP with intrinsic SSUC in quantized magnitude for a photon wave-corpuscle or particle in Eqs. (1.1) & (2.1). However, such appearance of $c \equiv \Delta v_c = 2.99792 \times 10^{10} cm \cdot sec^{-1}$ in Eq. (1.14) there would be no change in the basic principles of SRT. But the impacts of that quantized version of same postulate related to same inertial seed of light in SRT would have huge impact in current physics. In Eq. (2.5), the observer-independent (i.e. intrinsic) and absolute (i.e. quantized) values for inertial speed of light in Eq. (2.12), as Einsteinian Invariance under the principles of SRT, is one of the CIPs in one particle or IFR with SSUCs in quantized magnitude for specific scale of photons in the EMS.

Then, the inertial speed of light in SRT is one intrinsic (i.e. observer-independent) SSUC (i.e. absolute and constant) magnitude of inertial-motion $c \equiv \Delta v_c$ in Eq. (2.12). Therefore, the Eqs. (2.12) & (2.14) are also equally showing that the same $c \equiv \Delta v_c = 2.99792 \times 10^{10} cm\ sec^{-1}$ (in SRT) must have the corresponding scale specific quantized magnitude for its inertial mass-energy Δm_c. In same Eq. (2.12), that Δm_c, along with Δv_c, as CIPs are ultimately as two SSUCs on the same EMS beside other two corresponding SSUCs in magnitudes e.g. $\Delta \lambda_c$ & Δh_c. Hence, the c in SRT equations can also define as

$$c \equiv \Delta v_c = (k_2/\Delta m_c) = \Delta h_c/(\Delta m_c \cdot \Delta \lambda_c) \qquad (3.1)$$

from Eqs. (2.12) & (2.14). Since, all those luminous IFRs would have similar scale with $c \equiv \Delta v_c$ in SRT, with zero relative motions respect to each other, the Eq. (3.1) would also reveal one homogeneous inertial-quantized-state respect to $c \equiv \Delta v_c$ for a specific-scale of photons in the EMS. There would also have corresponding SSUCs in quantized magnitudes for all $\Delta m_c, \Delta \lambda_c$ & Δh_c equally observer-independent, absolute constancies in quantized magnitudes in same inertial state.

Therefore, the Sub-sections below would correspondingly redefine the universally invariant SRT equations in quantized manner in relevance of **(i)** relativistic equivalence of mass & energy, **(ii)** relativistic increments of masses, **(iii)** relativistic contractions of space, and (iv) relativistic dilation of time by assuming that $c \equiv \Delta v_c$ as one SSUC with universal independence of relative motions of all IFRs:

2.1. Relativistic Equivalence of Mass & Energy

The relativistic mass energy equivalence as defined in SRT, in terms of an equation respect to quantized inertial-motion of a particular scale of photons in EMS $c = \Delta v_c = 2.99792 \times 10^{10} cm\, sec^{-1}$, is

$$E = m_0 \cdot c^2. \tag{3.2}$$

In Eq. (3.2), it shows, a chunk of non-luminous initial non-relativistic mass m_0 could convert into an equivalent amount of energy E if multiplies by the square of $c = \Delta v_c = 2.99792 \times 10^{10} cm\, sec^{-1}$ in Eq. (3.1). From Eqs. (1.1) & (3.1) as well as Planck's quantum relation there would be

$$E = hv = hc/\lambda, \tag{3.3}$$

where same E could not be continuous. But instead of that, the E will be a sum of discrete energy quanta correspond to the magnitude of frequency v for corresponding wavelength λ. Then, for one specific photon with corresponding discrete value $\Delta\lambda_c$ in Eq. (3.1) there must have a quantized mass-energy

$$E_c = h/(c.\Delta\lambda_c) = \Delta m_c. \tag{3.4}$$

From Eq. (2.1) & (2.14) that non-luminous initial mass m_0 in Eq. (3.2) as one CIP must have also any specific-scale with quantized magnitude correspond to any PSs in the universe (as BBCOU)

$$m_0 \equiv \Delta m_0 \tag{3.5}$$

where $c = \Delta v_c = 2.99792 \times 10^{10} cm\, sec^{-1}$ has also assumed as discrete magnitude and E as sum all discrete integer 'packets' of mass-energies in Eqs. (3.3) & (3.4). Therefore, in Eq. (3.2), for that scale-specific quantized magnitude of initial mass (or say for any non-luminous mass) $m_0 \equiv \Delta m_0$ in Eq. (3.5), and since that E could also assume as the integer sum of quantized micro mass-energies (those are seemed to 'free' from the binding forces in Δm_0) of further micro scale of PSs in same Eq. (3.2), we can define the same E from Eq. (3.4) as

$$E = n^* . \Delta m_c \qquad (3.6)$$

where the n^* is an integer. Then, that quantized initial mass Δm_0 in Eq. (3.5), if can consider as a sum of all integer Δm_c in Eqs. (3.4) & (3.6) with similar micro mass-energies (those are seemed to 'bind' by binding forces in Δm_0), would be

$$m_0 \equiv \Delta m_0 = n^* \cdot \Delta m_c = n^* . E_c \equiv E \qquad (3.7)$$

when that Δm_c should have the quantized magnitude Δv_c in Eq. (2.14) which is obviously an observer-independent constant magnitude as c.

In the Eq. (3.7), alternately, the initial mass Δm_0 can also be described from the Eq. (2.14) as one macro-scale of particle or system-of-particles with heavier mass would have inverse corresponding scale-specific quantized 'slower' inertial-motion

$$v_0 \equiv \Delta v_0 \qquad (3.8)$$

Subsequently, that macro-scale of particle or system-of-particles with such initial mass Δm_0 in Eq. (3.7) is as if eroding or melting from its present macro scale (earlier under binding forces) to integer n^* number of 'free' micro scale of PSs (out of that binding forces) as E with smaller quantized masses of Δm_c but scale-specific quantized higher (inverse) inertial-motion $c = \Delta v_c$ due to the same Eq. (2.14). As a result, the Eq. (3.2) of SRT can finally describe in terms of relativistic equivalence for all mass & energy respect to the $c = \Delta v_c$ in quantized manner

$$\Delta E = \Delta m_0 \cdot (c = \Delta v_c)^2. \qquad (3.9)$$

This seems, one with smaller amount of kinetic energy (due to slower quantized motion say Δv_0) in one inertial macro-scale-

PS Δm_0 could erode (through increments of kinetic energy) up to one maximum kinetic energy (for quantized motion $c = \Delta v_c$). When, the integer n^* number of discrete PSs of micro scale quantized mass-energies and inverse higher motions with higher kinetic energy are released out of the binding energies of Δm_0. Therefore, in Eq. (3.9), the so-called non-luminous quantized mass Δm_0 (with higher potential energy with lower kinetic energy) transforms into luminous quantized energies or integer number of smaller quantized mass-energies of micro scale of particles (with lower potential energy but higher kinetic energy) having quantized inertial-motions $c = \Delta v_c$. After the course of that transformation, the Δm_0 would erode into $n^* \times \Delta m_c$. But, continues to hold initial structure (i.e. initial scale) of same Δm_0 after such transformation an intervention of infinite amount of external forces would be needed in relevant equation for relativistic increments of masses in SRT (that will describe in next Sub-Section-2.2).

However, in Eq. (3.9), the conventional SRT equivalence of mass and energy, which has assumed as non-quantized in earlier Eq. (3.2), has defined in terms of quantized manner due to the Eqs. (2.2), (2.14), (1.17) & (1.21). But in doing so it has never harmed anything to basic assumptions of SRT.

2.2. Relativistic Increment of Mass

The relativistic increment of masses in any non-luminous IFRs and the relevant universal invariant equation in SRT is

$$m^* = m_0/\sqrt{(1 - v^*)^2/c^2}, \qquad (3.10)$$

where m^* and v^* are correspondingly relativistic increments in both relative-inertial-mass and relative-inertial-motion of any non-luminous material-body with *initial* relative-inertial-mass Δm_0. The same entity that was assumed in Eq. (3.9). If the value of v^* in Eq. (3.10) becomes equal to c then m^* would be infinite.

However, from the Eq. (3.1), again there must be the $c \equiv \Delta v_c$ in Eq. (3.10), and the Eq. (3.10) could re-write accordingly as

$$m^* = \Delta m_0 / \sqrt{(1 - v^*)^2 / (c = \Delta v_c)^2}, \qquad (3.11)$$

which would also be universally invariant to all relatively moving IFRs. But the $c \equiv \Delta v_c$ as a universal invariant constant with observer-independent value is one SSUC in quantized magnitude and not any UC irrespective of scales.

Subsequently, the relativistic increments of mass, that has defined in same Eq. (3.11) compare to the Eq. (3.10), would have also all SSUCs in quantized magnitudes for all relatively moving IFRs instead of any single UC magnitude for all scales of masses. Then the Eq. (3.11) is also compatible to same Eq. (3.10) in SRT.

There would have obvious fundamental differences in type of those two motions v^* & Δv_c in Eq. (3.11). The v^* is merely a relative difference in magnitude between two inertial-motions involve with two corresponding IFRs.

Whereas, the Δv_c is an intrinsic i.e. an observer-independent, absolute, SSUC in quantized magnitude for inertial-motion of a particular scale of photon particles in EMS.

The relativistic mass m^* in Eq. (3.11) could also be any SSUCs in differences of integer quantized magnitudes as the $m_0 \equiv \Delta m_0$ in Eq. (3.7). Therefore, any relativistic mass m^* in Eq. (3.11) would be

$$m^* \equiv (n_1^* . \Delta m_c) \qquad (3.12)$$

in Eq. (3.7) where n_1^* is an integer number of same Δm_c. Then, for any relativistic mass m^* in Eq. (3.12) compare to initial non-relativistic mass m_0 in Eq. (3.7) would have

$$n^* < n_1^*. \qquad (3.13)$$

2.3. Relativistic Contraction of Space

The measurements of space also as one relative thing in relatively moving IFRs is universally invariant respect to same c in SRT as

$$s^* = s_0 \cdot \sqrt{(1-v^*)^2/c^2}, \qquad (3.14)$$

where s^* is relative inertial space that has relativistic contraction respect to increments in relative motion v^* and the s_0 is initial relative inertial-space at the beginning. The Eq. (3.14) is stated universally invariant to all IFRs due to c in SRT.

But from the Eq. (3.1), there must have $c \equiv \Delta v_c$ in same Eq. (3.14), where the Δv_c would be one SSUC with respective quantized magnitude depends on specific scale of photon in EMS. Then, the Eq. (3.14) also can further write as

$$s^* = \Delta s_0 \cdot \sqrt{(1-v^*)^2/(c \equiv \Delta v_c)^2} \qquad (3.15)$$

where universal invariance of the above equation depends on merely the particular SSUC for quantized magnitude of inertial-motion for photons in EMS. Alternately, the parameter s^* in Eq. (3.15) could also assume as the scale-specific quantized conceptual space of the Δm_c in Eq. (3.7) with corresponding SSUC quantized magnitude of inertial-motion $c \equiv \Delta v_c$.

However, at the SSUC quantized inertial-motion $c \equiv \Delta v_c$ in same Eq. (3.15) there would have obvious relativistic contraction of space up to the magnitude of $s^* = 0$. But in Eqs. (1.17) & (2.23), that relativistic contraction of space would be also 'discrete' in type rather than any 'continuous'. As a result, the initial space s_0 in Eq. (3.15) can write

$$\Delta s_0 \equiv n^* . \Delta s_c \qquad (3.16)$$

where Δs_c is the SSUC quantized inertial-space of Δm_c in Eq. (3.7). As a result, when $s^* = 0$ by achieving after it's inertial-motion $v^* = c \equiv \Delta v_c$ there would be ultimately no relativistic space difference in between two IFRs having the two corresponding Δm_c for two SSUCs in quantized magnitudes $\Delta s_c = \Delta s_c$.

Subsequently, when the same relativistic inertial-space even reaches in that $s^* = 0$ that would actually have the corresponding SSUC in quantized magnitude of inertial-space equal to Δs_c for Δm_c in Eq. (3.7). Hence, the $s^* = 0$ at the inertial speed of $c \equiv \Delta v_c$ would have in actual term

$$s^* = 0 \ (\equiv \Delta s_c). \qquad (3.17)$$

2.4. Relativistic Dilation of Time

The time, which has also defined earlier in scale-specific quantized manner as Δt in all particles or systems-of-particles in Eq. (1.21) and shown to have SSUCs with quantized magnitudes due to all corresponding SSUCs quantized magnitudes of Δr in inertial condition. Obviously, that would include all sense of so-called non-luminous and luminous IFRs in SRT. Most importantly, in that Eq. (1.21), the unit of Δt could have in cm, i.e. in unit of distance, due to the units of Δr as like as Δs in Eq. (1.17), beside usual unit in sec.

But the time which is defined in SRT as relative, in all relatively moving IFRs, that has universalised respect to c. But in other side, the universe (BBCOU) that has described by Eq. (2.28) would have no detectable time with absolute void, no detectable space with same absolute void where corresponding radius of the same void conceptually $\Delta r = 0$. That would be beyond the exchange limit of any quantized signals with obvious $\Delta r \neq 0$. That is, the existence of both such (SSUCs) quantized CIPs like time Δt and radius Δr should have always any attachments with the specific LAR of a volume Δs for specific scale of particles or systems-of-particles. As a result, any inertial magnitudes for both Δt & Δr for any such scales of PSs in Eq. (1.17) must be the observer-independent CIPs with SSUCs in quantized magnitudes. As a result, the scale-specific quantized inertial-time Δt in Eq. (1.21), that is quantized time, is also compatible with the non-scale-specific time Δt in Eq. (1.20) with unit in *sec*.

However, that non-scale-specific time in Eq. (1.20) could measure in relative manner in any relatively moving IFRs but universally invariant respect to c in SRT as

$$t^* = t_0 \cdot \sqrt{(1-v^*)^2/c^2} \qquad (3.18)$$

where t^* is relative inertial-time. That dilates in relativistic manner respect to all relativistic changes in relative-motion v^*, and the t_0 is relative inertial-time at the beginning. The Eq. (3.18) is universally invariant respect to c in all scales of IFRs in SRT.

From the same Eq. (3.1), there must be the $c \equiv \Delta v_c$ in same Eq. (3.18), where the Δv_c is one SSUC quantized magnitude that depends on the scales of particles or systems-of-particles or IFRs. Then, the Eq. (3.18) of SRT also can write in quantized manner as

$$t^* = \Delta t_0 \cdot \sqrt{(1-v^*)^2/(c=\Delta v_c)^2}. \qquad (3.19)$$

However, at the SSUC quantized inertial-motion $c \equiv \Delta v_c$ in same Eq. (3.19) there would have obvious dilation in relativistic time up to the magnitude of $t^* = 0$. But in Eqs. (1.21) & (2.23), that relativistic dilation of time would be also discrete rather than continuous. As a result, the initial time t_0 in Eq. (3.19) can write as

$$\Delta t_0 \equiv n^* \cdot \Delta t_c \qquad (3.20)$$

where Δt_c is the SSUC quantized inertial-space of Δm_c in Eq. (3.7). As a result, when such initial time t_0 becomes the $t^* = 0$ by achieving relative motion $v^* = c \equiv \Delta v_c$, that would have no relativistic time differences in between two IFRs having two corresponding Δm_c in Eq. (3.7). But those two Δm_c with two IFRs would remain have the SSUCs in quantized magnitudes

$\Delta t_c = \Delta t_c$. Subsequently, when the same relativistic time would even achieve at the $t^* = 0$ that would actually have SSUC quantized magnitude of quantized time equal to Δt_c for two identical Δm_c of same scale of particles in Eq. (3.7). Hence, the $t^* = 0$ achieving inertial-motion of $c \equiv \Delta v_c$ would be actually

$$t^* = 0 \ (\equiv \Delta t_c). \qquad (3.21)$$

3. Quantized Speeds and Localized Special Relativity:

Factually, the *light*, not only has one inevitable position in the course of progresses in mechanical theories in physics in the previous centuries, but the same has equally re-presented as if one alien kind of entity in comparison of all other perceivable physical entities in the surrounding universe. It was appearing also with one unique higher magnitude of speed. Earlier such a speed was assumed to have infinity magnitude. Although, it is now finally a finite. Next was its way of propagation that imagined in rectilinear path. But it is now in curved paths, along the geodesic, due to curvature of universal spacetime. It has also elementary wave property. But later have emerged with wave-corpuscular duality. Consequently, the same 'light' simultaneously appears as a wave and a corpuscle. It makes our unique biological sense of perceptible visibility that no other non-luminous material can do. The *fire*, that causes same light to emit through the process of exothermic chemical reactions within so-called non-luminous substances, also once was assumed as one of the basic 'elements' in physical nature.

In 1638, Galileo was not only first to find the speed-of-light through any experimental basis, if not infinite, must have at least ten times higher magnitude than the speed-of-sound. Then highest measurable unit of anything in physical nature. But thirty-eight years later of it, while studying eclipses of Jupiter's moon *Io*, Romer was first estimated the magnitude for such finite speed of light. He had managed to calculate speed of light that takes about 22 minutes to travel a distance equal to a diameter of earth's orbit around the sun. From this, he inferred that speed of light travels about 2,20,000 kilometres per second (which was about 26% less compare to today's estimate 299,792.458 km/sec).

In 1637, one year before Galileo, natural philosopher René Descartes argued that the light is made of small discrete particles or 'corpuscles' which are travelled with finite speed along the rectilinear paths. But that particle theory of light was unable to explain the phenomena like refraction, reflection, interference and polarization. Afterward, during years of 1672, Newton was pioneered the corpuscular theory of light. He reasoned that a wave cannot travel in straight line. Then, that corpuscular doctrine of light was dominated in physics for next 100 years or so. But later, such corpuscular hypothesis became failed due to not having any satisfactory descriptions for the phenomena like interference, diffraction and polarization for light. Later years it was mostly abandoned. The 'Wave Theory of Light' was dominated in physic that was proposed once by Christiaan Huygens in 1678 who was also contemporary to Newton.

However, in the 1865, in his *A Dynamical Theory of the Electromagnetic Field*, Maxwell had uniquely defined same light not merely as an electromagnetic wave that creates out of the disturbances in electromagnetic fields but as wave that would propagate in a finite and constant speed c independent to the speeds of its source and receiver fields. This unique independent property in the speed of light was also gained indirect experimental support in 1887 through experiments of Michelson-Morley's. While they were trying to find out if there any absolute inertial-motion of earth compare to absolute state of rest for universally existing aether that was presumed over the previous centuries 'filling' the entire universal space.

Thereafter, another new episode related to same light was started with beginning of the last century. The corpuscular convention of light was revived again, beside its wave property, through black body radiation experiments of Planck in 1900. Due to this, the first decade of twentieth century not only had a great role in further illumination of entire physics of all previous centuries through light but simultaneously it was also initiated a split in physics of after decades through same light. The corpuscular property of light was well footed in term of 'quanta' for electromagnetic waves through works of Planck (in 1900) and 'photons' of same electromagnetic waves in explanation of Photoelectric Effects by Einstein (in 1905). The same light was parallelly assumed remain as wave with absolute constancy and observer-independence to the motions of its sender and receiver in foundations of SRT by Einstein (in 1905).

However, incorporation of such observer-independent absolute constancy c in foundation of SRT had changed the all absolute perceptions of all basic physical parameters like space, time and matter involving with every IFRs in physical nature during the same decade. But all those relative parameters have had universalized respect to such postulated universally invariant value c.

In addition, it was also in the first time appeared in history of physics in same decade through SRT that all relatively moving inertial material-bodies or IFRs in two basic categories: one is 'luminal' and other is 'non-luminal' in types. A luminal-IFR would have three very distinct behaviours in its *inertial* (i.e. in absence of affecting forces) magnitude; e.g. it would have **(i)** a universal constancy or observer-independence or intrinsic finite magnitude of inertial-motion, **(ii)** an absolute constancy or observer-independence or intrinsic (discrete like) magnitude - of inertial mass-energy, and **(iii)** an absolute constancy or observer-independence or intrinsic (discrete like) magnitude of inertial wavelength in EMS. In contrary, all non-luminal IFRs would have all corresponding relative inertial-motions respect to each other but finite, no absolute zero and sub-luminal type of speeds.

Another new thing that might be immerged in the equation for 'quanta' of Planck, for electromagnetic waves, would be an inverse relationship in-between quantized mass-energies and corresponding wavelengths in EMS.

But in 1924, that was introduced as universal inverse relationship in Eq. (1.1) as wave-corpuscular duality in all scales of particles or systems-of-particles by de Broglie. That would have a universal inverse constant k_1 in Eq. (2.1).

Those were in earlier foundation of SRT assumed as non-luminal-IFRs later appeared to have identical wave-corpuscular duality in all luminal IFRs through same de Broglie's equation. That is all luminous and non-luminous IFRs in SRT are as wave-corpuscular-phenomena.

Moreover, in various experiments of todays' particle physics, it is also comprehended that some fermions (i.e. the so-called non-luminous-IFRs in the range of micro scales have the scale-specific observer-independent constant magnitudes of inertial-motions) similar to that c for a photon (boson) in EMS in foundation of SRT.

However, as was stated in above, making any of our sense for biological 'vision' is also light specific. The notion for such sense-of-vision have had also many changes in theories over the centuries in the types from *extromission* to *intromission*. Initially for extromissonists, date back to Empedocles in fifth century BC, it was assumed that sense of vision or visual perception was due to something in the form of 'eye beams' that emits from the eye and reflected back from visual object to that eye again. But later, the intromissionists' notion about 'vision' has now dominated for the sense of vision for any visible object. It states sense of vision emerges especially due to entry of reflected light from visible object into the eye. Although, Newton and John Locke in seventieth century were notable modern propagators of such intromission theory but same was initiated by Aristotle and Galen in fourth century BC. Moreover, Newton had shown white light is composite and a sum of total seven different colours where black is as an absence of any such colours in light. Those different seven colours in light can also split separately through a prism in creation for sense of colours in human eyes. Now this theory has more established by describing that sense of vision by shifting from the eyes to the brain through all complex neuro-somatic processes.

Therefore, ultimately, the human eyes are biologically adopted to make such sense of visions through a very limited range of wavelengths e.g. VIBGYOR for corresponding photons in whole EMS. Hence, such range of our observations through wavelengths in EMS have varied through biological evolutions or by the inclusions of instruments. Consequently, there can be such different observers who may have capacities to observe same universe through their different range of wavelengths of light in EMS by intaking of different wave-corpuscles in their 'eyes'. Since all luminous and non-luminous wavelengths would be respective wave-corpuscles in de Broglie's Eqs. (1.1) & (2.1), that range of observational wavelengths could also be extended up to the non-luminous wavelengths beyond those luminous range in EMS as any observer. Hence, conceptually, observers in universe can have any corresponding scales of wave-corpuscles, irrespective of luminous or non-luminous in types, for the 'vision'.

Now the convention of light, in terms of its uniqueness in observer-independent magnitude for speed and wave-corpuscular duality, is involving all our senses of such vision in every relatively moving IFRs. Hence, conceptually one can have not only the indifference between a luminous and the non-luminous types of IFRs in terms of such quantized wavelengths and quantized mass-energies of same light but equally in terms of observer-independent scale specific inertial magnitudes for its quantized speed. As a result, there will be all observer-independent scale-specific constant inertial-motions Δv in Eq.

(2.14) and each of those Δv can replace that only one $c \equiv \Delta v_c$ or only one quantized speed of light in SRT in the Eqs. (3.9), (3.11), (3.13), (3.15) & (3.19). The Sub-Sections below would deduce all such different sets of SSUCs for different quantized values of Δv but similar universally invariant as were in SRT equations:

3.1. Respect to Inertial Speed of a Visible-Light photon

In SRT, the foundational postulate in relevance of observer and receiver relative motions' independence of constant inertial-motion c for light had basically three major observational basis: **(i)** the 'null' results, those were found while measuring anticipated relative motion of light respect to (to and from) motions of earth in orbit that conceptually moving through aether, in Michelson & Morley experiments; **(ii)** prediction of constancy in c, as source-motions' independent propagation for any electromagnetic wave in EMS through space by Maxwell's Equation; and **(iii)** such finite and constant magnitude of c was mostly had measured correspond to visible light in the EMS.

However, todays' overwhelming expansions in observational horizon of physics, compare to the last half of nineteenth century, may have more information regarding the same c. If so, such additional information can enhance indirect supports to comprehend more precisely about (intangible) inertial behaviours of both luminous & non-luminous IFRs.

For example, at present, the overwhelming quantized perception that seems to have within all those earlier material-bodies or IFRs (as any scale of particles or systems-of-particles) as one such inertial-behaviour for all luminous and non-luminous IFRs (after SRT). The de Broglie's wave-corpuscular universal inverse relationship within all those same (quantized) inertial material-bodies as described in Eqs. (1.1) & (2.2) would be another such inertial-behaviour for all luminous & non-luminous IFRs (after SRT).

Subsequently, from those quantized inertial-behaviour and inverse relationship between waves & corpuscles in all inertial material-bodies would have further – **(i)** a universal-invariance for physical laws in all IFRs, **(ii)** SSUCs in quantized magnitudes of all inertial matters irrespective of luminous & non-luminous IFRs in terms of SRT, and **(iii)** a wave-corpuscular duality in light as any scale of photon-particle in same EMS similar to one scale of any non-luminous PSs.

There can also be similar other new observational understandings in todays physics those might have scopes to emerge as new basis to draw more precise assumptions about inertial behaviours of all same luminous and non-luminous IFRs after outbreak of SRT.

The non-quantized electromagnetic field theory of Maxwell (as one of observational basis of SRT) is now totally replaced by the QED field theory where c has wave-corpuscular duality.

The 'null' results in Michelson & Morley's Experiments, that was assumed (in SRT) as an observer-independent constancy for inertial-motion of $c = 2.99792 \times 10^{10} \, cm \, sec^{-1}$ of light, had conducted under weak gravitational field strength of earth with very lower escape velocity about 11.2 km/sec (compare to such higher velocity of light), could see same type of null results or observer-independent inertial-motion if it can be conducted through exchange of a "bullet" from a gun under corresponding escape velocity any comparatively minor gravitational mass than earth. There the speed of such a bullet would appear similarly higher than its escape velocity in comparison of the speed of light for escape velocity of earth.

In Chapters 1 & 2, each of those wave-corpuscular PSs, irrespective of micro and macro scales or luminous and non-luminous IFRs, are postulated to have all SSUCs in discrete magnitudes for all 5+5 inverse CIPs along with total 7 numbers of UCs.

The inertial-motion Δv is one of such 10-CIPs in Eq. (2.14) with al SSUCs in quantized magnitudes inversely co-related with similar SSUCs if quantized magnitudes of Δm through inverse UC k_2. Even, different scales of photons in EMS correspond to specific wavelengths in inertial state seems to have such different intrinsic SSUCs in quantized magnitudes for inertial-motions in Eq. (1.14).

Therefore, the constant c in SRT that defines as an observer-independent absolute constant for inertial-motion of light in inertial state has assumed as one such SSUC for any specific scale of photons. As a result, in SRT that absolute constant has appeared only for the $c = 2.99792 \times 10^{10} cm\ sec^{-1} = \Delta v_c$ in Eq. (1.14). Hence, there could be all such different scale-specific SRT equations respect to all other SSUCs in quantized magnitudes of inertial-motions for all scales of PSs with $c = \Delta v$ including that $c = \Delta v_c = 2.99792 \times 10^{10} cm\ sec^{-1}$.

Moreover, from the Sub-Section of 1.3.1. in Chapter-1, there would be all different magnitudes of such observer-independent constants for the quantized inertial-motions, even in different scales of photons, depend on corresponding wavelengths or quantized mass-energies in same EMS. That starts from the wavelengths of gamma-rays in one end to the radio-waves in other.

In Sub-Section of 1.3.1. (iii) of same Chapter-1, that observer-independent constant magnitude for inertial-motion was postulated in SRT equations for that $c = \Delta v_c = 2.99792 \times 10^{10} cm\ sec^{-1}$ corresponds to one scale of photons could assume against corresponding wavelength expectedly exist anywhere within the range of yellow colours under visible light zone of wavelengths in EMS.

In the same EMS, there would be both slower and faster observer-independent SSUCs in quantized magnitudes of inertial-motions for some photons compare to those with such SSUCs in quantized magnitude of $c = \Delta v_c = 2.99792 \times 10^{10} cm\ sec^{-1}$. Therefore, on the same EMS that c, as in SRT, would have all different SSUCs in quantized observer-independent constant magnitudes for different photons respect to their intrinsic wavelengths.

Not only that, in Sub-Section-1.3.2. in same Chapter-1, we also have different scale-specific observer-independent constant (SSUCs) quantized magnitudes even for the different scales of fermions as different non-luminous-PSs as like as different scales of photons or bosons. In Sub-Section-1.3.3 in same Chapter-1, all scales of PSs are appeared to have different scale-specific observer-independent constant i.e. quantized magnitudes of inertial-motions as one CIP with SSUC.

As a consequence, it appears that all SSUCs in quantized values for $c = \Delta v$ as corresponding SRT-constants could have also all such similar range of scale-specific quantized or observer-independent constant magnitudes in all scales of PSs (including that magnitude of $c = \Delta v_c = 2.99792 \times 10^{10} cm\, sec^{-1}$).

Consequently, the corresponding set of SRT Eqs. (3.9), (3.11), (3.13), (3.15) & (3.19) respect to the $c = \Delta v_c = 2.99792 \times 10^{10} cm\, sec^{-1}$ for a photon with wavelength expectedly somewhere within the yellow colours under visible light zone in EMS as

$$E = m_0 \cdot (c = \Delta v_c)^2, \qquad (3.22)$$

$$m^* = m_0/\sqrt{(1-v^*)^2/(c = \Delta v_c)^2}, \qquad (3.23)$$

$$t^* = t_0 \cdot \sqrt{(1-v^*)^2/(c = \Delta v_c)^2}, \qquad (3.24)$$

$$s^* = s_0 \cdot \sqrt{(1-v^*)^2/(c = \Delta v_c)^2}. \qquad (3.25)$$

where (m^*, t^*, s^* and v^*) are corresponding relativistic mass-energy, time-dilation, contraction-of-space and motion respect to the ($c = \Delta v_c$) for any IFR. The (m_0, t_0 & s_0) are specific scales of quantized magnitudes for non-relativistic initial matter, time and space for same in Eqs. (3.22), (3.23), (3.24) & (3.25) respectively. All those values of $(m^*, t^*, s^* \& v^*)$ and $(m_0, t_0 \& s_0)$ in corresponding set of SRT Eqs. (3.22), (3.23), (3.24) & (3.25) are universally invariant to all relatively-moving IFRs exchangeable by signals of quantized constant observer-independent inertial-motion $c = \Delta v_c$.

3.2. Respect to Inertial Speed of a Radio-Wave Photon

Likewise, for the SSUCs of the observer-independent constancy in quantized magnitudes of inertial-motion for a scale of radio-wave photon in Eq. (1.14), there would be the $c = \Delta v_\eta$ in same set of above SRT Eqs. (3.22) - (3.25)

$$E = m_0 \cdot (c = \Delta v_\eta)^2, \tag{3.26}$$

$$m^* = m_0 / \sqrt{(1-v^*)^2/(c = \Delta v_\eta)^2}, \tag{3.27}$$

$$t^* = t_0 \cdot \sqrt{(1-v^*)^2/(c = \Delta v_\eta)^2}, \tag{3.28}$$

$$s^* = s_0 \cdot \sqrt{(1-v^*)^2/(c = \Delta v_\eta)^2}, \tag{3.29}$$

where obviously the corresponding magnitudes of relativistic parameters $E, m^*, t^* \& s^*$ must be different due to $c \equiv \Delta v_c \neq \Delta v_\eta$. But, respective magnitudes for all non-relativistic initial magnitudes $E_0, m_0, t_0 \& s_0$ may be same in those two sets of SRT Eqs. (3.22) - (3.25) and (3.26) - (3.29).

3.3. Respect to Inertial Speed of a Neutrino

In Sub-Section-1.3.2. of Chapter-1, it has also considered that the fermions of different scales, as different scales PSs with similar wave-corpuscular duality, would have different SSUCs in quantized magnitudes of inertial-motions. That is, different observer-independent constancies for all conceptual scale-specific quantized magnitudes. No matter those are yet measured or not. Therefore, the different SRT sets of equations could be derived in terms of such different SSUCs in quantized magnitudes of constant inertial-motions of PSs (with wave-corpuscular properties) in Eq. (2.2).

Subsequently, any such specific scale of wave-corpuscular neutrino, as one of such similar PSs, within the mass-energy spectrum of all neutrinos, say with scale specific quantized magnitude of mass-energy Δm_{n_1} will have the corresponding observer-independent SSUC inverse quantized magnitude of inertial-motion Δv_1 in the Eq. (2.14). Therefore, respect to that Δv_1 in the set of SRT Eqs. (3.22) - (3.25) can write as

$$E = m_0 \cdot (c = \Delta v_1)^2, \qquad (3.30)$$

$$m^* = m_0/\sqrt{(1-v^*)^2/(c = \Delta v_1)^2}, \qquad (3.31)$$

$$t^* = t_0 \cdot \sqrt{(1-v^*)^2/(c = \Delta v_1)^2}, \qquad (3.32)$$

$$s^* = s_0 \cdot \sqrt{(1-v^*)^2/(c = \Delta v_1)^2}, \qquad (3.33)$$

where the Eqs. (3.30), (3.31), (3.32) & (3.33) respect to SSUC quantized magnitudes of inertial-motion for a neutrino would be a set of SRT equations so-called non-luminous in type. Although, those non-relativistic initial parameters like E_0, m_0, t_0 & s_0 would have same magnitudes in beginning with those two concerned IFRs in two respective sets of SRT Eqs. (3.22) - (3.25) and (3.30) - (3.33) but due to corresponding differences in constant magnitudes of quantized inertial-motions $c \equiv \Delta v_c \neq \Delta v_1$ there would be all respective differences in relativistic magnitudes of parameters E, m^*, t^* & s^* in those same two sets of SRT Eqs. (3.22) - (3.25) and (3.30) - (3.33).

3.4. Respect to Inertial Speed of a Hydrogen Molecule

From the same Eq. (2.14), not only a fermion like neutrino but all other scales of fermions as PSs, would have also intrinsic scale-specific or SSUCs in quantized magnitudes of inertial-motions.

Even an atom, being such a specific scale of mass-energies as PSs, in the spectrum of all 105 atoms of periodic tables different chemical properties, would have also respective scale specific or SSUCs in quantized magnitudes of such inertial-motions. Although, inverse quantized magnitudes of inertial-motions for such an atom would be far smaller compare to a neutrino of $c = \Delta v_1$ in Eqs. (3.30) - (3.33) and a photon of $c = \Delta v_c$ in Eqs. (3.22)-(3.25) but inverse quantized magnitudes of mass-energies of same atom say Δm_a would be far heavier in Eq. (2.14).

Atoms are usually not in free motions except inert-gases except within deep astronomical void like spaces. The lightest atom, the ordinary hydrogen atom, can also be in free motions in the form of any ordinary hydrogen molecule by two such atoms.

If the quantized mass-energy for such an ordinary hydrogen molecule is say Δm_2 then its inverse quantized inertial-motion from Eq. (2.14) would be say Δv_2. That inertial-motion of such a hydrogen molecule would have also an observer-independent SSUC in discrete magnitude of Δm_2. Therefore, such an observer independent SSUC in quantized magnitude of inertial-motion must be another $c \equiv \Delta v_2$, although that $c \equiv \Delta v_2 \neq \Delta v_1 \neq \Delta v_c$ in corresponding sets SRT Eqs. (3.30)-(3.33) and (3.22)-(3.25).

Subsequently, there would be another set of SRT equations respect to such $c = \Delta v_2$ for the SSUC in quantized magnitude of constant inertial-motion for an ordinary free hydrogen molecule

$$E = m_0 \cdot (c = \Delta v_2)^2, \qquad (3.34)$$

$$m^* = m_0/\sqrt{(1 - v^*)^2/(c = \Delta v_2)^2}, \qquad (3.35)$$

$$t^* = t_0 \cdot \sqrt{(1 - v^*)^2/(c = \Delta v_2)^2}, \qquad (3.36)$$

$$s^* = s_0 \cdot \sqrt{(1 - v^*)^2/(c = \Delta v_2)^2}, \qquad (3.37)$$

where obviously the magnitudes of all corresponding relativistic parameters $E, m^*, t^* \& s^*$ would be different compare to the same in earlier two sets of SRT Eqs. (3.30)-(3.33) and (3.22)-(3.25).

But all initial non-relativistic parameters $E_0, m_0, t_0 \& s_0$ may have same magnitudes in the beginning for all those three respective IFRs with respective values of $c = \Delta v_2 \neq \Delta v_1 \neq \Delta v_c$ though in the end there would be all different relativistic magnitudes for parameters $E, m^*, t^* \& s^*$ for all those three sets of SRT Eqs. (3.30)-(3.33), (3.33)-(3.25) and (3.34)-(3.37).

3.5. Respect to Inertial Speed of any Scales of Particles

Now finally, for every scale of PSs, irrespective of micro to macro scales, boson and fermion particles or systems-of-particles, luminous and non-luminous IFRs with corresponding scale-specific quantized magnitudes of inertial mass-energies Δm as SSUCs have the inverse quantized magnitudes for inertial-motions Δv as SSUCs in Eq. (2.14).

Subsequently, each of those scale specific observer-independent constant SSUCs in discrete magnitudes of inertial-motions would have the all similar scale-specific equivalent magnitudes for $c \equiv \Delta v$, and respect to each of those scale-specific values of $c \equiv \Delta v$ there would be a universal set of SRT equations

$$E = m_0 \cdot (c = \Delta v)^2, \qquad (3.38)$$

$$m^* = m_0/\sqrt{(1-v^*)^2/(c = \Delta v)^2}, \qquad (3.39)$$

$$t^* = t_0 \cdot \sqrt{(1-v^*)^2/(c = \Delta v)^2}, \qquad (3.40)$$

$$s^* = s_0 \cdot \sqrt{(1-v^*)^2/(c = \Delta v)^2}. \qquad (3.41)$$

The magnitudes of relativistic parameters $(E, m^*, t^* \& s^*)$ in all those scale specific sets of SRT Eqs. (3.38)-(3.41) correspond to any scale-specific SSUCs quantized magnitudes of $c \equiv \Delta v$ will represent all the *scale-specific local* values.

All the initial parameters $m_0 = \Delta m_0, s_0 = \Delta s_0, t_0 = \Delta t_0$ beside E_0 may have same values in all respective local scale-specific IFRs with all scale-specific local sets of SRT in Eqs. (3.38)-(3.41).

Consequently, due to all scale specific local SSUCs observer-independent constant magnitudes of quantized inertial-motions $c \equiv \Delta v$ for all micro to macro PSs in Eq. (2.14), the set of Eqs. (3.38)-(3.41) actually depicts all scale-specific or local sets of SRT equations. The conventional set of SRT Eqs. (3.22)-(3.25) also has appeared as one such scale-specific / local set of SRT equations respect to the scale-specific SSUC observer-independent constant magnitude of quantized inertial-motion for a particular scale of visible light photons with the $c \equiv \Delta v_c$.

Hence, all the magnitudes of relativistic parameters E, m^*, t^* & s^* have appeared as local SSUC (but variable) in all those scale-specific local sets of SRT Eqs.(3.22)-(3.25), (3.26)-(3.29), (3.30)-(3.33), (3.37)-(3.37) and (3.38)-(3.41) as different IFRs due to all corresponding constant SSUCs magnitudes $c = \Delta v_c$, $c = \Delta v_\eta$, $c = \Delta v_1$, $c = \Delta v_2$ & $c = \Delta v$ although the initial non-relativistic magnitudes for all those corresponding relativistic parameters may have all $m_0 = \Delta m_0$, $s_0 = \Delta s_0$, $t_0 = \Delta t_0$ & E_0. Therefore, due to the different scale-specific quantized magnitudes of c in Eq. (2.14), the set of SRT Eqs. (3.22)-(3.25) one of all those scale-specific or local sets of SRT Eqs. (3.38)-(3.41).

4. Quantized Speeds and Invariant Special-Relativity:

In above SRT set of Eqs. (3.22)-(3.25), there inertial speed was as an observer-independent constant (or say as SRT-constant) for light $c \equiv \Delta v_c = 2.99792 \times 10^{10} cm\ sec^{-1}$ which had assumed as one SSUC in discrete magnitude for a specific-scale of photons in the EMS. Then, that $c \equiv \Delta v_c$ in SRT remains a universal-invariant constant but only as one of SSUCs in its quantized magnitudes in universe. In following paragraphs, all such SSUCs universally invariant SRT-constants correspond to every localized SRT sets of equations could be 'universalized' through intervention of the k_2 in Eq. (2.14) as one UC in the same SRT sets equations. Because, k_2 is involved with all those same inertial-motions (as SRT-constants) of particles or systems-of-particles or IFRs irrespective of the scales.

As a result, that would have also the universal consequences, in terms of all respective quantized changes in relativistic mass & energy equivalence, increments in mass, contraction of space and dilation of time $(E, m^*, t^* \& s^*)$ starting from all respective scale-specific values of initial parameters like mass, time & space $(E_0, m_0, t_0 \& s_0)$, respect to each localized or SSUC quantized valued SRT-constants $c = \Delta v$ for scale-specific PSs in all those localized sets of SRT-equations.

The Eqs. (3.38)-(3.41) are describing generally the SRT-equations for all scales of PSs irrespective of all SSUCs in quantized magnitudes as SRT-constants $c = \Delta v$ (irrespective of any luminous or non-luminous types of IFRs). Each of those SRT set of equations in the corresponding Eqs. (3.38)-(3.41) respect to $c = \Delta v$ as the scale-specific 'local' SRT-constant IFRs.

Because, all those SRT-constants $c = \Delta v$ including that $c = \Delta v_c = 2.99792 \times 10^{10} cm.\,sec^{-1}$ are merely SSUCs depending on respective scales of PSs unlike any of those constants like UCs independent of any scales. As a result, all those magnitudes for relativistic parameters $(E, m^*, s^* \& t^*)$ in corresponding Eqs. (3.38)-(3.41) respect to all SSUCs including discrete magnitudes of $c = \Delta v$ having local values for respective set of SRT-equations in all same relatively moving IFRs.

Although, initial magnitudes of parameters $(E_0, m_0, s_0 \& t_0)$ could have identical for all those localized relativistic parameters $(E, m^*, s^* \& t^*)$ in all scale specific SRT set of equations respect to all the different SSUCs values of SRT-constants $c = \Delta v$.

Hence, it requires any further universally invariant solution to overcome all those scale-specific localized relativistic magnitudes for parameters $(E, m^*, s^* \& t^*)$ to get a universal set of SRT-equations for the Eqs. (3.38)-(3.41).

In Eq. (2.14), both the CIPs like Δm & Δv are as SSUCs quantized magnitudes with such scale-specific localized observer-independences also inversely linked to each other through the UC like k_2 irrespective of scales (or IFRs) for PSs. Where localized SRT-constant for any photon-PSs have $c = \Delta v = \Delta v_c$. Therefore, during any quantized signal transformations, if magnitudes of any one 5+5 CIPs in Eqs. (2.35) and (2.36) can measure, then magnitude of Δm in Eq. (2.14) can also be calculated through Eq. (2.28) to universalize the all localized SRT sets of Eqs. (3.38)-(3.41).

Therefore, in Eqs. (3.38)-(3.41) we can write Eq. (3.1) from Eqs. (2.14) and (2.1) in terms of UC like k_2, where the k_2 is universally (inverse) invariant to all scales of IFRs correspond to all SSUCs of SRT equations.

As a result, due to the presence of k_2 as one of UCs in Eq. (3.1), the local SSUCs like $c = \Delta v$ can be universalized along with the corresponding localized relativistic magnitudes of $(E, m^*, t^* \& s^*)$ in all respective SSUCs of SRT sets of Eqs. (3.38)-(3.41).

If the scale-specific quantized SSUCs magnitudes of any one of the 5+5 inverse 10-CIPs e.g. $\Delta r, \Delta \lambda, \Delta m, \Delta s, \Delta t, \Delta s_u, \Delta t_u$ & Δh can directly be measured then the corresponding SSUCs in quantized magnitudes for $c = \Delta v$ can be automatically calculated through known values of any of those UCs like $k_1, k_2, k_3, k_4, k_5, k_6$ & k in respective Eqs. (2.1), (2.14), (2.17), (2.19), (2.21), (2.23) & (2.28).

As a result, all the localize IFRs or SRT sets of Eqs. (3.22)-(3.25), (3.26)-(3.29), (3.30)-(3.33), (3.34)-(3.37) and (3.38)-(3.41) correspond to all local SSUCs as observer-independent quantized constancies of inertial-motions like $c \equiv \Delta v_c, c \equiv \Delta v_\gamma, c \equiv \Delta v_\eta, c \equiv \Delta v_1, c \equiv \Delta v_2$ and also for every $c \equiv \Delta v$ there would be a universal invariant set of SRT-equations due to the Eq. (3.1)

$$E = m_0 \cdot (c = k_2/\Delta m)^2 = m_0 \cdot (\Delta h/\Delta m \cdot \Delta \lambda)^2, \qquad (3.42)$$

$$m^* = m_0/\sqrt{(1-v^*)^2/(k_2/\Delta m)^2} = m_0/\sqrt{(1-v^*)^2/(\Delta h/\Delta m \cdot \Delta \lambda)^2}, \qquad (3.43)$$

$$t^* = t_0 \times \sqrt{(1-v^*)^2/(k_2/\Delta m)^2} = t_0 \times \sqrt{(1-v^*)^2/(\Delta h/\Delta m \cdot \Delta \lambda)^2}, \qquad (3.44)$$

$$s^* = s_0 \times \sqrt{(1-v^*)^2/(k_2/\Delta m)^2} = s_0 \times \sqrt{(1-v^*)^2/(\Delta h/\Delta m \cdot \Delta \lambda)^2}. \qquad (3.45)$$

Then, Eqs. (3.42)-(3.45) would be a universally invariant set of SRT-equations for physical laws in all localized SSUCs of IFRs as SRT set of Eqs. (3.42)-(3.45) equations which would have identical or universally invariant magnitudes for relativistic parameters $(E, m^*, t^* \& s^*)$. Although in initial non-relativistic state those parameters might have same magnitudes of $m_0 = \Delta m_0, s_0 = \Delta s_0, t_0 = \Delta t_0$ and E_0 respect to $c = k_2/\Delta m$ irrespective of scale-specific magnitudes for any localized SRT-constant $c = \Delta v$ due to presence of the UC like k_2.

5. Consequences:

The universal invariance for quantized SRT-constant in SRT set of Eqs. (3.42)-(3.45), unlike all scale-specific localized sets of SRT Eqs. (3.26)-(3.29), (3.30)-(3.33), (3.34)-(3.37) & (3.38)-(3.41) with localized SRT-constants for IFRs, could also reveal many new consequences compare to the conventional localized SRT set of Eqs. (3.22)-(3.25) respect to only one of those SSUCs with the observer-independent SRT-constant and quantized magnitude for inertial-motion of 'light' $c = \Delta v_c$. The Universal Invariances in Eqs. (3.42), (3.43), (3.44) & (3.45) for all corresponding scale-specific localized relativistic magnitudes of parameters E^*, m^*, s^* & t^* would actually appear in all changes of scales from one to another starting from respective initial scale-specific quantized magnitudes for $E_0 = \Delta E_0$, $m_0 = \Delta m_0$, $s_0 = \Delta s_0$ & $t_0 = \Delta t_0$.

This has also described in corresponding Eqs. (3.9), (3.11), (3.15) & (3.19) respect to conversions of such scales of PSs as IFRs from a non-luminous to a luminous respect to one localized SRT set of equations for the 'light' $c = \Delta v_c$. The same thing can happen in all other correspondingly localized sets of SRT Eqs. (3.38)-(3.41) for every localized value of changes in scales from one to another in all relativistic magnitudes of parameters E^*, m^*, s^* & t^* respect to all localized scale-specific constant inertial-motions $c = \Delta v$. But all those localized relativistic magnitudes of parameters E^*, m^*, s^* & t^* would finally be universally invariant respect to k_2 in Eqs. (3.42)-(3.45).

However, such a universalized invariant set of SRT Eqs. (3.42)-(3.45) for different localized SRT sets of Eqs. (3.38)-(3.41) could have two major consequences are described in two Sub-Sections below in addition of all other known predictions of SRT set of Eqs. (3.2), (3.9), (3.22), (3.30), (3.34), (3.38) & (3.42) respect to $c = \Delta v_c$.

5.1. Energy as Sum of Discrete Smaller Masses

The equivalence in mass and energy of any material-body, that has defined in Eq. (3.2), has actually appeared as if an integer sum (n^*) for smaller masses with corresponding micro scale of PSs in Eqs. (3.6) & (3.7) can write from Eq. (2.14)

$$E = n^* \cdot \Delta m_c \cdot \Delta v_c^2 = n^* \cdot k_2 \cdot \Delta v_c \quad (3.46)$$

when mass m_0 in Eq. (3.7) would translate into an energy of particular scales of light as any boson or photon in EMS with quantized smaller mass-energies. The same energy E in Eq. (3.46) can also define through Eq. (3.2) by Planck's law as

$$E = h\nu = (\Delta h_c \cdot \Delta v_c)/\Delta \lambda_c \quad (3.47)$$

where ν is frequency of same light with wavelength $\Delta \lambda_c$ in EMS. By equating the Eqs. (3.46) & (3.47) there would be

$$n^* = \Delta h_c / k_2 \cdot \Delta \lambda_c \quad (3.48)$$

if Plank's constant Δh_c also as the SSUCs respect to c is with its conventional known value along with the calculated value (in Chapter-2) of k_2 as UC, then the magnitude of n^* can define from corresponding magnitude(s) of $\Delta \lambda_c$. If units of $h_c = gm \cdot cm^2 \cdot sec^{-1}$, $k_2 = gm \cdot cm$ & $\lambda_c = cm$ then there will be no unit for the integer n^*.

5.2. Equal Mass Can Release Different Energies

The Eq. (2.2) states that all the mass-energies in the universe must be quantized in scale-specific way. Therefore, the magnitude of all localized initial inertial masses m_0 in corresponding Eqs. (3.22), (3.26), (3.30) & (3.34) should also be intrinsically quantized with any specific scale of PSs. As a result, that $m_0 = \Delta m_0$ quantized in all those different localized sets of SRT Eqs. (3.22), (3.26), (3.30) & (3.34) correspond to different scale-specific quantized constant values for $c \equiv \Delta v_c$, $c \equiv \Delta v_\eta$, $c \equiv \Delta v_1$ & $c \equiv \Delta v_2$ become inverse universal invariant in Eq. (3.42).

However, for any given initial scale-specific quantized inertial mass say $m_0 = \Delta m_0$ is equal to all those localized Eqs. (3.22), (3.26), (3.30) & (3.34) would have all the different equivalent 'output' energies $E^* = E_c^*$, $E^* = E_\eta^*$, $E^* = E_{v_1}^*$ & $E^* = E_{v_2}^*$ for corresponding scale-specific different constant values of $c \equiv \Delta v_c$, $c \equiv \Delta v_\eta$, $c \equiv \Delta v_1$ & $c \equiv \Delta v_2$.

Since, there would be all scale-specific localized magnitudes $c \equiv \Delta v_2 < c \equiv \Delta v_1 < c \equiv \Delta v_c < c \equiv \Delta v_\eta$, from the corresponding Eqs. (3.34), (3.30), (3.22) & (3.26) there might be all different localized magnitudes for equivalent energies as well

$$\left(E^* = E^*_{v_2}\right) < \left(E^* = E^*_{v_1}\right) < \left(E^* = E^*_c\right) < \left(E^* = E^*_\eta\right). \quad (3.49)$$

Hence, the Eq. (3.49) has also revealed that the translated equivalent energies would be incremented if the corresponding initial quantized magnitudes of mass also be incremented with the increments of scale-specific localized quantized inertial-motions.

The Eqs. (3.42) & (3.49) also have revealed that not only from the luminous to non-luminous IFRs but in scales of non-luminous i.e. from one fermion to another fermion IFRs and even from one luminous to another luminous IFRs the similar kind of mass and energy conversions can be occurred as any wave-corpuscular phenomena.

That is, this whole process of mass to energy conversion in Eq. (3.49) through Eq. (3.42) actually appears as if one liberation of any sum of corresponding higher quantized inertial-motions associated with other micro scale of PSs from one lower quantized inertial-motion to another macro scale of PS. Consequently, Eq. (3.49) through Eq. (3.42) would ultimately be like a 'freedom' of higher quantized inertial-motions of micro scale PSs under binding energies within one lower quantized inertial-motions macro scale of PSs in scale specific ways.

5.3. Superluminal Quantized Motions

In localized set of SRT Eqs. (3.22)-(3.25) respect to $c = \Delta v_c = 2.99792 \times 10^{10} cm \cdot sec^{-1}$ it would not be possible to detect any inertial-motion which may greater than the same $c = \Delta v_c = 2.99792 \times 10^{10} cm \cdot sec^{-1}$. Similarly, in other localized sets of SRT Eqs. (3.26)-(3.29), (3.30)-(3.33), (3.34)-(3.37) and (3.38)-(3.41) with corresponding quantized magnitudes for SRT-constants $c \equiv \Delta v_\eta$, $c \equiv \Delta v_1$ & $c \equiv \Delta v_2$ would not have respective higher inertial-motions than $c \equiv \Delta v_\eta$, $c \equiv \Delta v_1$ & $c \equiv \Delta v_2$.

If there, $c \equiv \Delta v_2 < c \equiv \Delta v_1 < c \equiv \Delta v_c < c \equiv \Delta v_\eta$ then $c \equiv \Delta v_\eta$ for a radio photon would have higher inertial-motion than $c \equiv \Delta v_c$ in same SRT frame of Eqs. (3.26)-(3.29) although in set of Eqs. (3.22)-(3.25) respect to $c \equiv \Delta v_c$ such higher inertial-motion $c \equiv \Delta v_\eta$ never could be observed.

Same thing would happen in localized SRT set of Eqs. (3.30)-(3.33), where $c \equiv \Delta v_2 < c \equiv \Delta v_1$. Subsequently, that higher inertial-motion $c \equiv \Delta v_1$ cannot be found. But that could be observed in other sets of SRT Eqs. (3.34-.37), (3.22)-(3.25) and (3.26—(3.29) due to corresponding higher valued SRT-constants $c \equiv \Delta v_1 < c \equiv \Delta v_c < c \equiv \Delta v_\eta$.

However, all those different sets of SRT Eqs. (3.22)-(3.25), (3.26)-(3.29), (3.30)-(3.33) and (3.34)-(3.37) are defining that there are all corresponding upper limits of motions respect to all SRT-constants like $c \equiv \Delta v_c$, $c \equiv \Delta v_\eta$, $c \equiv \Delta v_1$ & $c \equiv \Delta v_2$. Similarly, there would be all corresponding superluminal upper limits of motions $c \equiv \Delta v_c < c \equiv \Delta v_\eta$ or $c \equiv \Delta v_c < c \equiv \Delta v$ if there corresponding SRT-constants $c \equiv \Delta v_c < c \equiv \Delta v$ in same SRT sets of equations.

Therefore, all those localized values of superluminal limits of motions in all corresponding SRT sets of equations respect to SRT-constants could be universalized in universally invariant quantized extension of SRT set of Eqs. (3.42)-(3.45) without violating the basic principles of SRT. Such a universalized superluminal motion could be observed through Eqs. (3.42)-(3.45) if and only if there would be detected any particle in universe having smaller quantized mass-energy than a photon-PS say $\Delta m < \Delta m_c$ in Eqs. (1.16) & (2.14). That is, such a PS needs to have further micro scale than the scale of photon-PS in SRT set of Eqs. (3.22)-(3.25) with respective SRT-constant $c \equiv \Delta v_c = 2.99792 \times 10^{10} cm.\,sec^{-1}$.

Any such superluminal PSs would accordingly include the micro scales not only to have the higher magnitudes of superluminal motion $\Delta v > \Delta v_c$ but also the higher magnitudes of wavelengths say $\Delta \lambda > \Delta \lambda_c$ of visible light in EMS. Those could be in realms of dark matters and/or dark energies if not within visible matters in universe. Even, a fermion PS if could found to have similar smaller quantized inertial mass-energy than Δm_c might have such superluminal motion in such universally invariant quantized extension of SRT set of Eqs. (3.42)-(3.45) without violating basic principles of SRT.

Summary

The Chapter-3 would be the first set of inferences based on postulates taken in foundation Chapters 1 & 2. Particularly, the newer universally invariant inverse relationship in Eq. (2.14) in-between two CIPs like Δm & Δv those are involving with every scale of particles or systems-of-particles and universal constant (UC) like k_2. The Special Relativity Theory, that has framed respect to one local or SSUC-magnitude like $c \equiv \Delta v_c = 2.99792 \times 10^{10} cm.\,sec^{-1}$ ultimately universalized irrespective of SSUCs of magnitudes correspond to all relatively moving inertial frames of reference in Eqs. (3.42)-(3.45). The same could also be considered as the universal 'quantized extension of the Special Relativity Theory' that has initially framed respect to SSUC-magnitude of inertial-motion of light as one wave-corpuscular phenomenon or photon-particle.

Moreover, in this Chapter, such universal quantized extension also has shown some additional predictions within the structure of Special Relativity Theory. Those are like the scopes of **(i)** higher output of energies from the same chunk of matter through Eqs. (3.42) & (3.49) if that matter can convert into the corresponding sum of particles or systems-of-particles having smaller quantized mass-energies compare to a photon-particle with $c \equiv \Delta v_c = 2.99792 \times 10^{10} cm.\,sec^{-1}$; and **(ii)** also the superluminal quantized inertial-motions in physical nature without violating the basic principles in Special Relativity Theory if corresponding quantized inertial mass-energy for any scale of particles or systems-of-particles would have inertial mass-energies $\Delta m < \Delta m_c$ (for mass-energy of particular photon-particles) Electromagnetic Spectrum with corresponding SSUC with corresponding value of $c \equiv \Delta v_c = 2.99792 \times 10^{10} cm.\,sec^{-1}$.

However, in next Chapters 4 & 5, there would be the universal non-inertial quantized extensions of same inverse invariance where correspondingly 'gravitational forces' and 'gauge fields of forces' will be included from those new set of postulates in Chapters 1 & 2.

CHAPTER-4:

QUANTIZATION OF GENERAL RELATIVITY
[Through Quantization of Einstein Field Equations]

> "The actual state of our knowledge is always provisional and ... there must be, beyond what is actually known, immense new regions to discover."   - Louis de Broglie

1. Two Basic Non-Inertial Assertions:	166
1.1. Quantization of Inertial Accelerations	170
1.2. Scale-specific Escape Velocities	171
2. Equivalence of Quantized Accelerations	177
3. Consequences:	180
3.1. Smallest-Bound Particles in Gravitational-Field	182
3.1.1. Convergence of Smallest-Bound Particles	182
3.1.2. Homogeneity of Smallest-Bound Particles	186
3.1.3. Proportionality in Convergence & Homogeneity	189
3.1.4. Infinitesimally Discrete Inertial States	189
3.2. Scale-Specific Einstein Field Equations	191
4. Inferences:	196
4.1. Scale-Specific Quantization of Gravitation	198
4.2. Non-invariant Einstein Field-Equations	203
4.3. Inversely Invariant Einstein Field-Equations	204
4.4. Variable Blackness in Black Holes	206
4.5. Simultaneous Gravitation & Anti-Gravitation	208
4.6. Non-Inertial Definition for Gravitating Bodies	211
4.7. Left & Right-handed Duality in Gravitating-Body	212
Summary	214

Present Chapter is the second set of inferences in a series of total eight on the basis of new postulates in Chapters 1 & 2. In previous Chapter-3, there are all scale-specific quantized inertial-motions (i.e. observer-independent absolute constancies in magnitudes) for all scales of particles or systems-of-particles and respect to which all space, time and matter have redefined universally invariant when conceptually forces are absent. That is quantization of Special Relativity Theory in terms of all quantized magnitudes of inertial-motions in physical nature.

The Chapter-4 would reveal non-inertial (in presence of forces) applications for same quantized inertial-motions of all those scales of particles or systems-of-particles, and that non-inertial applications would involve gravitation.

Any non-inertial motion as an inertial-acceleration or gravitational-acceleration would assume as the rate of changes in infinitesimal discrete magnitude for inertial-motions. Then, in each of those infinitesimally rate of changes in discrete values of inertial-motions, the gravitational field equations in General Relativity Theory would derive quantized for each of such same infinitesimal rate of discrete values of inertial-motions. In addition, the gravitational field as curved spacetime (if assume as scale-specific volume) of a gravitating-body or a gravitationally-shaped-body, for any one such infinitesimal discrete duration of rate of quantized inertial-motions, would be a sum of homogeneous smallest-bound discrete mass-energies with inverse homogeneous highest-bound discrete motions. As a consequence, in every such infinitesimal discrete durations for rate of changes in quantized inertial-motions, any gravitating-body or gravitationally-shaped-body would appear ultimately the sum of all homogeneous discrete spacetime (correspond to the quantized inertial mass-energies for the smallest-bound particles or systems-of-particles).

Therefore, the gravitation or curved spacetime of any scale-specific gravitating-bodies or gravitationally-shaped-bodies, as defined by Einstein Field Equations in General Relativity Theory, would become ultimately as quantized, i.e. the Quantization of General Relativity Theory along with few new predictions beside earlier.

In Section-1, such a quantized description of inertial-acceleration would derive as any infinitesimal rate of changes in non-inertial motions or inertial-acceleration of any particle or system-of-particles under influence or effect of forces would occur through any scale specific quantized values inertial-motions. That is, in any infinitesimal discrete moment of time an accelerated particle or system-of-particles would have infinitesimally scale-specific discrete inertial-motions.

The Section-2 would depict an equivalence for such a quantized inertial-acceleration with one gravitational-acceleration (under gravitational forces) of any particle or system-of-particles.

The Section-3 would have introduced two new important concepts like 'convergence' and 'homogeneity' of smallest-bound particles or systems-of-particles in all corresponding gravitational fields of every gravitating-body or gravitationally-shaped-bodies beside scale-specific deductions of the Einstein Field Equations in scale-specific manner. In Section-4, there would have seven new predictions from that quantized convention of gravitation.

1. Two Basic Non-Inertial Assertions

In non-inertial surrounding of universe, yet four types of fundamental forces seem to present and are believed to have unified origin. Out of those four, the three forces like electromagnetic, strong & weak nuclear origins are observed to interact through exchange of respective quantized mass-energy with all specific scales of particles or systems-of-particles. Such exchange particles or systems-of-particles with corresponding quantized mass-energies are defined through different gauge (quantum) fields in the Standard Model of Particle Physics of Quantum Mechanics. Moreover, all those three basic forces are found to dominate mostly within micro scales of particles or systems-of-particles in structures formation.

But the fourth type of fundamental force, the gravitation, that has defined as curvature of spacetime, in the Einstein Field Equation of General Relativity Theory in presence of mass-energy, are found to dominate over those three-gauge forces in scales of macro systems-of-particles mostly in astronomical structures formation. Although, the gravitation is now appearing universally effective to all micro to macro scales of particles or systems-of-particles in long distances but also has effect in short ranges within same particles or systems-of-particles in physical nature. In macro scales, particularly from a scale of planetesimals, the systems-of-particles have started to appear in the range of different scales of hydrostatically equilibrium astronomical-bodies by the dominating gravitational force over those three material forces. Therefore, all the micro to macro scales of particles or systems-of-particles are appearing under the effects of gravitational force as gravitational bodies (GBs) but all those same scales of particles or systems-of-particles are not necessarily always the gravitationally shaped bodies (GSBs) in universe.

In this context, such four fundamental forces in non-inertial universe along with the inertial postulates like 5+5 inverse 10-CIPs and 7 numbers of UCs in Chapters 1 & 2 would have two outcome assertions for all those scales of particles or systems-of-particles:

(i) whether the *inertial acceleration*, that may involve in every particle or system-of-particles (as has explained in Classical Mechanics and General Relativity Theory), would remain be the *continuous* rate of changes in inertial-motions along the direction of applying forces. Or whether it would actually be, from the Eq. (2.14), the *discrete* rate of changes in same inertial-motions with simultaneous inverse changes in corresponding discrete inertial mass-energies along the direction of same applying forces;

(ii) whether *escape velocity* for any scale-specific gravitational field, as outcome of the respective gravitational-field-strength of any such scale-specific material-body that has postulated as Δm in Eqs. (2.1) & (2.14) parallelly any scale-specific gravitating-body or gravitationally-shaped-body definable by the Einstein Field Equation of General Relativity Theory, would possess the scale-specific value irrespective of any gravitating-bodies or gravitationally-shaped-bodies.

In two sub-sections below, both of these two assertions would be described in scale-specific discrete or quantized terms compatible with present observational comprehensions in physics:

1.1. Quantization of Inertial Accelerations

In both Classical Mechanics and GRT, the concepts of a material-body as well as its subsequent accelerations have considered as non-quantized values and continuous rate of changes respectively. Although, the large-scales of the universe, which is still now better explained by the Classical Mechanics and GRT, but has comprised by all micro structures of material-bodies having scale-specific intrinsic quantum property in magnitudes.

Consequently, the Eqs. (2.1) & (2.14) have assumed all those quantized material-bodies to have any specific scale as any system-of-particles in this universe. Each of those same scales quantized material-bodies would have also any intrinsic scale-specific quantized magnitudes for associated mass-energy that inversely co-related to similar associated inertial-motion. As a result, any inertial acceleration for each of those scale-specific quantized material-body, unlike the definition in Classical Mechanics, would have the rate of changes in magnitude of motions during acceleration in the form of all scale-specific infinitesimal discrete magnitudes under influence of any forces or fundamental forces.

The Classical definition for the inertial acceleration **a** of any material-body of mass m on direction of applying force **f**

$$\mathbf{a} = \mathbf{f}/m, \qquad (4.1)$$

where scale-specific intrinsic property of quantization that would involve with the m never has precisely defined. The same m has defined in quantized manner as Δm in the Eqs. (2.1) & (2.14). Then, the m in Eq. (4.1) will have

(i) any scale specific intrinsic quantized magnitude in physical nature from the Eqs. (2.1) & (2.14), and

(ii) an inverse relationship with that **a** on the direction of constant intervention of **f**.

Then from Eq. (2.14), the Eq. (4.1) could further be write in such a way that through $m \rightarrow \Delta m$ that Δm would refer to have all scale-specific quantum changes

$$\mathbf{a} = \mathbf{f}/\Delta m \qquad (4.2)$$

where the **a** & Δm are still inversely co-related like in Eq. (4.1).

Then, in Eq. (4.2) we can write from Eq. (2.14), there will be all instantaneous and infinitesimal discrete changes in the magnitudes of the **a** for all simultaneous scale-specific *quantum changes in magnitudes* of $m = \Delta m$ for same material body as mentioned in Eq. (4.1), because of the universal inverse relationship between Δm & $\Delta \mathbf{v}$ in Eq. (2.14). That is, in Eq. (4.2), in the direction of **f**, for each of the instantaneous and infinitesimal quantum change in scale of, say, $\Delta m'$, there will be the corresponding instantaneous as well as infinitesimal inverse quantum change in scale of motion, say, $\Delta \mathbf{v}'$. Therefore, that instantaneous and infinitesimal quantum *change in motion* $\Delta \mathbf{v}'$ of Δm, on the direction of **f**, will be nothing but the instantaneous and infinitesimal inertial acceleration **a'** for that Δm.

Then, each of such instantaneous as well as infinitesimal scale-specific quantum changes in vector motions $\Delta \mathbf{v}'$ of Δm, on the direction of $\mathbf{f}$ can define in Eq. (4.2) through Eq. (2.14) as

$$\mathbf{a}' = \mathbf{f}/\Delta(m - m') = (\mathbf{f}/k_2) \cdot \Delta \mathbf{v}' \qquad (4.3)$$

where, for every quantum change in specific scales of $(\Delta m - \Delta m')$ on the direction of $\mathbf{f}$ for the $\mathbf{a}'$, we will have:

(i) the corresponding instantaneous and infinitesimal changes in scale-specific quantized magnitudes of $\Delta \mathbf{v}'$, and

(ii) that $\Delta \mathbf{v}'$ is direct proportional to the $\mathbf{a}'$ as well.

Therefore, Eq. (4.3) depicts that in the level of all instantaneous, as well as infinitesimal, rate of changes in motions $\Delta \mathbf{v}$ of the Δm on direction of the $\mathbf{f}$, there will ultimately be the all instantaneous and infinitesimal quantum magnitudes of ($\mathbf{a}' = \Delta \mathbf{a}$) in the Eq. (4.3) as

$$\Delta \mathbf{a}' = \mathbf{f}/\Delta(m - m') = (\mathbf{f}/k_2) \cdot \Delta \mathbf{v}' \qquad (4.4)$$

From Eq. (4.4), we also have quantized magnitude for the force:

$$\Delta \mathbf{a}/[\Delta(m - m')] = \Delta \mathbf{f}, \qquad (4.5)$$

where we have for such total influence of forces $\mathbf{f}$ on the m in Eq. (4.1)

$$\Delta \mathbf{f} = \Delta \mathbf{f}_1 + \Delta \mathbf{f}_2 + \cdots + \Delta \mathbf{f}_{n-1} + \Delta \mathbf{f}_n \qquad (4.6)$$

for any duration of time; say, $t = (\Delta t_1 + \Delta t_2 + \cdots + \Delta t_{n-1} + \Delta t_n)$ where the 'Δ' signifies all instantaneous and infinitesimal discrete changes in time. For the corresponding total quantized inertial acceleration $\mathbf{a}$ of Δm, we have from Eq. (4.1)

$$\mathbf{a} = \Delta \mathbf{a}_1 + \Delta \mathbf{a}_2 + \cdots + \Delta \mathbf{a}_{n-1} + \Delta \mathbf{a}_n \qquad (4.7)$$

for total scale-specific changes in quantized inertial motions (or velocities) **v** of m during the course of such inertial acceleration

$$\mathbf{v} = \Delta\mathbf{v}_0 + \Delta\mathbf{v}'_1 + \Delta\mathbf{v}'_2 + \cdots + \Delta\mathbf{v}'_{n-1} + \Delta\mathbf{v}'_n \qquad (4.8)$$

with corresponding total inverse changes in quantized magnitudes of m itself in Eq. (4.1) via Eq. (2.14) as

$$k_2/m = \; k_2[1/\Delta m_0 - 1/\Delta m'_1 - 1/\Delta m'_2 - \cdots - 1/\Delta m'_{n-1} - 1/\Delta m'_n] \qquad (4.9)$$

where Δm_0 and $\Delta\mathbf{v}_0$ are the respective initial scale-specific quantized inertial mass and motion of the classical material body m in Eq. (4.1). That inertial acceleration increments in Eq. (4.7), from $\Delta\mathbf{a}_1$ to $\Delta\mathbf{a}_n$, will sum up all instantaneous and infinitesimal changes in quantized inertial-motions from $\Delta\mathbf{v}'_1$ to $\Delta\mathbf{v}'_n$ in Eq. (4.8) of the material body with all corresponding changes in quantized magnitudes from $\Delta m'_1$ to $\Delta m'_n$ in Eq. (4.9) on the direction of affecting force(s) from $\Delta\mathbf{f}_1$ to $\Delta\mathbf{f}_n$ in Eq. (4.6), and *vice versa*.

Therefore, in Classical Mechanical as well as GRT concepts of inertial acceleration, the **a** being the rate of changes in vector motions in Eq. (4.1). That would be at infinitesimal level not as continuous but one discrete process of all quantized changes in Eq. (4.7) through corresponding quantized changes in vector inertial-motions or velocities in Eq. (4.8) on the direction of instantaneous quantized application of forces in Eq. (4.6) with all scale-specific quantized inverse changes of inertial-mass in Eq. (4.9) due to Eq. (2.14). That is, any inertial acceleration ultimately appears as quantized at its most infinitesimal level for each instantaneous rate of changes due to Eq. (2.14).

1.2. Scale-Specific Escape Velocities

However, like the micro-scales of particles or systems-of-particles, the macro-scales of system-of-particles mostly in astronomical range of universe are also appearing with incrementing scale-specific magnitudes of mass-energies. In such macro-scales, in convention of 'material-bodies' in Classical Physics, are either configured by above molecular scales through dominating electromagnetic forces (i.e. form of viscosity, surface tension etc. in such macro astronomical-systems) or by dominating gravitational force.

In current astrophysical conventions, now it has also appeared that internal gravitation of those macro material-bodies as system-of-particles starts to dominate over other non-gravitational forces from a scale of planetesimal like astronomical objects when mass appears to increase greater than 10^{12} kg [4]. Thereafter, with all gradual increments in scales of astronomical-systems would be gravitation-dominated and shaped as massive-bodies up to the macro-most scale of whole universe (as BBCOU). It would be continued to all corresponding scales of such incrementing mass-energies starting from likely a scale planetesimal to dwarf planets. A dwarf planet to the range of scales like different bigger solid planets. A solid planet to the range of scales for different mass-energies of giant gaseous planets. A gaseous giant plant to the brown dwarfs. A brown dwarf to the scales of sub-solar stars. A sub-solar star to the range of scales of different stars, A solar sized star to the scales of thousand million times heavier giant-stars'. A star to a binary system of stars up to the scales of many-stars' constellations. A stars' constellation to the galaxies. A galaxy to the clusters of galaxies. A galactic cluster to the superclusters of galaxies and finally to the macro-most scale of universe (as BBCOU).

All these gravitationally dominated massive bodies in macro scales have shaped by gravitation are also perceived with similar scale-specific range of mass-energies. Oppositely, when proceeds onward micro scales of GBs, there are all other dominating natural forces than gravitation e.g. electromagnetic, strong & weak nuclear forces those have shaped all those same.

But each of that micro to macro scales of particles or systems-of-particles also as GBs or GSBs are influenced by the universal gravitation. Where other three natural forces are no more similar universally influencing. But all those GBs which are influenced by such universal gravitation not necessarily could shape every scale. Instead, only the certain macro scales, starting from a scale of planetesimals up to the macro-most scale, could shape as any GSBs.

However, all of such GBs or GSBs would have corresponding scale-specific magnitudes of mass-energies Δm in Eq. (2.14). But for convenience, to avoid any misunderstandings in proceeding mathematical equations, the scale-specific magnitudes mass-energies only for GSBs would be stated as ΔM (in places of Δm universally for all scales of GBs) in same Eq. (2.14).

Conceptually, due to all scale-specific differences in values of mass-energies Δm for GBs or ΔM for GSBs there would be all scale-specific different gravitational field strengths (GFSs).

Subsequently, there might have all scale-specific lower to higher magnitudes of 'escape-velocities' say v_e depending on such scale-specific different GFSs in all those corresponding scales of GBs or GSBs. Such scale-specific differences in escape-velocities as Δv_e are now also observed in different macro scales of Astronomical Objects as GSBs due to such scale-specific magnitudes of mass-energies ΔM and GFSs.

Also, from Eq. (1.17), all those GBs or GSBs [1] with respective GFSs would have scale-specific radii Δr, if considers those as scale-specific spherical volumes Δs or $\Delta S = ¾(\pi \Delta r^3)$ with increments onward scale-specific incrementing values along with corresponding values of Δm or ΔM.

If we imagine two parallel moving particles, having same scale (i.e. with identical inertial mass-energy) with corresponding equal magnitude of quantized inertial-motions which are *just*

---

1. In proceeding paragraphs the GSBs could mean for both GBs and GSBs for convenience.

below, say v_{e-1}, the scale-specific escape-velocity (v_e), toward the center-of-mass of any of those particular scale of GSBs. Then due to infinitesimally rate of discrete changes in quantized inertial-motions, in the course of gravitational acceleration in particular gravitational field of GSB, would converge at the center-of-mass.

But, if those two parallelly moving identical scale of particles have quantized inertial-motions further less than that v_{e-1}, then both of those parallel moving particles would converge even before reaching to the same center-of-mass.

Alternately, if the quantized inertial-motions of those two parallel particles are greater than that v_{e-1}, both of those parallelly moving same particles would just *fly by* that center-of-mass.

Subsequently, there would a *highest limit of convergence* of quantized inertial-motions at the center-of-mass for every scale-specific GSBs in universe. That would be equal to that v_{e-1} for any gravitationally in-bound smallest scale of particles within any particular GSBs. This can also assume as the highest limit of convergence in curvature of spacetime (as assumed in in GRT) that is involving with specific GSBs.

Since, those escape-velocities are appearing to have different scale-specific magnitudes, due to different scale-specific ΔM & GFSs for all specific scales of GSBs, there would also be all different scale-specific *maximum limits of convergence* for respective highest-bound quantized (inverse) inertial-motion v_{e-1} (for all respective parallelly moving two similar scale of particles) at the center-of-masses of same GSBs. This type of all scale-specific maximum limits of convergence in different scales of GSBs also in otherwise depicting the existences of all scale-specific curvatures in spacetime enfolding those scale-specific GSBs.

But due to the inverse relation in the Eq. (2.14), each of those highest-bound quantized inertial-motions would have also the scale-specific quantized inertial-mass-energies. Therefore, due to the same inverse relation in Eq. (2.14), simultaneously, each of those GSBs having 'highest limit' of scale-specific 'convergence' would have the scale-specific *maximum limits of homogeneity* of all those smallest-bound unescaped particles of quantized inertial-mass-energies say m_{e-1}. Subsequently, each of those scale-specific GSBs would conceptually be the scale-specific integer sum of all those smallest-bound quantized inertial-mass-energies Δm_{e-1} with all homogeneous highest-bound inverse quantized inertial-motions Δv_{e-1}.

Each of those smallest-bound particles would have *just a heavier or macro scale* than one escapable scale of particles having the corresponding quantized inertial-mass-energies say m_e (for the corresponding quantized inverse inertial-motions Δv_e) to escape from that specific scale GSBs.

Then, in above paragraphs, due to appearance of such scale-specific quantization in magnitudes of escape-velocity Δv_e in each of such specific scales of GSBs or Astronomical Structures with respective GFSs, there would be also the scale-specific maximum limits for highest-bound *inescapable particle motions*, say Δv_{e-1}. Moreover, we can also imagine that Δv_{e-1}, as any 'message' [1], instead of a smallest-bound particle with highest-bound quantized inertial-motion. That massage, just has missed out or cannot be escaped out through the corresponding GFS of scale-specific GSB, due to its critical scale-specific lower quantized magnitude within same GFS compare to another specific scale of escapable messages with just higher quantized inertial-motions Δv_e would be

$$\Delta v_{e-1} < \Delta v_e \qquad (4.10)$$

and never capable to reach in external surroundings of specific scale of the GSB where observers may wait forever to see it. We could imagine there three categories of observers in that external surrounding of the corresponding GSB. In Eq. (4.10),

---

1. intrinsic to such unescaped smallest-bound quantized mass-energy.

there would be one category of observers those have say capabilities in receiving and analyzing only any messages from the GSB if possess quantized magnitude for inertial-motions up to maximum range of say

$$\Delta v_1 < \Delta v_{e-1}, \qquad (4.11)$$

and another with maximum range up to the quantized magnitude of inertial-motions say

$$\Delta v_2 = \Delta v_{e-1} < \Delta v_e, \qquad (4.12)$$

and a third one with range beyond the Δv_{e-1} or say the $\leq \Delta v_3$

$$\Delta v_3 > \Delta v_{e-1}. \qquad (4.13)$$

Then, the first two classes of observers, those with limitations of receiving any messages from particular GSB beyond Δv_1 and/or Δv_2, will like to have corresponding *Event Horizon* type situation. That would actually make no sense whether or not that GSB is a black hole *i.e.* $\Delta v_e > c$. Consequently, that would prohibit them to know anything about particular GSB within the event-horizon due to no transmission of any messages having quantized inertial-motions ($\Delta v_2 = \Delta v_{e-1}$) < Δv_e and/or $\Delta v_1 \leq \Delta v_{e-1}$ from 'inside'.

Likewise, this can also be possible in cases of all other micro to macro scales of GSBs with all subsequent scale-specific magnitudes of ΔM and GFSs in physical nature. Because, for each of those different scale-specific magnitudes for Δv_{e-1} and Δv_e in Eq. (4.10), there will be also all corresponding maximum limits of inescapable motions ($\Delta v_2 = \Delta v_{e-1}$) < Δv_e and/or $\Delta v_1 \leq \Delta v_{e-1}$ of messages beside the third category of observers with corresponding limitations to receive $\Delta v_3 > \Delta v_{e-1}$ in Eq. (4.13).

Then, obviously, there would have all scales specific event horizons as well those are associated in every GBs or GSBs with all those three categories of observers with respective limitations in universe. Since all those scales of GSBs would have scale-specific magnitudes of ΔM (with GFS & event horizon), the respective magnitudes of Δv_{e-1} & Δv_e in Eq. (4.10) will be directly proportional to magnitudes of ΔM, GFS and event horizon (EH).

2. Equivalence of Quantized Accelerations

As per basic propositions of the GRT, all types of accelerations, under influences of either non-gravitational forces (for inertial accelerations) or gravitational forces (for gravitational accelerations), must be equivalent [8]. Where an observer cannot differentiate in-between an inertial acceleration and a gravitational acceleration.

In Sub-Section-1.1, such an inertial acceleration has described as discrete in type. Because, at its most infinitesimal scale of rate of changes has assumed equal to any quantized magnitude of inertial-motions by substituting the earlier continuous rate of changes in motions of Classical Mechanics. Consequently, in following paragraphs of this Section, there would be also a similar Equivalence Principal of inertial and gravitational accelerations but assuming all those accelerations as infinitesimally discrete rate of changes in scale-specific quantized magnitudes of inertial-motions of accelerating particles under influences of both non-gravitational and gravitational forces).

Therefore, due to Eq. (2.14), the discrete rate of changes for 'inertial acceleration' $\Delta \mathbf{a}$ of any object Δm in Eq. (4.4) is infinitesimally quantized under influence and direction of the force $\Delta \mathbf{f}$ in Eqs. (4.5) & (4.6), where that $\Delta \mathbf{f}$ represents any non-gravitational force, or say as *inertial force(s)*. Then, there would be an obvious question whether the gravitational force(s) say $\Delta \mathbf{f_g}$ of GSBs [1] with same corresponding ΔM equivalent to the

---

1. Conceptually that could also be similar in all GBs having corresponding values of Δm in places of ΔM for GSBs.

inertial force(s) $\Delta \mathbf{f}$ in Eq. (4.5). Consequently, whether same $\Delta \mathbf{f_g}$ in its similar infinitesimal rate of changes for gravitational acceleration in GSBs, say $\Delta \mathbf{g}$, would also be quantized like $\Delta \mathbf{a}$ in same Eq. (4.5).

To reach in any such equivalence for any infinitesimal discrete rate of changes between $\Delta \mathbf{a}_1$ and $\Delta \mathbf{g}_1$ in Eq. (4.7), we can start from earlier example of accelerated 'Lift' [9] in GRT, that can occur alternately under influences of both inertial forces say $\mathbf{f}_i$ and gravitational forces $\mathbf{f}_g$ with few essential modifications. That would assess whether there will be any differences in feelings of inside observer in 'Lift' once under influence of infinitesimal inertial acceleration as $\Delta \mathbf{a}_1$ and next under influence of infinitesimal gravitational acceleration as $\Delta \mathbf{g}_1$ in terms of corresponding 'upward thrusts' from floor.

Suppose, one very 'smart device' we have and that has installed on the inner vertical wall of that 'Lift', say at point A in Figure-1 below. However, our device is so fine-tuned that it could able to spontaneously emit a photon-signal with speed $c = \Delta v_c$ in a straight line to B on opposite wall as soon as the lift starts to move in velocity $\Delta \mathbf{v}'_1$ in Eq. (4.8) reverse to the direction of applied force as $\Delta \mathbf{f}_1$ in Eq. (4.6) that would create an instantaneous acceleration $\Delta \mathbf{a}_1$ (or $\Delta \mathbf{g}_1$) in Eq. (4.7).

In GRT, the observer who stands on the floor of that 'Lift' would feel an upward thrust from his floor when the 'Lift' accelerates at $\Delta \mathbf{a}_1$ (or $\Delta \mathbf{g}_1$) in Eq. (4.7).

If the 'Lift', in Figure-1 would place conveniently in the gravitational field of Earth, the inside observer could not recognize whether the acceleration of the 'Lift' has occurred due to gravitational or inertial forces from the feelings of an upward thrusts out of the floor.

If the same 'Lift' would place somewhere in deep astronomical space, where impact of gravitation is negligible, and under influence of some instant inertial forces that might create equal infinitesimal acceleration that was in earth (but in opposite direction), the same inside observer could not identify

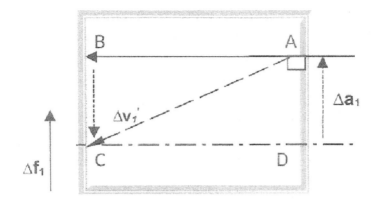

Figure-1. For $\Delta t_1 = 1$ sec, quantized infinitesimal acceleration of 'Lift'= $\Delta t_1 \Delta \mathbf{a}_1 = \Delta \mathbf{a}_1 = \Delta \mathbf{v}_1'$ shifting of a light ray inside BC=DA distance traveled by lift outside under instant influence of instantaneous force $\Delta \mathbf{f}_1$.

whether that acceleration has occurred due to gravitational or inertial forces from his feelings of outward thrust out of the floor.

But the inside observer could measure that amount of infinitesimal change in motion of '*Lift*' from the distance of shifting in photon signal's path on the opposite wall, say as a distance BC $\equiv \Delta \mathbf{v}_1'$, and instant acceleration of that '*Lift*' would be either as

$$\Delta \mathbf{v}_1' \equiv \Delta \mathbf{a}_1 \qquad (4.14)$$

or

$$\Delta \mathbf{v}_1' \equiv \Delta \mathbf{g}_1 \qquad (4.15)$$

after duration of time Δt_1. Since in Eq. (4.7), for an infinitesimal discrete rate change in quantized acceleration of same '*Lift*' could be $\Delta \mathbf{a}_1 = \Delta \mathbf{g}_1$, subsequently from the Eqs. (4.14) & (4.15) there would be an equal to infinitesimal quantized inertial-motion $\Delta \mathbf{v}'_1$ of the same '*Lift*' in Eq. (4.8).

Even from those Eqs. (4.14) & (4.15) respectively, the inside observer could calculate the amount of such infinitesimal force has influenced from outside on the '*Lift*' as $\Delta \mathbf{f}_1$ or $\Delta \mathbf{f}_{g_1}$ from Eqs. (4.4) & (4.6).

But that observer could not differentiate from his inside whether such infinitesimal force is inertial $\Delta \mathbf{f}_1$ or gravitational $\Delta \mathbf{f}_{g_1}$, or such infinitesimal acceleration is inertial $\Delta \mathbf{a}_1$ or gravitational $\Delta \mathbf{g}_1$. Though the observer could realize that all those infinitesimal parameters would possess quantized magnitudes.

This not only illustrates an Equivalence in inertial and gravitational accelerations as was in GRT moreover depicts an Equivalence up to the infinitesimal scales of discrete rate of changes for same inertial and gravitational accelerations.

However, instead of one photon messaging particles, due to Eq. (2.14), one can use any other scales of messaging or signal particles inside of the '*Lift*'. But that could show only the differences in quantized magnitudes for all parameters in Eqs. (4.14) & (4.15). But the feeling of same inside observer would remain be unchanged.

3. Consequences

In previous Sections-1 & 2, both inertial and gravitational sourced accelerations for a particle have defined as equivalent in magnitudes of all infinitesimal discrete rate of changes in quantized inertial-motions. That is, every such equivalent gravitational acceleration of any particle or system-of-particles would have also the similar infinitesimal discrete rate of changes, equal to its all discrete changes in magnitudes of inertial-motions, under the influence of gravitational force.

Moreover, in above Sub-Section-1.1, every 'material-body' or 'massive body', as was assumed in GRT, as non-scale-specific, would have any specific scale of GBs or GSBs as per growing observational understandings of astrophysics.

Therefore, two such basic reflections, such as **(i)** the scale-specific appearances of all earlier non-scale-specific 'material-bodies' in basic assumptions of GRT as GBs or GSBs, and **(ii)** also the appearances of all earlier equivalences of inertial & gravitational accelerations in GRT as equivalent but quantized due to all infinitesimal discrete rate of changes in scale-specific quantized inertial-motions for any accelerated particles or systems-of-particles.

Those two new assertions in relevance of gravitation would have huge importance in further understanding of gravitation as extension of GRT. Consequently, the gravitation, that has revealed through the Einstein Field Equations of GRT, as curvature of spacetime [1] in presence of imprecise 'material-bodies' can derive instead as scale-specific curvature of spacetime that would involve to each of those precise GBs or GSBs. Since, the intrinsic magnitudes of mass-energies ΔM have assumed as scale-specific in all those GBs or GSBs, the corresponding scales of GBs or GSBs would also have the intrinsic scale-specific curvatures of spacetime from Eq. (2.28). In such way, that scale-specific curvature of spacetime i.e. gravitation has described in Einstein Field Equations of GRT would be extended as scale-specific.

In below Sub-Section-3.1, as already have depicted in above Section-2, the same gravitational field for any scale-specific GSBs would derive equally as the scale-specific *convergence* for all respective Δv_{e-1} in Eq. (4.12) and simultaneously as the scale-specific integer sum of all *homogenous* smallest-bound particles with quantized inertial-mass-energy Δm_{e-1} due to $\Delta m_{e-1}=(k_2/\Delta v_{e-1})$ from Eq. (2.14). Those with intrinsic quantized inertial-motions Δv_{e-1} just below the respective escape-velocity Δv_e of the particular gravitational field.

---

1. That seems omnipresent universally without 'material-bodies'

These new outcomes within each of the scale-specific gravitational field of GSBs (or GBs) would have enormous consequences. Those are not merely to define Einstein Field Equation of GRT in scale-specific manner in proceeding Sub-Section-3.2 but also to equate same scale-specific Einstein Field Equations (for curved spacetime) to Supersymmetric quantized fields in Standard Model of Particle Physics that could associate with every homogeneous smallest-bound particle (Δm_{e-1}) within such gravitational field of GSBs in next Chapter-5:

3.1. Smallest-Bound Particles in Gravitational-Field

Above mentioned '*Lift*, that could place in any scales of GSBs, actually has revealed two basic things within each of those GSBs. In one, the GSBs would have scale-specific sums equal to total integer number of homogeneous smallest-bound quantized mass-energies Δm_{e-1} within itself, and another would be its embedded scale-specific gravitational field due to such total mass-energies that creates scale-specific convergence as scale-specific curvature of spacetime.

The Sub-Sections 3.1.1. to 3.1.4. below would describe the same gravitation (curved spacetime) as scale-specific along with corresponding scale-specific magnitudes of mass-energies in all scales of GSBs. However, in doing that such scale-specific magnitude of mass-energies would appear as an integer sum of all smallest-bound inescapable homogeneous particles within scale-specific convergence of spacetime from the Eq. (2.14):

3.1.1. Convergence of Smallest-Bound Particles

In Figure-2 below, for convenience, there have fixed two more similar kind of 'devices' on the outside of bottom wall of same '*Lift*' in earlier Figure-1, as those two devices can simultaneously eject two identical signal-particles on downward vertical direction. In that arrangement, it assumes that those two new devices would spontaneously eject two signal-particles of same scale along with simultaneous ejection of another identical signal-particle from the earlier device which has already fixed on inside wall of '*Lift*'.

Those three identical quantized magnitudes say Δv could be imagined equal to any infinitesimal rate of change in the course of any quantized gravitational acceleration in Eq. (4.5) for of the said *Lift* under respective gravitational force(s) of any specific scale of GSB.

Then, during the moment of simultaneous ejections from all those two outside bottom's devices, the two signal-particles would start to move parallel to each other. But after that, under the influences of gravitational force(s) of particular GSB, both of signal-particles would gradually lean from the mutual parallel path, and finally converge onwards the center-of-mass for respective GSB.

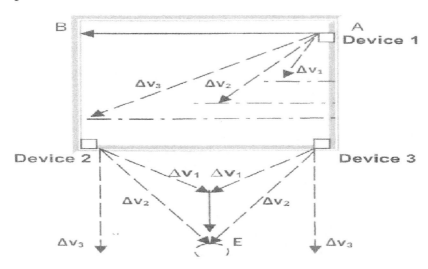

Figure-2. Convergence of bound particles with quantized motions at the center of mass E of different scales of gravitating bodies.

If that '*Lift*' in Figure-2 could place on the Event Horizon of such specific scale of GSB, and:

    (i) there it would have an infinitesimal acceleration for discrete rate of change in quantized motions equal to say $\Delta \mathbf{g}_1 = \Delta \mathbf{a}_1$ in Eq. (4.7), and those two parallel moving quantized signal-particles could eject simultaneously from the outside bottom devices 2 & 3 in Figure-2 having two quantized identical motions Δv_1 & Δv_1 also in Eq. (4.11) will converge before reaching at E (center-of-mass);

(ii) but when that '*Lift*' would have infinitesimal acceleration for discrete rate of change in quantized motions equal to say $\Delta \mathbf{g}_2 = \Delta \mathbf{a}_2$ in Eq. (4.7), and those two parallel moving quantized signal-particles could eject simultaneously from same bottom devices 2 & 3 in Figure-2 having also two quantized identical motions Δv_2 & Δv_2 in Eq. (4.12) will converge at E;

(iii) and finally, when the same '*Lift*' would have infinitesimal acceleration for discrete rate of change in quantized motions equal to say $\Delta \mathbf{g}_3 = \Delta \mathbf{a}_3$ in Eq. (4.7), those two parallel moving quantized signal-particles could eject simultaneously from bottom of those two devices 2 & 3 having quantized identical motions Δv_3 & Δv_3 in Eq. (4.13) will never converge but just fly by to the same E.

Since, there could be all scale-specific Event Horizons for all scale-specific GSBs along with all scale-specific magnitudes of $\Delta v_1, \Delta v_2$, and Δv_3 in corresponding Eqs. (4.11), (4.12) & (4.13), there it could also be different scale-specific convergences for all those three types of signal-particles.

Therefore, if any one of those signal-particle appears to converged in different scales of GSBs, then the space, that was defined in Eq. (1.17), would be not only the curved but also simultaneously be the scale-specific quantized in all those different scales of GSBs.

Moreover, the Δv_2 in Eq. (4.12), would be the scale-specific highest magnitude of quantized signal-particles those have just lower quantized motion to escape out through respective Event

Horizon of GSB. Therefore, such scale-specific appearance in curvature of space for particular scales of GSBs, could be defined by such scale-specific convergence of that highest quantized signal-particle motion Δv_2 in Eq. (4.12) within same GSBs. Since the magnitudes of that Δv_2 in Eq. (4.12) would be always intrinsically quantized in Eq. (2.14), its corresponding magnitude of *convergence* at the respective center-of-mass E or curved space of every GSBs would also have the similar scale-specific magnitudes.

Again, we have also $\Delta t = 2\pi \Delta r$ for a quantized time scale in Eq. (1.21), a quantized mass Δm in Eqs. (2.1) & (2.14), and a quantized volume of space Δs in Eqs. (1.17) & (2.19) as respective CIPs for all scales of particles or systems-of-particles in universe, and those CIPs never could be separated also within all those GSBs (ultimately also as any GBs). Then the intrinsic quantized magnitudes of Δs, Δt & Δm for every scale of GSBs in Eq. (2.28) as CIPs are also inseparable. Therefore, from Eqs. (2.1), (2.14), (2.17), (2.19), (2.21), (2.23) & (2.28), all those same scales of GBs or GSBs in universe would have scale-specific quantized convergence (curvature) of corresponding spacetime say

$$\Delta p = \Delta s \cdot \Delta t, \qquad (4.16)$$

where $\Delta t = 2\pi \cdot \Delta r$ & $\Delta s = \frac{3}{4}(\pi \cdot \Delta r^3)$ have defined in respective Eqs. (1.21) & (1.17)). Therefore, from Eqs. (2.19) & (2.21), the Eq. (4.16) can re-write as

$$\Delta p = \Delta s \cdot \Delta t = \left(\tfrac{3}{4}\pi \cdot \Delta r^3\right) \cdot (2\pi \cdot \Delta r) = \tfrac{3}{2}\pi^2 \cdot \Delta r^4, \qquad (4.17)$$

where radius Δr possesses scale-specific quantized magnitude of radius for that scale-specific convergence of spacetime Δp. Then, from Eq. (2.23) further there may have in Eq. (4.17)

$$\Delta p = \tfrac{3}{2}\pi^2 \cdot \Delta r^4 = \tfrac{3}{2}\pi^2 \cdot k_6^4/\Delta V^4, \qquad (4.18)$$

where for convenience, the capital letter symbol ΔV has used to define the scale-specific quantized magnitude of motions in all macro scales of GBs as GSBs. Although, the same smaller letter symbol Δv in Eq. (2.14) has already used for general to define for all scales of particles or systems-of-particles in universe. Then, each of such scale-specific quantized convergence of spacetime in all scales of GSBs in universe respect to the highest-bound inescapable quantized motions Δv_{e-1} from Eq. (4.12) as well as from Eq. (2.14) in Eq. (4.18) would be

$$\Delta p_{e-1} = \tfrac{3}{2}\pi^2 \cdot k_6^4/\Delta V^4 = \left(\tfrac{3}{2}\pi^2 \cdot k_6^4/k_2^4\right) \cdot \Delta M^4 \qquad (4.19)$$

where Δp_{e-1} is also the usual quantized (convergence of) spacetime for such scale-specific curvature of same scale-specific GSB, and we can also consider the same as respective normal hydrostatic-equilibrium (HSE) state [1] for the respective ΔM of that GSB under influence of gravitation.

However, for our further conveniences in following text, we could define such scale-specific total mass-energies or simply mass of any GSB (or GBs) as ΔM instead of the generalize symbol Δm in Eqs. (2.1) & (2.14) for total mass-energy of any micro to macro scales of PSs (as same GBs or GSBs) in universe.

3.1.2. Homogeneity of Smallest-Bound Particles

There already have the scale-specific highest-bound convergence of the discrete inertial-motions for signal-particles in Figure-2 at the respective center-of-mass E of any relevant GSB as $\Delta v_2 = \Delta v_{e-1}$ in Eq. (4.12). Then from the Eq. (2.14), in same Eq. (4.12), there would be for such highest-bound converging discrete inertial-motion corresponding inverse

---

1. that occurs when external forces like gravitation are balanced by the internal pressure-gradient force/s like other basic natural forces in all scales of GBs or GSBs.

smallest-bound discrete inertial-mass-energy

$$\Delta v_2 = \Delta v_{e-1} = k_2/\Delta m_{e-1} = k_2/m_2 \qquad (4.20)$$

which has just missed to escape through respective Event Horizon of the GSB.

Alternately, same converging $\Delta m_2 = \Delta m_{e-1}$, as unescaped smallest-bound discrete inertial-mass-energies in corresponding (scale of) GSB, having just next heavier scale than all those particles capable to escape through respective Event Horizon of same GSB. Consequently, any scale of such GSBs, with maximum scale-specific HSE, having the ΔM would be ultimately a sum for all those smallest-bound homogeneous quantized inertial-mass-energies $\Delta m_2 = \Delta m_{e-1}$ in Eq. (4.20) with inverse highest-bound quantized inertial-motions $\Delta v_2 = \Delta v_{e-1}$ in Eq. (4.12).

Furthermore, from the Eq. (4.11), there can also be other scales of heavier signal particles or systems-of-particles with lower quantized inertial-motions $\Delta v_1 < \Delta v_2 = \Delta v_{e-1}$. From Eq. (2.14) there would be also the quantized mass-energy (say Δm_1) for same as

$$\Delta v_1 = k_2/(\Delta m_1 > \Delta m_{e-1}), \qquad (4.21)$$

which could further converge (due to their corresponding lower quantized inertial-motions) before reaching the same center-of-mass E of same GSB (in Figure-2), and also those could be the integer multiples of smaller scale particles like $\Delta m_2 = \Delta m_{e-1}$ in Eq. (4.21) within same GSB.

Moreover, there could be also other signal-particles with scale-specific higher quantized inertial-motions $\Delta v_3 > \Delta v_2 (= \Delta v_{e-1})$ within the same GSB, and those may just fly by without converging on the same center-of-mass E of GSB from Eqs. (4.13). Therefore from Eq. (2.14), we will have inverse quantized mass-energy say Δm_3 for that

$$\Delta v_3 = k_2/(\Delta m_3 < \Delta m_{e-1}), \qquad (4.22)$$

and such signal-particles having corresponding Δm_3 and $\Delta v_3 \geq \Delta v_e$ greater than respective escape velocity would escape out through respective Event Horizon of GSB. As a result, such signal-particles with Δm_3 in Eq. (4.21) will never be any integrated part of so-called homogeneity of all smallest-bound quantized mass-energies within the same scale of GSBs.

Then in Eq. (4.20), we can assume that, in every respective scale of GSBs, there could be all corresponding *optimum homogeneity* for all those homogeneous respective smallest-bound signal-particles of quantized mass-energy $\Delta m_2 = \Delta m_{e-1}$. Those smallest-bound particles just have missed to escape through the corresponding Event Horizon in Eq. (4.12) exceeding escape velocity of specific Gravitational Field Strength. Where obviously the corresponding scale-specific total mass-energy ΔM of GSB would conceptually be defined by the any integer (say n) multiple of that Δm_{e-1} of that specific scale of GSB. Therefore, such corresponding scale-specific *optimum homogeneity* of respective scale of smallest-bound signal particles or systems-of-particles in any GSB, from Eq. (4.20) will be

$$\Delta q_{e-1} = \Delta M = \Delta n \cdot \Delta m_{e-1}. \qquad (4.23)$$

Since in the Eq. (4.23), both of those parameters ΔM & Δm_{e-1} are scale-specific variables and seem to have all scale-specific quantized magnitudes in all different scales of GSBs, the Δn will have all different scale-specific magnitudes. But integer in type due to discrete values of all smallest-bound particles within every GBS. From Eq. (2.14), the Eq. (4.23) can further write as

$$\Delta q_{e-1} = \Delta n \cdot \Delta m_{e-1} = \Delta M = k_2/\Delta V \qquad (4.24)$$

where capital letter ΔV as symbol has used exclusively for the scale-specific inertial-motions of any GSB only.

3.1.3 Proportionality in Convergence & Homogeneity

For same Eq. (4.24), the Eq. (4.19) could further write through Eq. (2.14) as

$$\Delta p_{e-1} = \left(\tfrac{3}{2}\pi^2 \cdot k_6^4/k_2^4\right) \cdot (\Delta q_{e-1})^4. \qquad (4.25)$$

Since in Eq. (4.25) the $\left(\tfrac{3}{2}\pi^2 \cdot k_6^4/k_2^4\right)$ has appeared as one proportionality constant, therefore Δp_{e-1} would be *directly proportional* to $(\Delta q_{e-1})^4$ in every micro to macro scales of GBs or GSBs. Then in Eq. (4.25), for all GSBs, if the scale-specific quantized magnitudes of mass-energy of $\Delta q_{e-1} = \Delta M_{e-1}$ in Eqs. (4.23) & (4.24) becomes smaller and smaller, then the scale-specific curvature of spacetime Δp_{e-1} for all corresponding scale of GSBs would be proportionally decremented or vice versa.

3.1.4. Infinitesimally Discrete Inertial States

Every scale of PSs has one common definition in Eq. (2.28) in inertial states. That is, when those are conceptually in absence of the effects of all external forces on them. Those would have also the scale-specific quantized inertial-mass-energies Δm & inertial-motions Δv. Each of those scale-specific quantized magnitudes of Δv like other CIPs intrinsic to any PSs would have SSUCs in quantized magnitudes. That is, each of those PSs can travel with constant observer-independent inertial-motions and fixed inertial-mass-energies along with all other corresponding CIPs like Δr, $\Delta \lambda$, Δs, Δs_u, Δt & Δt_u which have also simultaneous all unchangeable quantized inertial-magnitudes.

But that unchangeable scenario would instantaneously become changed as soon as one such conceptually isolated inertial or free particle or system-of-particles enters in a surrounding of forces, like gravitation or/and non-gravitation, through exchange of respective force-carriers or interactional signals start with those free particles or systems-of-particles in such non-inertial surrounding.

Such a conceptual non-inertial entry of one isolated inertial particle or system-of-particles would create an acceleration (also deceleration) of the same.

That would instantaneously change the scale or in all 5+5 inverse SSUCs in quantized magnitudes including corresponding inertial constancy of magnitude and direction for inertial-motion of such an inertially isolated particle or system-of-particles.

Such acceleration in inertial particles or systems-of-particles causes due to exchange of the force carrying particles with other similar non-inertial particles or systems-of-particles in non-inertial surrounding. If it is through exchange of force carrying particles of any non-gravitational forces, then for conveniences, such an acceleration of same particles or systems-of-particles would say as *inertial acceleration*. If the acceleration of same particles or systems-of-particles under the influence of gravitational force would say as *gravitational acceleration*.

But, unlike all other non-gravitational forces, the gravitational force having both long and short ranges effects irrespective all scales of mass-energies for GBs or GSBs universally which could be imagined to envelope all those scale-specific mass-energies as scale-specific curved spacetime in universe. As a result, such gravitational force and gravitational acceleration would be universal. Then, the universal gravitation would be either could have effects in micro scales of GBs those are dominated by non-gravitational forces or could internally shape all macro scales of GSBs by dominating over all same non-gravitational forces.

In Eq. (4.4), it has also appeared that inertial acceleration of every non-inertial PSs has infinitesimally discrete rate of changes in quantized inertial-motions for every discrete durations of quantized-time.

Consequently, for every such discrete duration time, for every infinitesimal discrete rate of changes in inertial acceleration, every non-inertial PSs as GBs or GSBs would fundamentally be also an inertial-PSs. Therefore, for every such infinitesimally discrete duration of time for corresponding discrete infinitesimal non-inertial changes under influences of forces every scale of same non-inertial-PSs as any GBs or GSBs would be in isolated inertial states.

Further, any such inertial-acceleration as well as gravitational-acceleration for any non-inertial PSs, are also equivalent in Section-2. Not only that, that also have shown, one infinitesimal rate of change in quantized inertial-motion under that gravitational-acceleration of any non-inertial PSs (as GBs or GSBs) would be also equivalent to one infinitesimal rate of change in quantized inertial-motion under any inertial-acceleration.

As a result, not only the non-inertial PSs with inertial-accelerations but also the GBs or GSBs with gravitational-accelerations, in most infinitesimal rate of changes in corresponding quantized inertial-motions, would have discrete magnitudes of time.

That is, in most instantaneous discrete duration of quantized magnitude for every scale-specific time and in every infinitesimal rate of changes in scale-specific quantized inertial-motions, all non-inertial entities in the universe would be in inertial state.

3.2. Scale-Specific Einstein Field Equations

The Eq. (4.25) depicts a direct proportionality in between scale-specific quantized convergence (or quantized curved spacetime) Δp_{e-1} and fourth square of scale-specific quantized homogeneity of smallest-bound inescapable particles $(\Delta q_{e-1})^4$ in all scales of GSBs. That seems to be identical up to some extent with the Einstein Field Equations of GRT which express gravitation of the same GSBs (and GBs) but in non-quantized manner. This Sub-Section will describe those field equations of GRT as scale-specific.

In GRT, the Einstein Field Equations (EFE's) [10] that equate curvature of local spacetime (expressed by the Einstein Tensor) with local energy and momentum within that spacetime (expressed by Stress-energy-momentum Tensor) as

$$R_{\mu\nu} - \tfrac{1}{2}Rg_{\mu\nu} + g_{\mu\nu}\Lambda = (8\pi G/c^4) \cdot T_{\mu\nu} \qquad (4.26)$$

where $R_{\mu\nu}$ is the Ricci curvature tensor, R is the scalar curvature, $g_{\mu\nu}$ is the metric tensor, G is gravitational constant, Λ is the cosmological constant, c is inertial speed of light, and $T_{\mu\nu}$ is the stress-energy tensor or stress-energy-momentum tensor. If the Einstein tensor in Eq. (4.26) as

$$G_{\mu\nu} = R_{\mu\nu} - \tfrac{1}{2}Rg_{\mu\nu}, \qquad (4.27)$$

a symmetric second-rank tensor which is a function of the metric. Subsequently, the Einstein Field Equation will be

$$G_{\mu\nu} + g_{\mu\nu}\Lambda = (8\pi G/c^4) \cdot T_{\mu\nu} \qquad (4.28)$$

in more compact form. However, by using geometrized units for $G = c = 1$, the Eq. (4.28) can be in further reduced to

$$G_{\mu\nu} + g_{\mu\nu}\Lambda = 8\pi T_{\mu\nu}, \qquad (4.29)$$

where the left side stands for the curvature of spacetime by the metric and the right side for the mass-energy-momentum contents within that curved spacetime. Then the Einstein Field Equation ultimately appears as if a set of equations that defines how mass-energy-momentum curves the spacetime.

However, in Eqs. (4.28) & (4.29), we have a *direct proportional relationship* in between such spacetime curvature $(G_{\mu\nu} + g_{\mu\nu}\Lambda)$ can be for one infinitesimal duration of time and the $T_{\mu\nu}$. As if we have a similar direct proportional relationship in between quantized $\Delta(s \cdot t)$ and Δm in Eq. (2.28), and between Δp_{e-1} and

$(\Delta q_{e-1})^4$ in Eq. (4.25). Although, in Eqs. (4.28) & (4.29), those same parameters would appear different as $G_{\mu\nu} + g_{\mu\nu}\Lambda$ and $T_{\mu\nu}$ respectively for the same GBs or GSBs.

However, through Eq. (2.14), there will also be $c = \Delta v_c$ in Eq. (4.28) as one of SSUC's among the all scales of quantized magnitudes of inertial-motions which is unlike the UCs 8π & G as well as $k_1, k_2, k_3, k_4, k_5, k_6$ & k in corresponding Eqs. (2.10), (2.14), (2.17), (2.19), (2.21), (2.23) & (2.28), those will remain unchanged in all scales of GBs or GSBs. Then from Eq. (4.28) we obtain

$$G_{\mu\nu} + g_{\mu\nu}\Lambda = (8\pi G/c^4) \cdot T_{\mu\nu} = (8\pi G/k_2^4) \cdot \Delta m_c^4 \cdot T_{\mu\nu} \quad (4.30)$$

where $c = \Delta v_c = k_2/\Delta m_c$ from Eq. (4.3) and the Δm_c is scale-specific quantized magnitude of inertial mass for corresponding photon. Therefore, the Eq. (4.30), respect to $c = \Delta v_c = k_2/\Delta m_c$ as one of SSUCs, appears as a *local* constant to define the gravitational characteristics of the particular scale(s) of GBs or GSBs in universe. Then Eq. (4.30) can be universalized irrespective of any scales of GBs or GSBs in universe as

$$G_{\mu\nu} + g_{\mu\nu}\Lambda = (8\pi G/c^4) \cdot T_{\mu\nu} = (8\pi G/k_2^4) \cdot \Delta m^4 \cdot T_{\mu\nu} \quad (4.31)$$

where $\Delta v = k_2/\Delta m$ in Eq. (2.14) for any scales of PSs in universe. That $\Delta v = \Delta v_c = (\Delta v_c)_{e-1}$ would be the maximum speed that just cannot escape through the Event Horizon of a particular black hole (as one GSB) in Eq. (4.12). Therefore, for that particular black hole, the Eq. (4.31) would be like:

$$G_{\mu\nu} + g_{\mu\nu}\Lambda = \{8\pi G/[(\Delta v_c)_{e-1}]^4\} \cdot T_{\mu\nu} =$$
$$(8\pi G/k_2^4) \cdot [(\Delta m_c)_{e-1}]^4 \cdot T_{\mu\nu} \quad (4.32)$$

because Eq. (2.14) makes $(\Delta v_c)_{e-1} = k_2/(\Delta m_c)_{e-1}$ in Eq. (4.20) of particular scale of photons in that black hole. Then, the Eq. (4.32) can further universalize for any scales of GBs or GSBs with corresponding maximum scales of unescapable particles through respective Event Horizons

$$G_{\mu\nu} + g_{\mu\nu}\Lambda = \left(\frac{8\pi G}{\Delta v_{e-1}^4}\right) \cdot T_{\mu\nu} = (8\pi G/k_2^4) \cdot \Delta m_{e-1}^4 \cdot T_{\mu\nu} \qquad (4.33)$$

inclusive of all scales of black hole as well. However, in Eqs. (4.23) & (4.24), the same total mass-energy in every micro to macro scales of GBs or GSBs in universe from Eqs. (4.26) & (4.258) has considered as scale-specifically quantized in magnitudes i.e. ΔM. Therefore, in Eq. (4.33), the parameter $T_{\mu\nu}$ would have also scale-specific magnitudes due to the scale-specific magnitudes of such GBs or GSBs. That is in Eq. (4.33) there will be

$$\Delta M = T_{\mu\nu} \qquad (4.34)$$

for all corresponding scales of GSBs in universe. Since in the right-hand side of Eq. (4.33), the $(8\pi G/k_2^4)$ would be one similar UC irrespective of all micro to macro scales of GBs or GSBs in universe where other parameters like Δm_{e-1}^4 & $\Delta M = T_{\mu\nu}$ would have SSUC-magnitudes. Then obviously, all the parameters in left-hand side in same Eq. (4.33) will also ultimately have the SSUC-magnitudes; and can define for the same GBs or GSBs in scale-specific way as

$$G_{\mu\nu} + g_{\mu\nu}\Lambda \equiv \Delta p_{e-1} \qquad (4.35)$$

from Eqs. (4.18) and (4.21). Therefore, Eq. (4.33) can further re-write for any specific scale of GBs or GSBs as

$$\Delta p_{e-1} = (8\pi G/k_2^4) \cdot \left(T_{\mu\nu}\right)_{e-1} \cdot (\Delta m_{e-1})^4 \qquad (4.36)$$

and then from Eqs. (4.19), (4.24) & (4.25) we can also write for the same GBs or GSBs as

$$\Delta p_{e-1} = \left(\tfrac{3}{2}\pi^2 \cdot k_6^4/k_2^4\right) \cdot (\Delta n)^4 \cdot \Delta m_{e-1}^4 \,. \qquad (4.37)$$

Since in Eq. (4.35) for the same GBs or GSBs in Eqs. (4.36) & (4.37) there would be

$$\Delta p_{e-1} = \Delta s \cdot \Delta t = (G_{\mu\nu} + g_{\mu\nu}\Lambda)_{e-1}, \quad (4.38)$$

and from the same Eqs. (4.36) & (4.37) also there will be the

$$(8\pi G/k_2^4) \cdot (T_{\mu\nu})_{e-1} \cdot (\Delta m_{e-1})^4 =$$
$$\left(\tfrac{3}{2}\pi^2 \cdot k_6^4/k_2^4\right) \cdot (\Delta n)^4 \cdot \Delta m_{e-1}^4. \quad (4.39)$$

Therefore, the quantized magnitude for the total mass-energy in Einstein Field Equations of GRT would be

$$(T_{\mu\nu})_{e-1} = [3\pi k_6^4 \cdot (\Delta n)^4]/16G = [(3\pi k_6^4)/16G] \cdot (\Delta n)^4 \quad (4.40)$$

for any corresponding integer magnitude of Δn to count the scale-specific Δm_{e-1} in a GB or GSB. Hence, from Eq. (4.38), the Eq. (4.37) can further write as

$$p_{e-1} = (G_{\mu\nu} + g_{\mu\nu}\Lambda)_{e-1}$$
$$= \left(\tfrac{8\pi G}{k_2^4}\right) \cdot (T_{\mu\nu})_{e-1} \cdot (\Delta m_{e-1})^4 = \epsilon \cdot \Delta q_{e-1}^4, \quad (4.41)$$

where proportionality constant $\epsilon = \left(\tfrac{3}{2}\pi^2 \cdot k_6^4/k_2^4\right)$; and the Eqs. (4.37) & (4.41) from Eq. (4.25) are identical to define the corresponding Einstein Field Equation of the same GB or GSB which is scale-specific quantized.

Furthermore, the Eq. (4.41) also depicts an enfolded but scale-specific quantized curvature of spacetime (Δp_{e-1}) in every micro to macro scales of GBs or GSBs. That same (Δp_{e-1}) is directly proportional to the fourth square of optimum homogeneity (Δq_{e-1}) of the respective scale of constituent particles with smallest-bound quantized mass-energy (Δm_{e-1}) which can be totaled by respective integer Δn^4 in Eq. (4.40). Then we will have all the scale-specific quantizations for Einstein Field Equation in GRT for Eqs. (4.26) & (4.29) in every scale of GBs or GSBs in universe through the Eqs. (4.25) & (4.35) but finally in the Eq. (4.41).

4. Inferences

The Sub-Section-3.2, along with the Sub-Section-3.1, is appearing as scale-specific extension of GRT, where Einstein Field Equation for the gravitational field in reference of any GBs or GSBs (which are now appearing in all scale-specific range of magnitudes for the corresponding gravitational mass-energies). Subsequently, the corresponding curvature of spacetime are also appearing in all similar kind of scale-specific range of magnitudes in all those same GBs or GSBs.

It is also found that every such scale-specific gravitational mass-energy is also the scale-specific integer sum of all homogeneous smallest-bound quantized mass-energies of particles which have just missed to escape through the respective escape-velocity of that corresponding gravitational field or curved spacetime. Each of those smallest-bound quantized mass-energies from the Eq. (2.14) would have also the inverse quantized motion with highest-bound quantized inertial-motion in the same gravitational field or curved spacetime of those scale-specific gravitational mass-energies.

Subsequently, all such extended Einstein Field Equations to describe all GBs and GSBs in scale-specific manner also depict the same scale-specific gravitational fields as integer sum for homogeneous units of smallest-bound quantized mass-energies would also be an integer sum of all simultaneous homogeneous units of quantized gravitational fields. Therefore, every such gravitational fields of all GBs or GSBs are not only appeared as scale-specific but equally as a sum of all homogeneous units, i.e. as quantized gravitational fields.

However, this can further clue, whether each of those smallest-bound quantized mass-energies, along with highest-bound quantized-motions involving with those unescaped particles through escape-velocity, could be considered as any specific quantum field [11] of non-gravitational forces in Standard Model of Particle Physics. If so, the same GBs or GSBs as any scale-specific integer sum of all those homogeneous units of quantized gravitational fields could also equally be the scale-specific integer sum of all homogeneous units of gauge quantum-fields of Standard Model of Particle Physics. Those two could be equated. Subsequently, the scale-specific curved spacetime of extended Einstein Field Equation in GRT (involves with the gravitational mass-energies of GBs or GSBs) could emerge as equivalent to integer sum of all homogeneous units of gauge fields in Standard Model of Particle Physics in all same GBs or GSBs in universe. Because, each of those GBs or GSBs, through such scale-specifically extended Einstein Field Equation in Eq. (4.41), would ultimately be a scale-specific sum of total gravitational mass-energies also as a scale-specific integer sum of all homogeneous smallest-bound discrete particles.

The inertial speed of light as stated $c = \Delta v_c$ with a SSUC scale-specific quantized magnitude of inertial-motion $c = 2.9792 \times 10^{10} cm.\,sec^{-1}$ for a particular scale of photons in EMS. Any such SSUC-magnitudes of Δv, including that $c = \Delta v_c$, has derived inversely co-related to all corresponding scale-specific quantized inertial-mass-energy Δm (including Δm_c) of a particle or system-of-particles with an outcome inverse-UC k_2. The conventional Einstein Field Equation in GRT that has stated as invariant respect to that USSC like $c = \Delta v_c$ must be also a USSC (or non-invariant to all scales of frames of references as any GBs or GSBs) in universe.

Therefore, the gravitational field or curved spacetime involving with non-inertial activities in one GB or GSB that appears through such Einstein Field Equation correspond to that $c = \Delta v_c$ must emerge with different values to other scales of GBs or GSBs with corresponding scale-specific Einstein Field Equations where $c = \Delta v \neq \Delta v_c$. But, by involving the UC like $k_2 = (\Delta v \cdot \Delta m)$ in all those variant scale-specific Einstein Field Equations will appear invariant again.

In proceeding Sub-Sections 4.1 to 4.8 there will be all new outcome inferences regarding the gravitation or gravitational fields on the basis of previous Sections 1 to 3. Those inferences would state that both gravitation as well as conventional Einstein Field Equation in GRT are 'local' or scale-specific quantized invariant respect to $c = \Delta v_c$, the 'blackness' of one black hole could be varied depending on the capacities of the observers to see it, and more beyond a gravitational field of GBs & GSBs. As a consequence, there would emerge the simultaneous *anti-gravitational* fields with every corresponding scale of GBs and GSBs. Finally, in non-inertial state, every GB and GSB would appear to have mutual mirror-images in terms of directions for such corresponding gravitational and anti-gravitational fields i.e. in non-inertial state alike the mutual mirror-images for all scales of inertial PSs have defined in the Eqs. (2.29) & (2.30).

4.1. Scale-Specific Quantization of Gravitation

In Newtonian convention, the same gravitation has assumed as 'force' of attraction between two 'material-bodies'. Where scale/s of those two material-bodies, which are separated in an omni-present continuous space, was not precisely considered. Although, the gravitational intensity has appeared to depend on 'quantity of masses' as well as 'inverse square of distances' in-between the two material-bodies.

There were also 'universal invariances' (in Newtonian Classical Mechanics) for both space and time due to conventions of signal exchanges with infinite speeds, i.e. with no SRT like restrictions in upper limits of speeds, in-between any two of inertial frames of reference. Therefore, in similar way of possible infinite exchange of forces, the gravitational material-bodies would also be the 'universal invariant'. Hence, the Newtonian definition of any gravitational interaction was universally invariant in such way.

But, the inertial speed of light c not only has been assumed later as an absolute invariant quantity compare to all other relative quantities like matter, space and time involved with non-luminous inertial frames of reference, but it has also appeared in SRT as upper most possible motion in physical nature. As a consequence, in GRT as Eq. (4.26), not only the definition of gravitation which has associated with all such relative or non-invariant matter, space and time in all non-luminous frames of reference have emerged as invariant in reference of invariant inertial-motion of light, but ultimately the upper speed of propagation for any gravitational interaction also become limited due the convention of such maximum upper limit of light. Though same Eq. (4.26) the gravitation also has appeared as any curvature of space (also with time) in presence of matter. Also, in such basic conventions of GRT, the matters of material-bodies of same GBs or GSBs also have no precise mentions about the scales.

However, the gravitation in Eq. (4.26) has defined by GRT ultimately not as any force but an amount of curvature in spacetime that enfolds corresponding whole mass-energies or matter of any GBs or GSBs (with no precise scales). Consequently, all those non-invariant or relative quantities like mass-energies, space and time which are associated with a gravitational frame of reference could define universally invariant respect to that c.

Since, it is conceptually impossible to exchange any space and time without involving with any kind of mass-energy, such curvature of spacetime can also be considered as corresponding curved spacetime of any mass-energies like Δm_1, $\Delta m_2 = \Delta m_{e-1}$ and Δm_3 in above Eqs. (4.21), (4.20) and (4.22).

Moreover, all the mass-energy as ΔM in universe are affected by universal gravitation, consequently, each of such mass-energies would have the own intrinsic curvature of spacetime.

Hence, conceptually, there would not have any non-curved (i.e. absolute flat) spacetime in universe (BBCOU), neither as a part of the whole curved spacetime of same universe itself nor as any presence of 'gravitationally unaffected' mass-energies there as any so-called GBs or GSBs.

Although, the GRT still considers that space and time or spacetime primarily would have universal continuity everywhere as (an omnipresent) separate fabric. But it has not clear, how one can practically exchange with such spacetime, through its only option of any non-zero & non-infinite quantized signals, if that is an absolute kind of void or having no curvature or a flat fabric of any such continuous spacetime.

The convention of infinite speed of exchangeable signals, particularly in Newtonian way of defining gravitation, can also have another problem. In Chapters 1 & 2, as postulated in Eq. (2.14), any quantized signal if moves with infinity speed must have a zero magnitude for its mass-energies. That is, any such exchangeable signal with zero and infinite magnitudes for two CIPs like inertial-mass-energy and inertial-motion respectively cannot be exchanged with anything that possesses non-zero & non-infinity magnitudes of inertial-mass-energies and inertial-motions. Hence, any exchangeable signal must be of finite-motions as like as finite motion of the light c in SRT & GRT. But Eq. (2.14) also states that not only the $c = \Delta v_c$ correspond to a scale of photons in EMS but also all other boson and fermion scales of PSs in universe could have the scale-specific finite and quantized magnitudes of inertial-motions. Those inertial-motions are also the corresponding infinitesimal discrete rate of changes in any such accelerations in non-inertial surroundings as derived in Eqs. (4.4), (4.14) & (4.15), due to all non-gravitational as well as gravitational influences of forces.

The Eq. (4.41), has extended Einstein Field Equation in Eq. (4.26). As a consequence, the Eq. (4.25) that has revealed gravitation in scale-specific manner depending on UCs (unlike earlier SSUC like $c = \Delta v_c$) as new inverse invariances in all corresponding Eqs. (2.1), (2.14), (2.17), (2.19), (2.21) & (2.23).

The gravitation, that has as derived in Eq. (4.41), is also a curvature of spacetime but such curvature is precisely in scale-specific values of spacetime. Because, that Eq. (4.41) also has defined the magnitudes of mass-energies involved with any GBs or GSBs in precise scales. It presumes that universe, as macro-most scale or BBCOU, is constituted only by all those scale-specific GBs or GSBs. As a result, the whole universe would conceptually have only an integer sum of all scale-specific magnitudes of total mass-energies for all GBs or GSBs.

Consequently, there would be a precise sum for all scale-specific curvatures of spacetime of those GBs or GSBs. This states that there would have no such exchangeable 'flat fabric' or absolute 'void' of so-called spacetime in the universe. Even, if there exists any such flat fabric of spacetime in anyway in anywhere in the current universe assumed as BBCOU that would be beyond exchangeable edge of us through any of ours available non-zero & non-infinite quantized type of signals.

Because, a flat fabric of spacetime could have any zero & infinite types of magnitudes in its mass-energies beyond the communicable limits of our conventional non-zero & finite values of signals.

Hence, any scale scale-specific non-zero & non-infinity GBs or GSBs as defined in Eq. (4.41) could have any such scale-specific magnitudes of gravitation due to similar kind of all non-zero & finite scale-specific magnitudes of mass-energies within each of those corresponding GBs or GSBs.

Moreover, the Eq. (4.26) that have defined gravitation in GRT respect to SSUC quantized magnitude of $c = \Delta v_c$ would also appear as SSUC-invariant or scale-specific-invariant or merely any local-invariant in same Eq. (4.41) rather than any UC-invariant. Because, if that scale-specific magnitude of Δv_c correspond to the specific photon in EMS, all the output results of Eq. (4.26) would be different. But the same would be always universally-invariant irrespective of changes in such scales as any UC-invariant (in next Sub-Sections 4.2 & 4.3) by the intervention of UC like k_2 in Eq. (2.14).

Another notable outcome of the same Eq. (4.41) is that, beside scale-specific curved spacetime of GBs or GSBs, it has one parameter like scale-specific sum of Δn^4 for all homogeneous smallest-bound unescaped particles or systems-of-particles Δm_{e-1} due to the corresponding highest-bound inverse quantized speed Δv_{e-1} in Eq. (4.12). This reveals any GBs or GSBs in Eq. (4.41) are not merely possessing any scale-specific curvature of spacetime but its sum of total mass-energies would also have a scale-specific sum (Δn^4) of homogeneous smallest-bound discrete mass-energies Δm_{e-1}. That is, the gravitation, that has appeared in Eq. (4.26) is also ultimately a scale-specific quantized in Eq. (4.41). This would have immense importance in unification of gravitation and all other non-gravitational forces (in next Chapter-5). Particularly, the parameter Δm_{e-1} in Eq. (4.41) could assume as any Supersymmetric quantum field of forces in Standard Model of Particles Physics.

However, the extended scale-specific quantized gravitation in Eq. (4.41) starting from the of Eq. (4.26) in GRT can finally be described in following sentence as:

> an invariant scale-specific curvature of spacetime (Δp_{e-1}) of GBs or GSBs which is proportional to the scale-specific total mass-energies (Δq_{e-1}) of the same, and such total mass-energies is ultimately a scale-specific integer sum (Δn^4) of all smallest-bound homogeneous quantized mass-energies (Δm_{e-1}) which have highest-bound inverse quantized inertial-motions (Δv_{e-1}) just lower the corresponding scale-

specific escape-velocity to escape out from the concerned GBs or GSBs.

4.2. Non-invariant Einstein Field-Equations

So, in above Sub-Section-4.1, it has inferred gravitation as invariant in Eq. (4.26) respect to the constant inertial speed of light c, and the same speed of light has also assumed in Eq. (2.14) as SSUC-invariant or local-invariant due to $c = \Delta v_c$ with quantized magnitude respect to a photon-particle in EMS.

Consequently, the (conventional) Einstein Field Equations Eqs. (4.26) & (4.28) in GRT would be a similar kind of SSUC-invariant respect to that scale-specific quantized magnitude of $\Delta v_{e-1} = \Delta v_c = c$ in Eq. (4.12) (for same photon-particle) with its corresponding inverse quantized SSUC-invariant magnitude of $\Delta m_{e-1} = \Delta m_c = k_2/c$ in Eq. (4.20). Then, all the predictions through the Einstein Field Equation of GRT (respect to that $\Delta v_{e-1} = \Delta v_c = c$) in Eq. (4.26) would be same in Eq. (4.33). As a result, the same Eq. (4.26) would be a 'local' rather than any 'universal' due to the localized SSUC discrete constant value for $c = \Delta v_c$ in respective Eqs. (4.28) & (4.33) unlike any universalized value of UCs.

Therefore, all the corresponding SSUCs-observers (say, those are stationed in different scales of PSs), having all different SSUCs-magnitudes of Δv in Eq. (2.14), would make the same usual predictions through that Eq. (4.26) in GRT (respect to that $c = \Delta v_c$) but with all different scale-specific values for same Δv.

As a consequence, the Eq. (4.26), respect to that $c = \Delta v_c$, must be a 'local' or SSUC-invariant equation where all other scale-specific constants would have $\Delta v \neq \Delta v_c$ as localized values other than constant inertial-motion of a photon in EMS from the Eq. (2.14). So, the conventional GRT predictions respect to that inertial-motion of light $\Delta v = \Delta v_c = c$ would have local values rather than any universal irrespective of all scales.

4.3. Inversely Invariant Einstein Field-Equations

Then, the gravitation, that is appearing SSUC-invariant respect to $c = \Delta v_c$ for inertial quantized speed of light in GRT for Eq. (4.26) also needs to define as universally invariant irrespective of all scales of PSs with all different magnitudes of quantized inertial-motions for Δv. Because, every individual magnitude of Δv including Δv_c are merely as SSUCs of local-invariant constant values unlike the universal-invariant value for UC k_2 in Eq. (2.14).

There are all inverse constants like k_1, k_2, k_3, k_4, k_5, k_6 & k which are UCs irrespective of all possible scales for PSs in universe having all respective inverse relationships for all 5+5 CIPs having SSUCs local-invariances but $k_1 = (\Delta m \cdot \Delta \lambda)$, $k_2 = (\Delta m \cdot \Delta v)$, $k_3 = (\Delta r \cdot \Delta \lambda)$, $k_4 = (\Delta s \cdot \Delta s_u)$, $k_5 = (\Delta t \cdot \Delta t_u)$, $k_6 = (\Delta v \cdot \Delta r)$ & $k = [(\Delta m \cdot \Delta s \cdot \Delta t) \cdot (\Delta v \cdot \Delta s_u \cdot \Delta t_u)]$ in all those corresponding Eqs. (2.1), (2.14), (2.17), (2.19), (2.21), (2.23) & (2.28) in Chapters 1 & 2.

The quantized inertial space $\Delta s = (\frac{3}{4}\pi \Delta r^3)$ in Eq. (2.19) being one of such SSUCs is not only appearing as scale-specific quantized but is also scale-specific curved due to scale-specific quantization in magnitudes of Δr. Similarly, the inertial time $\Delta t = (2\pi \cdot \Delta r)$ in Eq. (2.21) that emerges from the left-handed axial rotation of the same Δs in Eq. (2.19) and which is also inseparable from such curved and quantized inertial space is also associating with every scale of inertial particles or systems-of-particles. That is, in inertial state, every particle or system-of-particles, with scale-specific quantized magnitudes of mass-energies Δm in Eqs. (2.1) & (2.14), has also inseparable association with same scale-specific quantized and curved space & time or spacetime.

In Chapter-3, it has also defined that the physical laws in Eqs. (3.2), (3.10), (3.26) & (3.42) respect to SSUC $c = \Delta v_c$ are inverse-invariant correspondingly in the Eqs. (3.42)-(3.45) respect to k_2. That k_2 as UC in inertial state, irrespective of scales of particles or systems-of-particles, is universally invariant correspond to any other SSUCs in quantized magnitudes for the $c = \Delta v$. Therefore, the question is whether the physical laws in presence of gravitation in Eq. (4.26) respect to same SSUC $c = \Delta v_c$ would remain be an invariant in Eq. (4.41).

Actually, the Eq. (4.41) has already emerged as universally invariant due to interventions of UCs like k_1, k_2, k_5 & k_6 in presence of gravitation in non-inertial states of GBs or GSBs in universe. This can explain in the following ways.

In inertial states each of the particles or systems-of-particles irrespective of scales possesses only curved and quantized spacetime along with quantized mass-energies and quantized motion; and acceleration in Eqs. (4.5) & (4.7) is an infinitesimal discrete rate of changes in all quantized inertial-motions.

Then, for any one given infinitesimal discrete rate of changes in quantized inertial-motion for one corresponding infinitesimal discrete duration of time involving with each of those scales of GBs or GSBs would be ultimately any corresponding scale of quantized inertial particle or system-of-particles. That is, in every infinitesimal discrete duration of time, the same Eq. (4.41) for any scale of GBs or GSBs should be the universal invariance respect to any scale-specific quantized magnitudes of inertial-motions $c = \Delta v$ in Eq. (2.14).

However, all those physical 'events' which have predicted through the GRT in Eq. (4.26) will be as usual or remain identical only when those are starting respect to the SSUC of the $c = \Delta v = \Delta v_c = 2.9792 \times 10^{10}\ cm.sec^{-1}$. But, when $c = \Delta v \neq 2.9792 \times 10^{10} cm.sec^{-1}$ is scale-specific different, the same physical 'events' would yield the results those must have different SSUC-invariance with the different scale-specific quantized values of $c = \Delta v$. Although, through Eq. (4.41) such differences must be appeared universally invariant.

4.4. Variable Blackness in Black Holes

The *blackness* [1] of a black hole is one of such physical events which are the results of the Eq. (4.26) in previous Sub-Section-4.3, where $c = \Delta v = \Delta v_c = 2.9792 \times 10^{10}\ cm.sec^{-1}$ is the highest possible inertial-motion in the universe due to the Eqs. (3.2), (3.10), (3.26) & (3.42). Because, through such SRT equations respect to $c = \Delta v = \Delta v_c = 2.9792 \times 10^{10} cm.sec^{-1}$, it is really impossible to trace out if there any other inertial-motions are greater than the same Δv_e from the respective gravitational field of a black-hole for the $c = \Delta v = \Delta v_c$.

As a result, to any biologically-limited [2] as well as SRT-limited [3] observer like us, it is really impossible to find out whether there is any signal particles which can really escape from a black hole with $(c = \Delta v) > (c = \Delta v_c = 2.9792 \times 10^{10}\ cm.sec^{-1})$. Because, in Eq. (2.14) that $c = \Delta v_c = 2.9792 \times 10^{10}\ cm.sec^{-1}$ emerges merely as one of SSUCs in quantized magnitudes instead of any UCs like k_2 etc. in physical nature.

However, the Eq. (4.41) states, there will be all scale-specific but universally invariant Einstein Field Equation in GRT for all scales of GBs or GSBs, respect to every different SSUC-magnitude of highest-bound quantized inertial-motions as corresponding Δv_{e-1} having inverse smallest-bound- quantized inertial SSUC-mass-energies Δm_{e-1} of particles (in gravitational field or curved spacetime) with just below the respective escape velocity Δv_e in Eq. (4.12).

In Eq. (4.12), if any GSB (as a black hole) possesses its $\Delta v_2 = \Delta v_{e-1} = c$, then obviously in Eq. (4.13) it would have escape-velocity i.e. $\Delta v_e = \Delta v_3 > c$. That could be possible through inverse relation in Eq. (2.14) for any other superluminal scale-

---

1. The corresponding scale-specific quantized magnitude of inertial-motion for Δv_{e-1} just below the escape-velocity.
2. Conceptually that neuro-vision process that is limited with particular range of wavelengths to see anything beyond.
3. The corresponding higher limitations of quantized inertial-motions respect any particular SSUC-constant in respective SRT set of equations.

specific quantized inertial-motions in extended invariant SRT Eqs. (3.42)-(3.45), if and only if, such superluminal speeds would have scale-specific quantized inertial mass-energies $\Delta m < (\Delta m_{e-1} = \Delta m_c)$.

Then, the GSB as one black hole, to observer like us, with $\Delta v_2 = \Delta v_{e-1} = c$ in Eq. (4.12). But if there any other observer unlike us, would see the same black hole differently, from outside of its respective Event Horizon say with the correspondingly separate biological as well as SRT limitations beyond that $(c = \Delta v) > (c = \Delta v_c = 2.9792 \times 10^{10}\ cm.sec^{-1})$ to receive and analyze signal messages $\Delta v_3 > c$. The same GSB would not be a black hole any more to that superluminally viewing-capacity-observer in same universe. Therefore, the blackness of that GSB as black hole must become 'diminished' to that superluminal viewing capability observer.

Moreover, if we have the Schwarzschild radius [12] for any such GBs or GSBs in physical nature is say $r = 2G \cdot M/c^2$, where, due to Eqs. (2.14), (2.17) & (2.23) all the CIPs e.g. r, M & c would be also as SSUCs. Then, for those Eqs. (2.14), (2.17) & (2.23), the Schwarzschild radius of any scales of GB's would be

$$\Delta r = 2G \cdot \Delta M/c^2 \qquad (4.42)$$

and in Eq. (4.12) if $\Delta v_2 = \Delta v_{e-1} = c$ in generalize way for every GBs then Eq. (4.42) can be re-write as

$$\Delta r = 2G \cdot \Delta M/\Delta v_{e-1}^2 \qquad (4.43)$$

and there would be all scale-specific different magnitudes of blackness' of the GBs or GSBs. As a result, through Eq. (4.43), even for those some corresponding Biological as well as SRT limited observers, with lower capacity of receiving and analyzing the escaped signals out of any smaller scales of GSBs, would appear as black hole. An observer with 'biological' & 'SRT' limitations to receive and analyze the escaped signals below the escape-velocity of the planet Earth, then same Earth could appear to him also as one black hole beyond its similar kind of 'Event Horizon' say outside of the geostationary orbit.

Therefore, 'blackness' would be a scale-specific invariant convention rooted into every specific scale of GBs and GSBs depending on simultaneous Biological as well as SRT limitations of one observer who could receive and analyze the escaped signal-particles with corresponding speeds greater than Δv_{e-1} out of the same GBs and GSBs in this universe.

However, such scale-specific invariances in blacknesses of the black holes could be universally defined as invariant in Eq. (4.41) through the Eq. (2.14).

4.5. Simultaneous Gravitation & Anti-Gravitation

In the Eqs. (4.25), (4.37) & (4.41)), those SSUC-parameters like Δp_{e-1}, Δq_{e-1} and $(T_{\mu\nu})_{e-1}$ are left-handed due to intrinsic left-handedness of CIPs; e.g. Δs, Δt, Δv_{e-1}, ΔM and Δm_{e-1} in Eq. (2.28) for any common expression in every scale of particle-systems including all scales of GBs or GSBs in universe. Therefore, the extended gravitational field equations of GRT, as we have in Eq. (4.41), to define scale-specific optimum convergence (or curvature of spacetime) & homogeneity of respective smallest-bound particles would be intrinsically left-handed.

In Eqs. (2.19) & (2.20), conceptually there we have simultaneous anti-space Δs_u & anti-time Δt_u as right-handed mirror images of CIPs Δs & Δt respectively for every GB or GSB. Subsequently, there would be from Eq. (2.19) the anti-space $\Delta s_u = (\frac{3}{4}\pi \cdot \Delta\lambda^3)$ and from Eq. (2.20) the anti-time as $\Delta t_u = 2\pi \cdot \Delta\lambda$ for all same micro to macro scales of GBs or GSBs in universe.

Then from Eq. (4.16), there would be also a simultaneous *scale-specific quantized mirror-imaged convergence* of such quantized anti-spacetime in every same micro to macro scales of GBs or GSBs through Eqs. (2.19) & (2.20) as

$$(\Delta p_u)_{e-1} = (\Delta s_u \cdot \Delta t_u) \tag{4.44}$$

and from Eqs. (2.19) & (2.20), such mirror image convergences in Eq. (4.44) by Eqs. (2.1) & (2.14) can define as

$$(\Delta p_u)_{e-1} = \tfrac{3}{2}\pi^2 \cdot \Delta \lambda^4 = \tfrac{3}{2}\pi^2 \cdot (k_1^4/\Delta M^4)$$
$$= \tfrac{3}{2}\pi^2 \cdot k_1^4/k_2^4 \cdot \Delta v_{e-1}^4/\Delta n^4 \tag{4.45}$$

simultaneously in all scales-of GBs or GSBs in physical nature.

In Eq. (4.23), the different scale-specific homogeneities of respective smallest scale of inertial-mass Δm_{e-1} which has just cannot escaped through the particular magnitude of corresponding Event Horizons of curved spacetime Δq_{e-1} (that enfolds the specific mass ΔM of GBs or GSBs).

Therefore, conversely, it can consider the same as if a count of all similar scale homogeneous highest-bound quantized inertial-motions Δv_{e-1} in Eq. (4.12) of signal-particles with inverse smallest-bound quantized mass-energies Δm_{e-1} in Eq. (4.20), which are all similar just cannot have escaped through the corresponding Event Horizons of those GBs or GSBs. If $(\Delta q_u)_{e-1}$ is such homogeneity of Δv_{e-1} in Eq. (4.12) in any same GB or GSB, then it could define the same (GB or GSB) from Eqs. (4.23), (4.24) & (4.25)

$$\Delta q_{e-1} = \Delta M = \Delta n \cdot \Delta m = k_2 \cdot \Delta n / \Delta v_{e-1} = k_2 / (\Delta q_u)_{e-1} \tag{4.46}$$

and from Eq. (4.46) we can rewrite the Eq. (4.45) as

$$(\Delta p_u)_{e-1} = \tfrac{3}{2}\pi^2 \cdot \Delta \lambda^4 = \left[\frac{3\pi^2 k_1^4}{2 k_2^4}\right] \cdot \left[\frac{(\Delta v_{e-1})^4}{\Delta n^4}\right] = \epsilon_u \cdot (\Delta q_u)_{e-1} \tag{4.47}$$

where $\epsilon_u = (3\pi^2 k_1^4)/(2 k_2^4)$ is the mirror-imaged proportionality constant; and the Eq. (4.47) defines all right-handed mirror-imaged convergence of anti-spacetime and homogeneity of just not escaped highest-bound scale of quantized motion through Event Horizon of the respective GB or GSB.

Therefore, the Eq. (4.47) is also the simultaneous right-handed mirror-imaged field for Eqs. (4.25) & (4.41) of the same GB or GSB. Since, the Eq. (4.41) defines extension of gravitation for every scale GBs or GSBs in scale-specific quantized manner, therefore the Eq. (4.47) would be for simultaneous existence of *anti-gravitation* in each of those same GBs or GSBs with similar scale-specific quantized manner. That is, every scale of GBs or GSBs, those are seemingly possessed two simultaneous but mutually mirror-imaged fields: its gravitation which has defined in Eq. (4.41) and simultaneous inverse anti-gravitation that has defined in (4.47). While the former would be the convergence towards the center-of-mass of GBs or GSBs, then the later would be a simultaneous divergence from the same center-of-mass of respective GBs or GSBs. This would be due to corresponding inverse magnitudes of Δm_{e-1} (for such inertial mass) and Δv_{e-1} (for inertial motions) of Eq. (2.14) in corresponding Eqs. (4.41) & (4.47).

In gravitation, the gradual heavier quantized magnitudes of corresponding Δm_2 *to* Δm_1 in Eqs. (4.12) & (4.11) would appear to have more and more convergences onward center-of-mass ΔM of GBs or GSBs in Figure-2. But alternately for the simultaneous anti-gravitation within each of those same GBs or GSBs would appear to have as if more and more divergences from same center-of-mass ΔM for all inversely higher and higher quantized values from Δv_2 to Δv_3 in same Figure-2.

Not only that, for such anti-gravitation within each of those same GBs or GSBs, the left-handed parameters $(\Delta p_u)_{e-1}$ would also be directly proportional to fourth square of right-handed parameters $(\Delta q_u)_{e-1}$ in Eq. (4.47) as like as direct proportional relation between left-handed parameters Δp_{e-1} and fourth square of right-handed parameters Δq_{e-1} in Eq. (4.25) for gravitation.

4.6. Non-Inertial Definition for Gravitating Bodies

The Eq. (2.28), as defined one common inertial expression for all scales of particles or systems-of-particles, has accommodated both left and right-handed 5+5 all inverse 10-CIPs within same scales of particles or systems-of-particles. In left-hand side of Eq. (2.28), there are all left-handed 5-CIPs ($\Delta m, \Delta s, \Delta t$ & Δr), and in right-hand side right-handed 5-CIPs ($\Delta v, \Delta s_u, \Delta t_u$ & $\Delta \lambda$).

The above Sub-Section 4.3 has also inferred that every such scale of inertial particles or systems-of-particles in could be equivalent to any infinitesimal discrete rate of changes towards application of force/s as any respective accelerating entity for one instantaneous discrete duration of time. Since, any such discrete acceleration as well as accelerating entities would be equivalent under applications of both inertial-forces as well as gravitational-forces, each of those accelerating or non-inertial entities could be considered as GBs or GSBs due to universal effects of gravitation. As a consequence, all those same inertial particles or systems-of-particles as described in Eq. (2.28) could also be described fundamentally as any GBs or GSBs through Eqs. (4.41) & (4.47) or vice versa for such instantaneous discrete duration of time.

The Eqs. (4.25) & (4.41) for scale-specific gravitational fields have also all those same left-handed 5-CIPs within any scale-specific GBs or GSBs like $\Delta m, \Delta s, \Delta t$ & Δr. Conversely, the Eqs. (4.45), (4.46) & (4.47) also have depicted simultaneous existence of scale-specific right-handed mirror-imaged gravitational fields or anti-gravitational fields for corresponding right-handed 5-CIPs *e.g.*, $\Delta v, \Delta s_u, \Delta t_u$ & $\Delta \lambda$.

Then, the simultaneous existence of such left-handed and right-handed gravitational fields within any specific scale of

GBs or GSBs could be defined from the earlier inertial Eq. (2.28) in Eqs. (4.25) & (4.41) and Eqs. (4.45), (4.46) & (4.47)

$$[\Delta p_{e-1} = \in \cdot \Delta q_{e-1}^4] = k / [(\Delta p_u)_{e-1} = \in_u \cdot (\Delta q_u)_{e-1}^4] \ . \quad (4.48)$$

Therefore, the Eq. (4.48) would be a non-inertial common definition for all scales of non-inertial PSs as any GBs or GSBs inclusive of all gravitational & anti-gravitational forces in universe.

4.7. Left & Right-handed Duality in Gravitating-Body

The Eq. (4.48) has defined that all micro to macro scales of GBs or GSBs in universe are nothing but to achieve as if corresponding hydrostatic equilibria of simultaneous scale-specific quantized left-handed (gravitational) and right-handed (anti-gravitational) fields. The inverse equilibrium between a left-handed gravitational field as the result of Δp_{e-1} & Δq_{e-1} in Eq. (4.41) and an anti-gravitational field as the result of $(\Delta p_u)_{e-1}$ & $(\Delta q_u)_{e-1}$ in Eq. (4.47) would ultimate be a form of all scales of GBs or GSBs in universe from Eq. (4.48).

Then, the Δp_{e-1} & $(\Delta p_u)_{e-1}$ also show the simultaneous scale-specific quantized convergences or curvatures of spacetime and anti-spacetime in all those same scales of GBs or GSBs, respectively.

Moreover, the Eqs. (1.27) and (2.31) already have depicted one conceptual unified equation in inertial state for all scales of particles or systems-of-particles in universe. Since all those non-inertial scale-specific GBs or GSBs could be infinitesimally equivalent to any inertial scale-specific particles or systems-of-particles for any instantaneous duration of discrete quantized time, it seems that those same unified Eqs. (1.27) and (2.31) could be extended further to accommodate above Eq. (4.48) universally.

The same Eq. (4.48) could be also observed by those two conceptually opposite or mirror-imaged observers in BBCOU also might have opposite realizations as there were during deductions of the Eqs. (1.27) and (2.31).

Subsequently, the LE-C observer would see the Eq. (4.48) having the scale-specific gravitational fields with curvatures in spacetime for the GBs or GSBs. That is, onward LE-C observation, all GBs or GSBs would appear as products of left-handedly curved spacetime and simultaneous left-handedly flattened anti-spacetime.

Conversely, the same Eq. (4.48) from conceptual RE-C would appear to have all scale-specific anti-gravitational fields with 'flattening' [1] in anti-spacetime for the same GBs or GSBs. Subsequently, onward RE-C observation, all same GBs GSBs would appear as products of right-handedly curved anti-spacetime and simultaneously flattened spacetime.

Conveniently, from that LE-C direction through same Eq. (4.48) all those GBs or GSBs would appear to have all increasing curvatures in scale-specific spacetime and simultaneously all decreasing in curvatures of anti-spacetime. Alternately, from the RE-C direction in same Eq. (4.48) conceptually there might have all decreasing in curvatures of anti-spacetime for those GBs or GSBs and simultaneously all increasing in curvatures of spacetime. Then same could write for both of those two mutual mirror-images LE-C & RE-C directions of gravitation and anti-gravitation for any GBs or GSB through Eq. (4.48) from respective Eqs. (1.27) & (2.31)

$$\begin{cases} \{k/[(\Delta p_u)_{e-1} = \epsilon_u \cdot (\Delta q_u)^4_{e-1}]\} = [\Delta p_{e-1} = \epsilon \cdot \Delta q^4_{e-1}] \xrightarrow{LH-C} \\ \xleftarrow{RE-C} \approx [(\Delta p_u)_{e-1} = \epsilon_u \cdot (\Delta q_u)^4_{e-1}] = \{k/[\Delta p_{e-1} = \epsilon \cdot \Delta q^4_{e-1}]\} \end{cases}$$

(4.49)

---

1. 'flattening' of anti-spacetime from LE-C direction can also be imagined as curvatures of anti-space from RE-C direction.

Therefore, for every scale of GBs or GSBs in universe through Eq. (4.49) there will have always a simultaneous left-handed and right-handed mutual mirror-imaged pair of observations for corresponding LE-C & RE-C directions. If the scale-specific (SSUC) magnitudes of any one out of total 5+5 inversely related internal-common-parameters or 10-CIPs within the pair will change, then spontaneously the (SSUC) magnitudes of all other inversely related 9-CIPs within same GBs or GSBs will change. There will no need for any exchange of signals for such spontaneous changes in 9-CIPs within GBs or GSBs by facilitating ultimate changes in scales from one to another.

Summary

Present Chapter has revealed any non-inertial-motions or accelerations/decelerations of any micro to macro scales of particles or systems-of-particles in presence of forces, in Classical Mechanics, in quantized manner. That has defined any non-inertial-motion of same particles or systems-of-particles as infinitesimal discrete rate of changes in all respective quantized inertial-motions as described in the previous Chapters 1, 2 & 3. Furthermore, such quantized accelerations/decelerations also have shown equivalent to the gravitational accelerations/decelerations of same particles or systems-of-particles.

Consequently, in Chapter-4, the gravitation as a force or as a spacetime curvature of those same scales of micro to macro particles or systems-of-particles in universe further has revealed in all quantized manner. Not merely such equivalent gravitational accelerations/decelerations of same particles or systems-of-particles having the similar infinitesimal discrete rate of changes in scale-specific quantized inertial-motions but equally each of those particles or systems-of-particles seem to be in respective quantized inertial-motions for every infinitesimal duration of time. Consequently, the whole universe along with all its comprising particles or systems-of-particles in non-inertial state would ultimately be in any such infinitesimal discrete duration of quantized time must be in a state of its inertia.

However, the gravitation in this Chapter-4 has emerged in following forms:

i) gravitation creates the quantized accelerations / decelerations for particles or systems-of-particles in presence of any gravitating bodies;

ii) curved spacetime of gravitating bodies, that defines through Einstein Field Equations respect to $c = \Delta v_c =$ scale-specific quantized inertial speed of light which has appeared as one of SSUCs, consequently the same Einstein Field Equation appears a local and then has universalized through Eq. (2.14) by $c = \Delta v_c = k_2/\Delta m_c$;

iii) since those gravitating bodies are appearing in scale specific manner, the same gravitating bodies would have also the scale specific curved spacetime;

iv) most importantly each of those gravitating bodies under curved spacetime are also appeared as the scale-specific integer sums for the homogeneous discrete smallest-bound (just have hailed to escape) particles, and consequently that curved spacetime of those scale-specific gravitating bodies are in turn emerging as the sum of homogeneous discrete curved spacetime for all the corresponding smallest-bound particles;

v) as a consequence, the gravitation as curved spacetime of any scale-specific gravitating bodies has not only appeared as the scale-specific 'quantized curvatures of spacetime' but also simultaneously is the scale-specific integer sums for 'quantized curvatures of spacetime' for all corresponding homogenous smallest-bound particles;

vi) since every discrete mass-energy of smallest-bound particle from Eq. (2.14) inversely would have discrete highest-bound inertial-motion for same particles then in each infinitesimal duration of time Einstein Field Equation for respective scale of gravitating-body must be in inertial state, and it would be same for whole universe;

vii) there would be all usual predictions for gravitation respect to $c \equiv \Delta v_c$ in General Relativity Theory along with other few new predictions, and one of those about simultaneous universal existence of anti-gravitation;

viii) finally, one universally common unified non-inertial equation for every scales of gravitating-body including universe as macro-most scale in Eq. (4.49).

This new comprehension, about such scale-specific quantized conventions of gravitation, gravitating-bodies, gravitating mass-energies (in same GBs or GSBs) and smallest-bound quantized particles along with corresponding smallest-bound quantized curved spacetime through scale-specific quantized extension of Einstein Field Equations in General Relativity Theory, would further offer a new scope for equating same gravitation as curved spacetime to all other non-gravitational electromagnetic, weak and strong basic forces for unification of physics in next Chapters 5 & 6.

CHAPTER-5:

UNIFICATION OF ALL BASIC FORCES
[Equating Curved Spacetime to all Gauge Fields]

> "It is often assumed that science starts from facts and eschews counter-factual theories. Nothing could be further from the truth. What is one of the basic assumptions of the scientific world-view? That the variety of events that surrounds us is held together by a deeper unity."
>
> — Paul Feyerabend

1. Two New Perceptions About A Gravitating-Body:	219
1.1. Scale-Specificness in Gravitating-Bodies:	222
1.1.1. A Curved Spacetime and Mass-energies	222
1.1.2. An Integer Sum of Smallest-Bound-Particles	225
1.2. Quantum-Properties in Gravitating-Bodies:	226
1.2.1. Intrinsic Quanta of Smallest-Bound-Particles	227
1.2.2. Scale-specificness in Smallest-Bound-Particles	228
1.2.3. Scale can Change in Gravitational Collapse	229
2. A Smallest-Bound-Particle as Gauge-Field	231
3. Consequences:	233
3.1. Curved Spacetime Equivalence of Gauge-Fields	235
3.2. Unification of Gravitation with 3 Basic Forces	238
3.3. Unification of Gravitation and All Basic Forces	240
Summary	244

Gravitation has revealed in General Relativity Theory as curved spacetime. Further the same gravitation has appeared in Chapter-4 as a scale-specific (quantized) curvatures for the same spacetime involving the respective sum of all smallest-bound homogeneous particles or systems-of-particles constitute any gravitating-body or gravitationally shaped body. Consequently, such a scale-specific new interpretation of gravitation not merely exposes same gravitation as a scale-specific convergence or quantized curvature of spacetime with every gravitating-body or gravitationally-shaped-body but also finds itself as one scale-specific sum of homogeneous composition of smallest-bound quantized mass-energies or particles.

Each of such corresponding smallest-bound unescaped particles as constituents of any such scale-specific gravitating-body or gravitationally-shaped-body can also be assumed as any specific type of Quantum-Field or Supersymmetric Gauge Field that has already defined in Standard Model of Particle Physics under Quantum Mechanics.

Therefore, the Chapter-5, as a third set of consequences in the series from Chapters 1 & 2, would unify such quantized gravitation or curved spacetime to any gauge-fields of other basic forces like electromagnetic, strong & weak forces. Hence any such curvatures of spacetime for scale-specific gravitating-body or gravitationally-shaped-body has defined by Einstein Field Equations in General Relativity Theory (in Chapter-4) can also equate to the respective integer sum of any Supersymmetric Quantum-Fields in Standard Model of Particle Physics in Quantum Mechanics. That would be also a unification of all basic natural forces in physical nature or unification of General relativity Theory and Quantum Mechanics through a common equation.

The Section-1 would presume two basic things about any such gravitating-body or gravitationally-shaped-body from the Chapter-4. That is, any such gravitating-body or gravitationally-shaped-body would have

(a) the scale-specific curvature or converged spacetime, and that interlinks to

(b) the 'scale-specific integer sum of all homogeneous smallest-bound particles.

Next, in Section-2, any such smallest-bound particles under gravitational acceleration within the gravitating-body would have further

(i) '5+5 inverse ten Common-Internal-Parameters', and

(ii) each of Common-Internal-Parameters would have 'infinitesimally discrete rate of changes in scale-specific quantized magnitudes along gravitational acceleration', and

(iii) any smallest-bound particle can 'crush' under gravitational collapses.

The Section-3 would postulate every such crushing smallest-bound particle as Supersymmetric gauge-fields of Standard Model in Particle Physics. Finally, the Section-4 would have three different inferences and most importantly that would equate the sum of all homogeneous Supersymmetric (non-gravitational) gauge-fields to the converging or curved spacetime (gravitation) within all scales of gravitating-bodies or gravitationally-shaped-bodies as one unified definition for all basic forces in physical nature.

1. Two New Perceptions About A Gravitating-Body:

In one conceptual inertial state, all micro to macro scales of particles or systems-of-particles have all scale-specific quantized magnitudes in all 5+5 CIPs and each of those 10-CIPs are universally co-related in corresponding Eqs. (2.1), (2.14), (2.17), (2.19), (2.21), (2.23) and (2.24).

There are also different types of non-inertial fundamental forces which could influence all those same inertial particles or systems-of-particles through exchange of force-carrying quantized particles. Subsequently, those fundamental forces are observed to fabricate all 'intra' and 'inter' configurations of all same particles or systems-of-particles irrespective of scales.

Then, the issue is whether all those different types of fundamental forces in universe would be basically quantized as well as unified.

Three fundamental forces e.g. electromagnetic, weak and strong forces have already appeared unified as respective Quantum Fields or Gauge Fields of Forces in Standard Model of Particle Physics. Such unification has a common background assumption that intrinsic component like 'matter' or 'mass-energy' involving within every corresponding scale (from micro to macro) of particles or systems-of-particles are quantized in magnitudes (beside other relevant assumptions).

But another force i.e. gravitation, which has defined as curvatures in one universally pre-existing or omnipresent type of 'flat' spacetime in presence of material-bodies in GRT unlike those quantized Gauge Fields of Forces.

Although, in Eq. (4.41), such a GRT definition for gravitation as curved spacetime in presence of material-bodies, has revealed instead as the scale-specific curvatures of spacetime that is occupied by the total mass-energies of the corresponding scale-specific GBs or GSBs instead of (earlier non-specific scales of any material-bodies in GRT).

The present convention of universe, that conceptually appears as BBCOU, would be ultimately a composition of all such micro to macro scale-specific GBs or GSBs. Total spacetime of such BBCOU would be assumed further as the sum of the scale-specific curvatures of all such scale-specific spacetime of same GBs or GSBs. So, there would not be any such pre-exiting or omnipresent 'flat' or 'absolute void' spacetime in same universe as BBCOU.

Even, if there would be any such existence of flat void spacetime anywhere in same BBCOU, there would have no option for us to locate it other than exchanging through any quantized-signals. Because, without having any material-existences within such conceptual flat and absolute void spacetime it would be practically impossible to exchange through conventional quantized-signals.

However, through the GRT definition of such gravitation, particularly which has defined through the Einstein Field Equation in Eq. (4.26), basically could assume:

(i) a direct proportionality in-between the amount of curvatures in spacetime and the amount of total mass-energies within the respective gravitating-bodies;

(ii) an equilibrium between an inward pressure of curvature or collapse of that spacetime and an outward counter forces of total matters in respective gravitating-bodies; and finally

(iii) a non-scale-specific version of the same curved spacetime as well as total mass-energies involved with the respective gravitating-bodies.

But the same non-scale-specific description of gravitation in Eq. (4.26) has extended further to scale-specific depictions in Eq. (4.41). Then, the Eq. (4.41) actually has described the same gravitation in following scale-specific manner unlike the above mentioned three major assumptions of GRT:

(i) it would remain have a direct proportionality in-between the amount of curvatures in spacetime and the amount of total mass-energies with respective gravitating-bodies;

(ii) it would remain have an equilibrium between an inward pressure of curvature or collapse of that spacetime and outward pressure of counter forces of total matters in respective gravitating-bodies; and ultimately

(iii) it will have any scale-specific value for that curved spacetime enfolded the total mass-energies of respective gravitating-bodies where same total mass-energies would have any specific scale, and such scale-specific total mass-energies would be also a corresponding sum of all homogeneous smallest-bound-particles those just have missed to escape out through over taking specific escape-velocity of gravitating-bodies.

The Section-1 would elaborate such scale-specific gravitation having two components: a scale-specific curvature of spacetime involving with any gravitating-body, and also a scale-specific total sum of all unescaped homogeneous smallest-bound-particles equal to total scale-specific mass-energies of same gravitating-body. Therefore, it would require to have the brief descriptions from the previous Chapters regarding any earlier material-body now as GBs or GSBs in the universe or BBCOU:

1.1. Scale-Specificness in Gravitating-Bodies:

As assumed in previous Chapter-4, since gravitational effect on every material-body in BBCOU is universal, all those same material-bodies are universally as the gravitating-bodies. Moreover, all those gravitating-bodies could be categorized in two broad groups. In one group (for GBs) there would be all those material-bodies which have internally shaped by dominated non-gravitational forces and in another (for GSBs) there would be all those have internally shaped by the dominating gravitational forces. Although, all the GBs and GSBs would have specific scales. Subsequently, each of those gravitating-bodies could be defined by the Einstein Field Equation of GRT in Eq. (4.26) as well as in Eq. (4.41) irrespective of all GBs and GSBs as well as non-scales-specific and scale-specific manners respectively. In addition, in previous Chapter-4, as mentioned in above paragraphs, two important features have perceived in each of those specific scales of GBs:

1.1.1. A Curved Spacetime and Mass-energies

Although, all those micro to macro scales of GBs are affected by the gravitation, but at least within the domain of all 'visible matters' in universe, the gravitation being a force observers to dominate over all other three Gauge Fields of Forces in those GBs when scale-specific magnitudes are acquired total mass-energies $\Delta M \geq 10^{12} kg$ (approx.) and radius $\Delta r \geq 0.5 km$ (approx..). That is through formations of GBs equivalent to or above the scales of planetesimals as (macro) scales of GSBs.

Therefore, in practical sense, such a scale of planetesimals would be smallest scale of GSBs in universe those have shaped first time by dominating gravitational forces for that critical mass-energies $\Delta M \geq 10^{12} kg$ exceeding the dominations of all other three Gauge Fields of Forces within the same compare to all other micro scales of GBs. Starting from such a smallest scale of GSBs in universe, there would be all incremented scales of bigger mass-energies GSBs like different magnitudes of sub-planetary objects. From those scales of rocky planets to giant gaseous planets, from brown dwarfs to sub-solar objects, from a sub-solar-star to all massive giant stars, from a binary system of stars to the huge constellations of stars, from such a constellation of stars to a galaxy, from a galaxy to the clusters of galaxies, from a super cluster of galaxies to the filaments, from filaments and huge voids to the whole universe.

Oppositely, onward micro scales from one such planetesimal, where three Gauge Fields of (material) forces are observed to dominate over the gravitational force. Those micro scales are in contrary shaped by those Gauge Fields of (material) forces with scale-specific quantized mass-energies. Although scale-specific quantization in micro scales are more precise compare to scale-specific quantization in macro scales within ours' range of observations.

But it is also evident that, all that same gravitation dominated macro scales of GSBs are the 'integer' sums of different micro scales of precisely quantized magnitude PSs as individual GBs. Subsequently, each of those GSBs would be always as any definite integer sum of all such quantized mass-energies of micro-scale PSs.

Since, in Eq. (2.28), every such micro PSs must have (5+5) ten numbers of CIPs, the composite macro GSBs would also have the same scale-specific (5+5) ten numbers of CIPs. Moreover, since each of those (5+5) ten CIPs in micro scales have SSUCs in quantized magnitudes, therefore in the macro scales there would have also similar SSUCs in quantized magnitudes for those same (5+5) ten CIPs. Therefore, the CIPs likes gravitation in every scale of GBs or GSBs would be identical but different in SSUCs in quantized magnitudes.

In Eqs. (4.17) & (4.35) such CIPs $\Delta s, \Delta t$ & Δm are linked to define gravitation of any GBs or GSBs in scale-specific ways. Finally, the scale-specific spacetime as $\Delta(s \cdot t)$ with GBs or GSBs becomes converged in scale-specific ways in Eq. (4.41).

From the Eq. (4.17), the Eq. (4.25) can deduce via Eqs. (4.19) & (4.23). Then, ultimately, the Eq. (4.41) would find through Eqs. (4.34), (4.35) & (4.39) for GBs or GSBs, where scale-specific spacetime appearing to enfold total scale-specific mass-energies $\Delta M \equiv T_{\mu\nu}$ of the same GBs or GSBs. The scale-specific curved spacetime $\Delta p_{e-1} = (G_{\mu\nu} + g_{\mu\nu}\Lambda)_{e-1}$ of same GBs or GSBs as there in Eq. (4.41)

$$\Delta p_{e-1} = (G_{\mu\nu} + g_{\mu\nu}\Lambda)_{e-1} = \varepsilon \cdot \Delta q_{e-1}^4 = \varepsilon \cdot \Delta(n \cdot m_{e-1})^4 \quad (5.1)$$

that enfolds scale-specific total mass-energies Δq_{e-1}^4 from Eqs. (4.39) & (4.23) and there will be

$$\Delta q_{e-1}^4 = \Delta M^4 \quad (5.2)$$

for the same GBs or GSBs. Therefore, the Eq. (5.1) has enfolded scale-specific spacetime Δp_{e-1} that ultimately has appeared to equate the fourth power of the scale-specific total mass-energies ΔM in all GBs or GSBs.

As a result, the gravitation that was described in GRT as curved spacetime not only has appeared as scale-specific curved spacetime but also observed equivalent to the scale-specific fourth power of corresponding total smallest-bound quantized mass-energies involving with all those GBs or GSBs.

Consequently, in Eq. (5.1), the same gravitation can also be defined in terms of total mass-energies of GBs or GSBs; but both of such parameters like curved spacetime as well as total mass-energies must be always with the scale-specific magnitudes which are unlikely in the previous GRT definitions of same gravitation.

1.1.2. An Integer Sum of Smallest-Bound-Particles

In Eq. (5.1), the new parameter Δm_{e-1}, that was not involved earlier in GRT, as a new outcome from the scale-specific description of the gravitation would have huge importance in proceeding Sections. So, it is pertinent to redefine the Δm_{e-1} in all those GBs or GSBs. The Eq. (5.2) can re-write from Eq. (5.1) as

$$\Delta p_{e-1} = \epsilon \cdot \Delta q_{e-1}^4 = \epsilon \cdot \Delta M^4 = \epsilon \cdot (\Delta n^4 \cdot \Delta m_{e-1}^4). \quad (5.3)$$

Then scale-specific total mass-energies Δq_{e-1}^4 of a GB or GSB in Eq. (5.1), that can consider to envelop within scale-specific spacetime Δp_{e-1}. That scale-specific mass-energies Δq_{e-1}^4 would be actually a scale-specific inverse sum Δn^4 for all homogeneously identical discrete particles with quantized smallest-bound mass-energy Δm_{e-1}^4 along with inversely highest-bound quantized motion which has quantized magnitude just below the corresponding escape-velocity of that GB or GSB.

Therefore, the Eqs. (5.1) & (5.3) define the conventional 'gravitational mass' in GRT but as a scale-specific integer sum of (Δn^4) for all homogeneous smallest-bound quantized mass-energies Δm_{e-1}. All those homogeneous smallest-bound Δm_{e-1} would have also specific scale and an integer sum Δn^4 of all those homogeneous smallest-bound Δm_{e-1} would be equal to the corresponding 'gravitational mass' ΔM^4 for relevant GB or GSB.

Because, all those Δm_{e-1} would have a homogeneous highest-bound inverse quantized magnitude of motion Δv_{e-1} from Eq. (2.14) which have just a lower SSUCs in discrete magnitudes than corresponding escape-velocity $\Delta v_e > \Delta v_{e-1}$ in Eq. (4.10).

1.2. Quantum-Properties in Gravitating-Bodies:

In Eqs. (4.4) & (4.5) of Chapter-4, any acceleration has defined as any infinitesimally discrete rate of changes in quantized inertial-motions for all corresponding instantaneous discrete moments of quantized time. Then, acceleration of any particle, including one such smallest-bound particle within a scale-specific gravitational field of any gravitating-body, would also be similarly quantized. This infinitesimally discrete acceleration under the effect of any force has basically assumed due to the intrinsic quantization in magnitudes of inertial-motions and inertial mass-energies of the concerned PSs (along with discrete magnitudes of time) irrespective of scales in Eq. (2.14).

Therefore, such quantized acceleration in Eq. (4.4) actually has defined all infinitesimal changes in scales of accelerating PSs onward the direction of influencing force including gravitation (since both of the discrete inertial and gravitational accelerations have appeared as equivalent in Chapter 4). However, for each of those accelerating particles irrespective of scales in any scale-specific gravitational field, from the Eq. (2.28), actually will be not merely all infinitesimal discrete changes in scale-specific intrinsic quantized magnitudes of two inverse CIPs like Δm & Δv out of total 5+5 CIPs but also the simultaneous instantaneous changes in all scale-specific quantized magnitudes of other CIPs e.g. $\Delta s, \Delta t, \Delta s_u$ & Δt_u within the same accelerating PSs.

Although, one smallest-bound-particle with scale-specific quantized mass-energies Δm_{e-1} in Eqs. (5.1) & (5.3) would have definite acceleration within a scale-specific GB or GSB, but that acceleration must be any infinitesimal discrete rate of changes in motions along the direction of affecting force on it. Hence, during that quantized acceleration in the corresponding PSs, not only the two quantized CIPs like Δm_{e-1} & Δv_{e-1} will change in discrete magnitudes, but from the Eq. (2.28), the simultaneous magnitudes of all other CIPs like (Δs_{e-1}, Δt_{e-1}) & ($\Delta s_{u_{e-1}}$, $\Delta t_{u_{e-1}}$) will also have automatic discrete changes onward the direction of same effecting force/s. That is, in 5+5 CIPs dimensional universe, anything that describable by the Eq. (2.28), must have the infinitesimally quantized accelerations no matter whether such acceleration would be any inertial or gravitational in type, due to equivalence of both. But simultaneously there would be all instantaneous automatic changes in quantized values of other nine-CIPs following every such discrete values for rate of changes in motion-CIP of the accelerating particle within GB or GSB.

In this Sub-section, there would be three basic considerations related to any such smallest-bound-particles having Δm_{e-1} within any gravitational field of GBs or GSBs from the Eqs. (2.14), (2.28), (5.1) & (5.3).

Those would be assumed as base for unification of all non-gravitational and gravitational forces within all same GBs or GSBs in next Section-2:

1.2.1. Intrinsic Quanta of Smallest-Bound-Particles

A smallest-bound-particle, while in discrete-acceleration within corresponding gravitational field of any GBs or GSBs, for an instantaneous duration of discrete time would be ultimately any scale-specific intrinsic quantized inertial-PSs. Then such a smallest-bound-particle must be a web of all 5+5 quantized CIPs or dimensions which have described through inverse relations like Eqs. (2.1), (2.14), (2.17), (2.19), (2.21), (2.23). Therefore, the Δm_{e-1} in the Eqs. (5.1) & (5.3) would have, for any instantaneous moment of time, all those 5+5 intrinsic CIPs.

There would be one for de Broglie Wavelength $\Delta\lambda \equiv \Delta r_u$ as anti-radius, one for the inertial-mass-energy Δm (for 1), one for the inertial-motion Δv, one for radius Δr, three for the space $\Delta s(\Delta x, \Delta y, \Delta z)$, three for the anti-space $\Delta s_u(\Delta x_u, \Delta y_u, \Delta z_u)$, one for the time Δt, and one for the anti-time Δt_u. But the two CIPs e.g. Δr & $\Delta\lambda = \Delta r_u$ seem more basic. Because all other CIPs could be derived [1] from those two.

As a result, there would be ultimately one for inertial-mass-energy, one for inertial-motion, three for space, three for anti-space, one for time and one for anti-time after exclusion of those radius and anti-radius from the list in any such PSs. Consequently, there would be total 5+5 CIPs for every smallest-bound-particle in any scale of GBs or GSBs. However, all those 5+5 CIPs should have intrinsic quantization in magnitudes correspond to the intrinsic quantized properties in magnitude of the same smallest-bound-particle for any infinitesimally discrete moment of time in respective gravitational field of all respective GBs or GSBs.

1.2.2. Scale-specificness in Smallest-Bound-Particles

Every PSs in inertial state, has defined in Eqs. (2.1), (2.14), (2.17), (2.19), (2.21), (2.23) and (2.28), possesses not only intrinsic quantization in magnitudes for all 5+5 CIPs or dimensions but such magnitudes are always scale-specific and universally-invariant SSUCs. Hence, any such PSs would also be any smallest-bound-particle, during every infinitesimally discrete duration of time in corresponding gravitational fields of GBs or GSBs, during its course of quantized acceleration for every infinitesimal discrete rate of changes in scales from macro to micro. Then, for any such infinitesimally discrete moment of time, any such smallest-bound-particle in Eqs. (5.1) & (5.3), just

---

1. Because, a radius can define a sphere, its other parameters like circumference, and its circumferential areas which are correspondingly deduced quantized space, time etc. Anti-radius similarly defines anti-space, anti-radius etc.

below the discrete inertial-motion of corresponding scale-specific escape-velocity of the GBs or GSBs, must be defined by same 5+5 CIPs or dimensions in Eqs. (2.1), (2.14), (2.17), (2.19), (2.21), (2.23) and (2.28) with all scale-specific quantized magnitudes. Each of such scale specific quantized magnitudes for all 5+5 CIPs or dimensions of any smallest-bound-particle would be also universally invariant SSUCs during every discrete moment of time in the course of quantized acceleration in respective gravitational field of GBs or GSBs. Therefore, any respective smallest-bound-particle in gravitational field would have scale-specific quantized magnitudes for each of its 5+5 CIPs or dimensions.

1.2.3. Scale can Change in Gravitational Collapse

In Eqs. (4.41), (5.1) & (5.3) a GSB is also ultimately a hydrostatic equilibrium of inward pressure for collapsing curved spacetime or gravitation and outward counter pressure for all non-gravitational or say *material* forces [1]. That outward non-gravitational forces or material forces, those also have described earlier as inertial-forces, can also be assumed as the scale-specific sum of all homogeneous 'quantum-fields of gauge forces' equivalent to such smallest-bound-particles with all homogeneous mass-energy Δm_{e-1} comprising every GSB.

The incrementing gravitational (inward) collapse pressures of curved spacetime of the GBSs starts to dominate over non-gravitational (outward) material forces with gradual increments of scale-specific values of mass-energies ΔM in Eqs. (4.41), (5.1) & (5.2). Those non-gravitational forces, as quantum-fields of forces, as total integer sum of all homogeneous smallest-bound particles Δm_{e-1} also have specific scales.

Then, with the increasing gravitational collapses in different phases of collapses in spacetime of one GSB would also cause the gradually compounding inward pressures on the same Δm_{e-1}. As a consequence, that Δm_{e-1} would also suffer

---

1. Collectively Electromagnetic force, Strong and Weak Nuclear forces.

changes from one earlier scale to another. Actually, its earlier one macro scale would break into some fragments of further micro scales of integer smallest-bound particles. This could happen again and again with incrementing inward collapsing pressures. It seems like one scale of a smallest-bound particle of Δm_{e-1} is gradually gaining its micro scales within a GSB along its gradually incrementing inward pressures of gravitational collapses with incrementing scale-specific total mass-energies ΔM in Eqs. (4.41), (5.1) & (5.2).

Same thing often observes in different scales of heavier mass-energy scales of astronomical objects like stars to galaxies or even within different phases of gravitational collapses in 'life' of a heavier star starting from its formation out of astronomical dusts to a neutron star or a black hole. In all subsequent stages of such a star's life, the respective 'scales' of the smallest-bound-particles Δm_{e-1}, in all corresponding stages, would also have relevant changes from one micro scale to another further micro scale. For example, from the scale of any one heavier fermion to a lighter fermion like neutron, and then to a boson as photon.

These changes in scales of Δm_{e-1} in all corresponding collapsing stages in life of a star or in any other scales of GBs or GSBs is something like a process of *crushing in the scales* of all those homogeneous smallest-bound-particles from one micro to further micro scales.

Then scales of one such smallest-bound-particle with Δm_{e-1} through all infinitesimal and instantaneous discrete rate of changes in an accelerating-PSs is gradually gaining micro and further micro scales within the GSB. That is also, as if like a crushing of Δm_{e-1} with gradually incrementing gravitational collapses of respective GSB simultaneously with its gradual increments in macro and further macro scales of total mass-energies ΔM in Eqs. (4.41), (5.1) & (5.2).

2. A Smallest-Bound-Particle as Gauge-Field

When inward gravitational collapse forces become dominated over all those outward three known basic non-gravitational or Electromagnetic, Weak Nuclear & Strong Nuclear forces in 'total mass-energies' of $\Delta q_{e-1} = \Delta M \equiv \Delta T_{\mu\nu}$ in Eqs. (4.41) & (5.1), in astronomical range of visible matters in universe, all gradually incrementing scales of macro GSBs are observed. That gradual incrementing scale-specific gravitational collapse forces in respective macro scales of GSBs can imagine in the role of breaking or crushing the corresponding micro scale of its smallest-bound-particles Δm_{e-1} from one particular scale to all further micro scales within incrementing scales of a single GSB. For instance, it could be a gravitational crushing for a scale of one smallest-bound hydrogen-atom say $\Delta m_{H_{e-1}}$ within a scale of White Dwarf star as GSB continued to the further micro scales like smallest-bound-particle like photon say $\Delta m_{c_{e-1}}$ in a GSB of Black Hole.

This would be also a similar phenomenon within a Particle Collider. Where such breaking or crushing of particular micro scale of PSs often occurs from one micro scale to further micro scales but under effects of equivalent non-gravitational forces. Then, the crushing of one micro scale of one smallest-bound-particle as Δm_{e-1} to further micro scales under increasing collapse of gravitational force as well as non-gravitational forces are appearing similar 'equivalent'. That is, along the gradual incrementing effects of any natural forces, irrespective of gravitational or non-gravitational in types but with equivalent strengths, any specific micro scale of smallest-bound-particle or colliding-particle would crush or break into further micro scales.

So, in thousands time heavier solar-masses star, as one such of GSBs, that could occur. Through its gradual incrementing internal gravitational collapses in different phases of 'life', starting from origin by once accumulation of astronomical dusts up to its appearance in scale of a huge collapsing strength black hole, through all corresponding phases of collapses there would be all micro scales-specific gravitational crushing of initial Δm_{e-1} to all further micro scales.

If initially, that Δm_{e-1} would have a scale of any heavier fermion subsequently crush into further micro scales of other lighter fermions up to heavier scales of bosons. Then in next, such heavier scales of bosons would further crush lighter scales of bosons up to the scales of photons in EMS. From a scale of heavier gamma-ray-photons with shortest wavelengths in EMS up to a scale of radio wave photons with longest possible wavelength (that seems to possess lightest possible quantized mass-energy among the visible matters) in same EMS. That is with all corresponding collapsing phases up to a huge black hole from such a collapsing heavier solar-star masses star GSB that seems never can be observed even through any radio-photons in any radio telescopes as those radio-photons would transform into the smallest-bound particles.

However, within visible range of matters, irrespective of the scales of PSs, any of those same smallest-bound-particles with corresponding Δm_{e-1} could assume further as any 'quantum-field of forces' in Standard Model of Particle Physics. Any such quantum-fields Δm_{e-1} should be also any respective Gauge-Fields for non-gravitational field of forces which have unified in Supersymmetric Model of Gauge-Fields as Grand Unified Theory in Particle Physics.

Then, in all heavier scales of GSBs within the range of visible matters, through all such gradual smaller and smaller "scale-specific crushing" of respective $\Delta m_{e-1} \equiv$ any Quantum Field among Gauge Fields of Forces (GFFs) would have also similar kind of Supersymmetric Gauge Unifications of the GFFs in SMPP. From one heavier scale of fermion fields to a lighter scale of gluon fields $\equiv SU(3)$. Then from one gluon fields to an electroweak & Higgs Boson fields $\equiv SU(2)$; and ultimately to the scales of Higgs boson & photon fields $\equiv U(1)$ up to a possible lightest radio wave photon in same electromagnetic spectrum). Say any of those Supersymmetric smallest-bound-particle in any GSB is

$$\Delta m_{e-1} \equiv \Delta[SU(3) \times SU(2) \times U(1)]. \quad (5.4)$$

Then Δm_{e-1} in Eqs. (4.41), (5.1) & (5.3) as smallest-bound-particle, which can postulate as smallest-bound-Supersymmetric-gauge-field in Eq. (5.4) under any gravitational collapsing field of GSBs for one infinitesimal discrete duration of time for corresponding infinitesimal discrete rate of changes in the course of gravitational acceleration.

The Eq. (5.4) for any such smallest-bound Supersymmetric Gauge Fields emergence of the Δm_{e-1} in any gravitational field of GSBs would have huge impacts for unification of physics. In one side there would be the gravitational force as curvature of spacetime in all corresponding scales of GBs or GSBs that seems enveloped the total mass-energies ΔM, in Eqs. (4.41), (5.51) & (5.52), as a sum of all such homogeneous smallest-bound Supersymmetric Δm_{e-1} in another side. In proceeding Section-3, that inward collapsing gravitational force being a scale-specific curved spacetime in previous Chapter-4 will be equated (or be unified) ultimately to the sum of all such outward homogeneous smallest-bound Supersymmetric Gauge Fields of Eq. (5.4) as non-gravitational material-forces within all corresponding scales of GBs or GSBs in universe as macro-most scale BBCOU.

3. Consequences:

The Eq. (5.1) in above Section-1 has extended further the gravitation in Eq. (4.41) as scale-specific 'convergence' of (curved) spacetime for scale-specific total mass-energies (comprising fourth power sum of all homogenous smallest-bound-particles) additional to ideas of gravitation through

Einstein Field Equation of GRT in Eq. (4.26). Then gravitation in Eqs. (5.1), (5.3) & (5.4) can be now assumed from the Eq. (4.41) in its extended form below. The gravitation would

(i) remain have a direct proportionality relationship in-between the amount of curvature in spacetime and the amount of total mass-energies within respective gravitating-bodies;

(ii) remain to have an equilibrium between inward pressure of curvature or collapse for that spacetime and outward counter material (gauge fields of) forces out of total matters in respective gravitating-bodies;

(iii) remain have curved spacetime but unlike GRT (a) possessed scale-specific magnitude that has occupied by total scale-specific matter, where (b) same total matter not only a fourth power sum of all homogeneous smallest-bound-particles those have quantized magnitude of motion just below the corresponding escape velocity of GB or GSB but (c) each of those smallest-bound-particles as any Supersymmetric-gauge-fields.

This revised description of gravitation, through the postulation of Eq. (5.4) in Eq. (4.41), can relate/equate the basic concept of curved spacetime (as gravitation) of GRT in Eq. (4.26) to all corresponding total mass-energies of smallest-bound-particles as Supersymmetric Gauge Fields in Standard Model of Particle Physics.

In below Sub-sections 3.1., 3.2. & 3.3., ultimately, two unified equations would be derived. First unified equation would equate gravitation as curved spacetime to all three-known material (e.g. electromagnetic, weak nuclear & strong nuclear) forces (already have unified as Supersymmetric gauge fields of forces) within every scale of GBs or GSBs in visible range of universe. Next unified equation would be an extension of the first to include every possible known or unknown

Supersymmetric gauge fields of forces if exists even beyond the current scopes of Standard Model of Particle Physics and may exist within same universe in realms of Dark Matters and Dark Energies:

3.1. Curved Spacetime Equivalence of Gauge-Fields

Due to Eq. (5.4), the Eqs. (4.41) & (5.1), the different macro structures which are in the realm of macro scales, as different visible matter GSBs, ultimately the results of equilibria in-between respective outward pressures from the sum of its all smallest-bound gauge fields [18] of forces versus inward collapsing force of gravitation from curved spacetime of same.

It could assume to occur in all phases in 'life' of all scales of those GSBs. In practical sense, same observes in life spans of every star and galaxy or even in other larger astronomical structures up to the macro-most scale as whole universe. So, all those GSBs are seemingly now engaged as if to reach in all corresponding equilibria in-between the respective sum of outward smallest-bound-Supersymmetric gauge-fields of forces versus inward collapsing curved spacetime.

Hence, there would have all different scale-specific 'equilibrium' for all those different scales of balanced GSBs. This reveals that degree of equilibrium in one planetesimal would be different compare to one gaseous planet or one 'solar' star or compare to one neutron star. Subsequently, the corresponding scale for Δm_{e-1} within a scale of planetesimal would be different compare to the scale for Δm_{e-1} within one macro-scale of GSB. For example, one smallest-bound-Supersymmetric-gauge-field Δm_{e-1} in Eq. (5.4) would have scale-specific lower quantized magnitude for its inverse Δv_{e-1} in a scale of one planetesimal (due to its lower gravitational field strength and subsequent lower magnitude of escape-velocity Δv_e) compare to another smallest-bound-Supersymmetric-gauge-field Δm_{e-1} with corresponding inverse Δv_{e-1} (due to higher gravitational field strength and subsequent higher magnitude of escape-velocity Δv_e) in a scale of one neutron star.

In this context, suppose the scale-specific equilibrium of one earlier GSB say now changed into another scale-specific equilibrium of another heavier GSB. For example, when earlier it was a planetesimal and later transformed into a conceptual scale of say gluon-star. Subsequently, the particular scale of Δm_{e-1} in that planetesimal through all stepwise gravitational crushing same smallest-bound-Supersymmetric-gauge-fields would be transformed accordingly from one 'earlier heavier fermion in a planetesimal' to another 'lighter gluon now in one gluon-star'. This can happen in every scale of GSBs in universe. As a result, all corresponding smallest-bound-Supersymmetric-gauge-fields Δm_{e-1} for any such gluon from Eq. (5.4) would be

$$\Delta m_{e-1} \equiv \Delta[SU(3)] \qquad (5.5)$$

and due to the Eq. (5.5), the Eq. (5.1) will appear for that gluon star as

$$\Delta p_{e-1} = \in \cdot \Delta q_{e-1}^4 = \in \cdot \Delta M^4 = \in \cdot \{\Delta n^4 \cdot \Delta[SU(3)]_{e-1}^4\}. \qquad (5.6)$$

That is, the Eq. (5.6) conceptually has equated the sum of all integer gluon fields to curved spacetime within respective scale of a gluon-star GSB where the corresponding curved spacetime of same gluon star $\Delta p_{e-1} = \left(G_{\mu\nu} + g_{\mu\nu}\Lambda\right)_{e-1}$ has appeared equal to the fourth powered sum of all homogeneous smallest-bound-gluon-fields $\in \cdot \Delta q_{e-1}^4 = \in \cdot \{\Delta n^4 \cdot \Delta[SU(3)]_{e-1}^4\}$.

In same way, there could be all similar kind of scale-specific equilibria in all further heavier scales of GSBs where the scale-specific smallest-bound-gauge-fields Δm_{e-1} would have also all further micro scale-specific sums of homogeneous smallest-bound-gluon-fields like that $\in \cdot \Delta q_{e-1}^4 = \in \cdot \{\Delta n^4 \cdot \Delta[SU(3)]_{e-1}^4\}$ in Eq. (5.6). Say it as any electroweak & Higgs Boson fields for $SU(2)$. Therefore, all those earlier homogeneous gauge-fields in

Eq. (5.5) as was in Eq. (5.6) will gravitationally crush to transform into further micro scales for

$$\Delta m_{e-1} \equiv \Delta[SU(3) \times SU(2)] \tag{5.7}$$

as corresponding smallest-bound-Supersymmetric-gauge-fields from the Eq. (5.6) as

$$\Delta p_{e-1} = \in \cdot \Delta q_{e-1}^4 = \in \cdot \Delta M^4$$
$$= \in \cdot \{\Delta n^4 \cdot \Delta[SU(3) \times SU(2)]_{e-1}^4\}. \tag{5.8}$$

Then Eq. (5.8) reveals the unification for fourth powered sum of all homogeneous smallest-bound-Supersymmetric gluon and electroweak fields of forces in Eq. (5.7) with gravitation as curved spacetime in specific scale of gluon-star GSBs. A conceptual electroweak star (before transforming into a black hole) can be one such scale of GSBs would have corresponding scale of smallest-bound-particles Δm_{e-1} in same Eq. (5.7).

In that way, conventionally, one exotic boson star or a black hole where $c = \Delta v_c = \Delta v_{e-1}$ in Eq. (4.12) would have ultimately the smallest-bound-Supersymmetric-gauge-fields for the corresponding Δm_{e-1} in Eq. (5.7) will similarly transform through Supersymmetric ways as the scales of Higgs bosons & photons fields = *U(1)* (up to a possible lightest scale of radio-wave photon in electromagnetic spectrum) within the entire range of visible matters

$$\Delta m_{e-1} \equiv \Delta[SU(3) \times SU(2) \times U(1)] \tag{5.9}$$

through further gravitational crushing. That exotic boson star or black hole as one such GSB can define by Eqs. (5.1) & (5.8)

$$\Delta p_{e-1} = \left(G_{\mu\nu} + g_{\mu\nu}\Lambda\right)_{e-1} = \in \cdot \Delta q_{e-1}^4 = \in \cdot \Delta M^4$$
$$= \in \cdot \{\Delta n^4 \cdot \Delta[SU(3) \times SU(2) \times U(1)]_{e-1}^4\}. \tag{5.10}$$

Then, Eq. (5.10) is finally equating any smallest-bound-Supersymmetric-gauge-fields to all three electromagnetic, weak nuclear and strong nuclear known forces to all the corresponding gravitational force as curved spacetime in visible range for any scales of GBs or GSBs in universe. Where ultimate smallest-bound-particle is appearing as micro-most scale of a radio-wave-photon (in EMS) in Eq. (5.9) with unescaped quantized motion through the respective 'Event Horizon'. As a consequence, any such concerned scale of GSBs with respective $\Delta m_{e-1} =$ to a radio-wave-photon would become totally invisible within the range of visible matters and would turn into a candidate of dark matters in universe.

However, finally, in Eq. (5.10), the gravitation as scale-specific curved spacetime has equated (unified) to fourth powered sum of all homogeneous smallest-bound-Supersymmetric gauge fields in GBs or GSBs.

3.2. Unification of Gravitation and 3 Basic Forces

For conveniences, the GSBs with maximum equilibria in above Sub-section-3.1. the respective smallest-bound-Supersymmetric-gauge-fields in Eq. (5.6), i.e. up to a gluon star or any other GSBs as $\Delta m_{e-1} \equiv \Delta[SU(3)]$, if categorizes as ΔM_1 GBS, then all those scales of GBs or GSBs can be defined by Eq. (5.6) would be under that category as

$$\Delta M_1 = \Delta p_{e-1} = \epsilon \cdot \Delta q_{e-1}^4 = \epsilon \cdot \{\Delta n^4 \cdot \Delta[SU(3)]_{e-1}^4\}. \quad (5.11)$$

Therefore, the Eq. (5.11) is a unified equation for a range of all Supersymmetric fermionic gluon fields and curved spacetime as gravitation in corresponding scales of GBs or GSBs in visible range of matters as ΔM_1.

Similarly, from the same Eq. (5.7), due to the smallest-bound-Supersymmetric all possible transformations in scales of the $\Delta m_{e-1} \equiv \Delta[SU(3) \times SU(2)]$, the corresponding scales for GBs or

GSBs can be categorized as another range, those can be defined for conveniences say ΔM_2. Where the ΔM_2 from the Eq. (5.8) can write accordingly as

$$\Delta M_2 = \Delta p_{e-1} = \epsilon \cdot \Delta q_{e-1}^4$$
$$= \epsilon \cdot \{\Delta n^4 \cdot \Delta[SU(3) \times SU(2)]_{e-1}^4\}. \quad (5.12)$$

Then, the Eq. (5.12) is also another unified equation for an extended range of all Supersymmetric fermionic gluon fields and electroweak fields with gravitation in corresponding scales of GBs or GSBs as ΔM_2 in visible range of matters. Those GBs or GSBs are included whole category of ΔM_1 in Eq. (5.11) in addition of all other scales of GSBs having electroweak range of scales for smallest-bound-particles.

Therefore, in Eq. (5.9), for the other micro scales of smallest-bound-Supersymmetric gauge-fields for every unescaped smallest-bound-particle, in all relevant scales of further macro scales of GSBs, as $\Delta m_{e-1} \equiv \Delta[SU(3) \times SU(2) \times U(1)]$ would actually define a whole range of scales in visible matter for all GBs or GSBs up to either an exotic boson star or a black hole with $c = \Delta v_c = \Delta v_{e-1}$ in Eq. (4.12).

For convenience, that can further be categorized as ΔM_3. Then the ΔM_3 would include all the scales of GBs or GSBs within whole range of visible matters in universe and from Eq. (5.10) we can write that

$$\Delta M_3 = \Delta p_{e-1} = \epsilon \cdot \Delta q_{e-1}^4$$
$$= \epsilon \cdot \{\Delta n^4 \cdot \Delta[SU(3) \times SU(2) \times U(1)]_{e-1}^4\}. \quad (5.13)$$

Finally, the Eq. (5.13) as ΔM_3 for whole range of scales as GBs or GSBs in visible matters in universe would have a unified equation for all Supersymmetric fermionic gluon fields, fermionic electroweak fields and bosonic fields with gravitation as respective curved spacetime.

Because, ΔM_3 in Eq. (5.13) not only has included the whole range of scales of GBs or GSBs that includes with ΔM_2, but in addition that also includes all those corresponding scales of GBs or GSBs in visible range of matters as smallest-bound-particles for Higg's Boson Fields and Photon Fields.

Therefore, the Eq. (5.13) is ultimately a unified equation for all known smallest-bound-Supersymmetric-gauge-fields of forces and gravitational forces as curved spacetime in every visible scales of GBs or GSBs in universe.

3.3. Unification of Gravitation and All Basic Forces

In Sub-section-3.2 through Eq. (5.13) all scales of visible matter as GBs or GSBs having all three known natural forces (electromagnetic, weak & strong) defined by Standard Model of Particle Physics have equated to gravitation defined by extended field equations of GRT.

But beyond the range of ΔM_3, it is not yet precisely known whether there is/are any other [1] smallest-bound-gauge-fields of forces would be existed.

Although the effect of gravitation is universally everywhere, and influences of gravitation as remains there as curved spacetime in all scales of the GSBs or GBs for all visible matters, dark matters and dark energies.

Then, if the smallest-bound possible scale of quantized mass-energy within the range of visible matters would be Δm_{e-1} for a radio-wave photon in EMS with longest wavelength in Eq. (5.13), then from the same equation it can further imagine that there might be all other critical macro scales where could have similar collapsing inward gravitational pressures but within the range of scales for dark matters & dark energies in universe.

Those GSBs would be beyond that category of all scale-specific total mass-energies for $\Delta M_3 = (\Delta n \cdot \Delta m)_{e-1}^4$, which can comprise a homogeneous sum of all unescapable smallest-

---

1. Conceptually, the material types of forces other than Electromagnetic, Strong & Weak nuclear forces.

bound-mass-energies smaller than a radio wave photon [1] within domain of visible matters.

But beyond that critical scale of GSBs as ΔM_3 in visible matters domain, there could be still other heavier macro scales of GSBs up to the scale of macro-most scale universe where rest of 95% as invisible matter-energy scales of GBs or GSBs but are still influenced by the gravitation. All those also could have capabilities through corresponding heavier gravitational collapses to crush further concerned smallest-bound-particles as any gauge fields Δm_{e-1} into smaller to further smaller scales beyond that scale of $\Delta m_{e-1}=$ a radio wave photon in ΔM_3 from Eq. (5.13).

Therefore, conceptually, such a smallest-bound radio wave photon of Δm_{e-1} say transforms into a further smaller scale of mass-energy for dark matter beyond visible matters. Because, beyond of that critical scale of ΔM_3 in Eq. (5.13) there could have still heavier scales of GSBs those can facilitate collapsing through the heavier scales dark matters say in a category of $\Delta M_4 > \Delta M_3$.

Further there could be even heavier scales of GSBs those might have more gravitational collapses in dark energies and could be categorized as $\Delta M_5 > \Delta M_4$ up to the macro-most scale of whole universe (BBCOU). Ultimately which might have the Δm_{e-1} as made of dark-energy entities. Present Sub-section-3.3 will extend the earlier Eq. (5.13) starting from the visible matters to dark matters and then dark energies comprising the macro-most scale universe as BBCOU.

Therefore, in Eq. (5.13), if there $\Delta m_{e-1} <$ a smallest-bound-particle like conceptual radio-wave photon/boson for a scale of GSBs with collapsing gravitational strength for total scale-specific mass-energies $\Delta M_4 < \Delta M_3$, and if the same ΔM_4 would

---

1. Since one radio wave photon seems to have longest possible wavelength among all wave-corpuscular phenomena or PSs in visible realm of matters in universe.

have total sum of unescaped homogeneous smallest-bound-particles = Δm_{e-1} for any unknown gauge fields of forces within the realm of dark matters say $XU(N_{DM})$, then in Eq. (5.9) such a smallest-bound-Supersymmetric-gauge-field would be

$$\Delta m_{e-1} \equiv \Delta[SU(3) \times SU(2) \times U(1) \times XU(N_{DM})] \qquad (5.14)$$

where X represents those yet unknown gauge fields, and N represents the matrix, and DM stands for dark matters. Then due to Eq. (5.14), from Eq. (5.13) the ΔM_4 can define as

$$\Delta M_4 = \Delta p_{e-1} = \in \cdot \Delta q_{e-1}^4$$
$$= \in \cdot \{\Delta n^4 \cdot \Delta[SU(3) \times SU(2) \times U(1) \times XU(N_{DM})]_{e-1}^4\} \qquad (5.15)$$

and Eq. (5.15) will be a scale-specific equality between scale-specific curvatures of spacetime and Super-symmetric unified gauge fields in some heavier scales of GSBs in domains of visible matter & dark matter. As a result, ΔM_4 represents actually a whole range of scales of GBs or GSBs in both visible **x** matters as well as dark matters those are correspondingly included all ΔM_3 and remaining are scales of dark matters.

Similarly, there can further heavier scales of GSBs than ΔM_4 in Eq. (5.15). Those GSBs can incorporate dark energies in universe. Such heavier scales of GSBs would have total mass-energies $\Delta M_5 > \Delta M_4$ beyond the scopes of Eqs. (5.14) & (5.15), and if there homogeneous smallest-bound-particles Δm_{e-1} would have further micro scale/s than smallest-bound mass-energies of dark matters in ΔM_4; then in corresponding scale of collapsing gravitational inward pressure within such GSB ΔM_5 its smallest-bound-particles Δm_{e-1} would be conceptually a candidate of dark energy particles. Consequently, if we consider the total numbers of such smallest-bound-particles as another type of unknown Supersymmetric gauge fields of forces in realm of dark energies say $YU(N_{DE})$, then for the same ΔM_5 in Eq. (5.14)

$$\Delta m_{e-1} \equiv \Delta[SU(3) \times SU(2) \times U(1) \times XU(N_{DM}) \times YU(N_{DE})]$$

(5.16)

where Y represents such a yet unknown gauge fields, and N is matrix and DE for dark energies. Then from Eq. (5.16), we can obtain that ΔM_5 from Eq. (5.15)

$$\Delta M_5 = \Delta p_{e-1} = \in \cdot \Delta q_{e-1}^4$$
$$= \in \cdot \{\Delta n^4 \cdot \Delta[SU(3) \times SU(2) \times U(1) \times XU(N_{DM}) \times YU(N_{DE})]_{e-1}^4\}$$

(5.17)

which will be ultimately a scale-specific equivalence of scale-specific curvatures in spacetime and Super-symmetric gauge fields those involving all scales of GSBs for visible-matters, dark-matters & dark-energies in macro-most scale universe.

Therefore, the Eq. (5.17) will also be a unified equation for all scale-specific curvatures of spacetime or gravitation and all known & unknown possible smallest-bound-Supersymmetric-gauge-fields of forces irrespective of the scales of GSBs in realms of visible matters, dark matters & dark energies of universe.

Also, in micro scales of GBs, some of relevant parameters as well as CIPs in Eq. (5.17) will have very smaller or of practically negligible values. Subsequently, the same Eq. (5.17), for all those micro scales of GSBs or for any further micro scales of GBs, can show the appropriate scale-specific expressions within the range of gauge fields of the Standard Model of Particle Physics or field equations of GRT for scales of visible matters respect to such practically observable values.

Therefore, Eq. (5.17) will be a unified non-inertial definition for all fundamental known and unknown smallest-bound-Supersymmetric gauge-fields of forces and gravitation as curved spacetime. The range of GSBs under category of ΔM_5

has included all category of GSBs ΔM_4 in addition to all other scales in dark energy GSBs under gravitational collapse up to the macro-most scale of whole universe. As a result, in ultimate scale of ΔM_5, in Eq. (5.17), defines the whole universe unifiedly.

Summary

In previous Chapter, a gravitating-body or gravitationally-shaped-body has revealed as scale-specific but having corresponding – **(i)** curvature of spacetime that equals to **(ii)** an integer sum of all comprising entrapped homogeneous smallest-bound-particles those have failed to overtake the respective escape-velocity.

In present Chapter, same gravitating-body is/are also scale-specific but with corresponding **(i)** curvature of spacetime that equals to **(ii)** an integer sum of all entrapped homogeneous smallest-bound-particles is/are also the smallest-bound-Supersymmetric-gauge-fields those have just below the corresponding escape-velocity. This is completely a new outcome and ultimately has emerged not only to equate all the known three non-gravitational (or material forces) Supersymmetric Gauge forces (like electromagnetic, weak and strong forces) to gravitational force as spacetime curvature involving within any corresponding gravitating-body. This also equates subsequently all those non-gravitational forces (now unknown) which may exist in the domains of dark matters and dark energies in BBCOU. The Chapter-5 has actually unified or equated all existing or known or other possible basic material or non-gravitational forces in Eq. (5.17) which are involved to BBCOU along the direction of LE-C in Eqs. (1.25) & (2.29). Same Eq. (5.17) would be also an ultimate unification for General Relativity Theory and Standard Model of Particle Physics through extensions of both.

Most importantly, another new convention in current Chapter-5 is that, each of those smallest-bound-particles are infinitesimally discrete quanta for the duration of scale-specific instantaneous discrete moment of time throughout quantized acceleration in specific gravitational field. Consequently, any smallest-bound-particle, throughout the course of its quantized acceleration, should retain its inertial intrinsic scale-specific quantized values as mentioned in Chapters 1 & 2.

However, the next Chapter would deduce a RE-C directional unified equation for Eq. (5.17) that would be a non-inertial extension of the Eq. (2.31) for every ingredient-PSs of BBCOU. That would be also the final unified equation compatible with the simultaneous left-handed and right-handed duality involves with entire realm of quantum-reality in physical nature.

It has also observed in the Chapter-5 that Gravitation or curved spacetime of any GBs or GSBs is actually unified to the sum of its all smallest-bound homogeneous particles as Supersymmetric Gauge Fields. But such unification has happened in total 5+5 unfolded ten 'inverse' dimensions (as 10-CIPs) in places of only available 3+1 four unfolded spacetime dimensions in General Relativity Theory and Quantum Mechanics.

Since, the Eq. (5.17) is also ultimately a unification for all non-zero & non-infinity (i.e. any real & quantized) values of those same 5+5 unfolded ten inverse dimensions (as 10-CIPs) in all same particles or systems-of-particles or GBs or GSBs, that entire range of such real & quantized unification would be also considered as the realm of a *unified quantum-reality* in physical nature.

CHAPTER-6:

DUALITY IN EVERY QUANTUM-REALITY
[Left & Right Handedness in Particles or Systems]

"All of physics is either impossible or trivial. It is impossible until you understand it, and then it becomes trivial."
— Ernest Rutherford

1. Anti-Gravitation and Anti-Gauge Forces	247
2. A Unified Equation for whole Quantum-Reality	253
3. A Mirror-Imaged Duality in Quantum-Reality	257
4. Consequences:	260
4.1. Symmetry in Quantity of Matter & Antimatter	260
4.2. Resolving Entanglement in E-P-R Paradox	264
Summary	267

There was an assertion for inertial unification of everything or every scale of particles or systems-of-particles in physical nature in Eqs. (2.29) & (2.31) in Chapter-2 when all 5+5 inverse unfolded 10-dimensions have only non-zero & non-infinity magnitudes.

Further, the Eq. (2.31) had shown an intrinsic left-handed and right-handed 'duality' in every such scale of particles or systems-of-particles in inertial-state.

In Chapter-5, there was a non-inertial unification for all those same scales of particles or systems-of-particles including all four known natural forces like three-gauge forces related to 'matter' and one-gravitational force related to curved spacetime of same 'matter' in Eq. (5.17).

That non-inertial unification also has unified all scales of particles or systems-of-particles inclusive of all GBs or GSBs as well as natural forces i.e. for the entire quantum-reality [1] in non-inertial-state compare to the unified quantum-reality in Eq. (2.28). The BBCOU assumes as the macro-most scale, comprising all other scales, for that unified entire quantum-reality.

The Chapter-6, would be a fourth set of consequences in the series and derive from the basic postulated foundations of Chapters 1 & 2. The same would actually reveal that entire unified quantum-reality to possess simultaneous Left-handed & Right-handed intrinsic *duality* in non-inertial-state as an extension of the Eq. (2.31).

The Sections 1 to 3 would deduce one ultimate non-inertial duality for unified equation for everything in the realm of quantum-reality within the range of magnitudes for the 5+5dimensional co-ordinates are non-zero & non-infinity.

That would have many consequences in resolving various inconsistencies in todays observational comprehensions.

The Section-4 will resolve the issues like E-P-R Paradox, asymmetries in quantities of matters over antimatters in present universe, etc. The Section-4 would deal with all these as Consequences.

1. Anti-Gravitation and Anti-Gauge Forces

In Chapter-5, the Eq. (5.17) has equated all known & unknown Supersymmetric gauge fields of forces, which are primarily attached with one quantized-CIP out of 5+5 CIPs i.e. total mass-energy ΔM as 'material-forces', in Standard Model of Particle Physics to curved spacetime as other two also combined the entire category of ΔM_5 for the range of all possible micro to macro scales of GBs or GSBs irrespective of

---

1. Since observers like us have intrinsic limitation to exchange through any signal with anything that have only non-zero & finite discrete values for all 5+5 dimensions, i.e. all our conventional observers, signals & observances are limited by the Quantum-Reality in physical nature.

Quantized-CIPs attached with gravitational-force has defined in Einstein Fields Equations of GRT. The same Eq. (5.17) has visible-matters, dark-matters & dark-energies entities influenced by gravitation.

The Section-1 will extend further the scope of same earlier Eq. (5.17). Beside equivalence of all known & unknown Supersymmetric gauge fields and gravitational forces within all those (visible matters, dark matters & dark energies) scales of GBs or GSBs, the Eq. (5.17) will further reveal a simultaneous mirror-imaged counterpart of it. That will be an equivalence of all mirror-imaged Supersymmetric antigauge [1] fields and antigravitational [2] forces within each of those same scales of GBs or GSBs in universe.

Moreover, the whole universe BBCOU, conceptually as macro-most scale, has contained all those scales of GBs or GSBs. Consequently, the same universe includes all those scales of particles or systems-of-particles which are also comprised the ΔM_5 in Eq. (5.17). Then, such a macro-most scale as universe (also as BBCOU) would be constituted by all those same scales of PSs as GBs and GSBs under category of ΔM_5 plus any conceptual micro-most scale of all PSs as GBs or GSBs. No matter whether such conceptual micro-most scale would be a candidate of dark energies, and also no matter whether same micro-most scale of PSs would be a singular entity as similar as the macro-most scale as whole universe in other hand.

Hence, for further convenience, everything can be categorized next as ΔM_6 inclusive of all such scales of micro-most & macro-most and ΔM_5 irrespective of scales for GBs or GSBs, and consequently Eq. (5.17) would appear accordingly

---

1. Due to simultaneous existence of quantized inertial-motion inverse to corresponding inertial-mass-energies as gauge-fields of forces.
2. Due to the simultaneous existence of quantized curvature of antispacetime inverse to corresponding curvature of spacetime as gravitation.

$$\Delta M_6 = \Delta p_{e-1} = \in \cdot \Delta q_{e-1}^4$$
$$= \in \cdot \Delta \{n^4 \cdot [SU(3) \times SU(2) \times SU(1) \times XY(N_{DM}) \times YU(N_{DE})]\}_{e-1}^4 \quad (6.1)$$

where total mass-energies would be obviously $\Delta M_6 > M_5$. For conveniences, if a smallest-bound Supersymmetric gauge-fields of forces for the scale-specific total mass-energies of that ΔM_6 in Eq. (6.1) in brief say the

$$\phi = [SU(3) \times SU(2) \times (SU(1) \times XU(N_{DM}) \times YU(N_{DE})] \quad (6.2)$$

then the Eq. (6.1) would appear in brief as

$$\Delta M_6 = \Delta p_{e-1} = \in \cdot \Delta q_{e-1}^4 = \in \cdot \Delta [n^4 \cdot (\Phi)]_{e-1}^4 \quad (6.3)$$

where Δp_{e-1} would be the scale-specific curved spacetime or universal geodesic for universal gravitation and Δq_{e-1} is scale-specific all smallest-bound Supersymmetric gauge-fields of forces, that is emerging out of the entire matters (total mass-energies) involving within the category of ΔM_6 in Eq. (6.3).

In Eq. (2.28), any GBs or GSBs, for any infinitesimal discrete rate of changes in quantized motion and discrete moment of time, would be one discrete inertial-PSs irrespective of scales with all those 5+5 inversely co-related 10-CIPs. That is, each of those non-inertial GBs or GSBs as any specific scale of inertial-PSs would be infinitesimally in inertial state.

Since every such scale of GBs or GSBs are infinitesimally as inertial-PSs for similar infinitesimal moment of time, therefore in Eq. (2.28) for M_6 would have $\Delta p_{e-1} = \in \cdot \Delta q_{e-1}^4$ in Eq. (6.3) for all its left-handed 5-CIPs like $\Delta m, \Delta s(\Delta x, \Delta y, \Delta z)$ & Δt in all corresponding GBs or GSBs irrespective of scales

$$[\Delta m, \Delta s(\Delta x, \Delta y, \Delta z), \Delta t] \equiv \{\Delta p_{e-1} = \in \cdot \Delta [n^4 \cdot (\Phi)]_{e-1}^4\}, \quad (6.4)$$

where Δp_{e-1} referring curved spacetime for gravitation as a left-handed gravitational field due to left-handedness in 4-CIPs like $[\Delta s(\Delta x, \Delta y, \Delta z) \& \Delta t]$, and also the Δp_{e-1} as the $\Delta[n^4 \cdot (\Phi)]_{e-1}^4$ is referring all smallest-bound Supersymmetric gauge-fields of forces for all 'material-forces' would also be the left-handed fields due to the left-handedness in rest 1-CIP Δm. Then, the Eq. (2.28), for every infinitesimal discrete moments of time, will ultimately appear for those two left-handed fields of forces in Eq. (6.4) as

$$\{\Delta p_{e-1} = \in \cdot \Delta[n^4 \cdot (\Phi)]_{e-1}^4 \}/k \cdot [\Delta v \cdot \Delta s_u \cdot \Delta t_u] . \qquad (6.5)$$

Consequently, the Eq. (6.5) would suggest, there might have corresponding two right-handed fields of forces. One for all those right-handed 4-CIPs like $[\Delta s_u(\Delta x_u, \Delta y_u, \Delta z_u) \& \Delta t_u]$ as right-handed counterpart of left-handed gravitation or curved spacetime, and there remaining one right-handed 1-CIP like all homogeneous highest-bound unescaped quantized inertial-motions Δv as right-handed counterpart of left-handed total muss-energies as a sum for all smallest-bound Supersymmetric gauge-fields of forces in every same scale of GBs or GSBs within category of ΔM_6. Therefore, all such corresponding right-handed gauge fields say Φ_u opposite to left-handed Φ in Eq. (6.2) would be

$$\Phi_u = [SU(3) \times SU(2) \times U(1) \times XU(N_{DM}) \times YU(N_{DE})]_u \qquad (6.6)$$

and that right-handed 1-CIP in Eq. (6.5) must be for corresponding all homogeneous highest-bound quantized inverse inertial-motions for every unescaped smallest-bound-particle as $\Delta v \equiv \Delta v_{e-1} = k_2/\Delta m_{e-1}$ in Eq. (4.20).

That is, the Φ_u would stand for all those oppositely right-handed gauge-fields of forces need to be comprehended in terms of the all unescaped homogeneous highest-bound quantized inertial-motions below the scale-specific escape-velocities equivalent to Δv_{e-1} instead of Δm_{e-1} within all same GBs or GSBs within the category of ΔM_6.

For conveniences, those right-handed highest-bound Supersymmetric gauge-fields of forces equivalent to highest-bound quantized inverse inertial-motions in every scale of such GBs or GSBs can term as 'antigauge-fields' of forces.

Those right-handed gauge-fields or antigauge-fields can also be assumed further as a kind of Supersymmetric antigauge fields for the inverse presence of Δv_{e-1} as the homogeneous highest-bound quantized inertial-motions.

That homogeneous highest-bound quantized inertial-motions Δv_{e-1} as homogeneous Supersymmetric-antigauge-fields opposite to simultaneous smallest-bound homogeneous quantized inertial-mass-energies Δm_{e-1} as Supersymmetric-gauge-fields have co-existences in every scale of GBs or GSBs within category of M_6.

However, every scale of those GBs or GSBs as PSs for infinitesimal duration of time have total inversely co-related (5+5) quantized 10-CIPs. In Eq. (2.28), out of all those 10-CIPs e.g. $\Delta s\ (\Delta x, \Delta y, \Delta z)$ & Δt within left-handed set through Eqs. (5.1) & (6.3) have equated as gravitational forces as scale-specific curvature of spacetime to scale-specific all material forces as $\Delta m = \Delta m_{e-1}$ for smallest-bound Supersymmetric gauge-fields of forces.

Then, for its all simultaneous inverse and right-handed 4-CIPs e.g. $\Delta s_u(\Delta x_u, \Delta y_u, \Delta z_u)$ & Δt_u, in every same scales of GBs or GSBs for infinitesimal duration of time as PSs in Eqs. (2.28) & (6.5), there would have an inverse and right-handed kind of gravitation or say 'antigravitation' due to the conceptual inverse curvatures of antispacetime $\Delta s_u(\Delta x_u, \Delta y_u, \Delta z_u)$ & Δt_u collectively as $\Delta(p_u)_{e-1}$, compare to left-handed spacetime $\Delta s(\Delta x, \Delta y, \Delta z)$ & Δt collectively as Δp_{e-1}.

Hence, that can write as a right-handed type or antigravitational field equation opposite to the Eq. (4.17) as

$$\Delta(p_u)_{e-1} = \frac{3}{2}\pi^2 \cdot \Delta\lambda^4$$

$$= \frac{3}{2}\pi^2 \cdot (k_1^4/k_2^4) \cdot [\Delta(v/n)_{e-1}^4]$$

$$= \epsilon_u \cdot \Delta(q_u)_{e-1}^4. \qquad (6.7)$$

In Eq. (6.7), the $\Delta(p_u)_{e-1}$ and $\Delta(q_u)_{e-1}^4$ are the right-handed convergence of anti-spacetime and homogeneity of unescaped particles with highest-bound quantized inertial-motion Δv_{e-1} respectively.

The ϵ_u in same Eq. (6.7) is simultaneous mirror-imaged proportionality constant of ϵ in Eq. (6.5) irrespective of scales for same GBs or GSBs.

Then Eq. (6.7) would be a simultaneous right-handed Equation of antigravitation opposite to Eq. (5.1) for same GBs or GSBs for every infinitesimal duration of time as any PSs in Eq. (2.28) for 5-CIPs e.g. $[\Delta v, \Delta s_u(\Delta x_u, \Delta y_u, \Delta z_u), \Delta t_u]$.

Due to such co-existences of simultaneous left-handed gravitation in Eq. (5.1) related to five left-handed 5-CIPs e.g. $[\Delta m, \Delta s(\Delta x, \Delta y, \Delta z) \ \& \ \Delta t]$ and right-handed antigravitation in Eq. (6.7) for rest of right-handed 5-CIPs e.g. $[\Delta v, \Delta s_u(\Delta x_u, \Delta y_u, \Delta z_u) \ \& \ \Delta t_u]$, as like as in inertial state in Eq. (2.28) for every infinitesimal duration of discrete time, there will be ultimately

$$[\Delta p_{e-1} = \epsilon \cdot \Delta(q)_{e-1}^4] = k/[\Delta(p_u)_{e-1} = \epsilon_u \cdot \Delta(q_u)_{e-1}^4] \qquad (6.8)$$

and that Eq. (6.8) could be considered as the non-inertial definition for all scales of GSBs or GBs within the category of M_6. The Eq. (6.8) also depicts that every same scale of GSBs or GBs are nothing but the inverse symmetry of simultaneous scale-specific gravitational and anti-gravitational forces [3].

However, the Eq. (6.8) further could re-write correspondingly from Eqs. (5.1) & (6.7) by considering $\Delta m_{e-1} \equiv \Delta\Phi$ in Eq. (6.2) and $\Delta v_{e-1} \equiv \Delta\Phi_u$ in Eq. (6.6) as

$$[\Delta p_{e-1} = \epsilon \cdot \Delta(n \cdot m_{e-1})^4 = \epsilon \cdot \Delta(n \cdot \Phi)_{e-1}^4]$$
$$= k/\{\Delta(p_u)_{e-1} = \epsilon_u \cdot \Delta(v/n)_{e-1}^4 = \epsilon_u \cdot \Delta[\Phi_u/n]_{e-1}^4\} \quad (6.9)$$

and from the Eq. (2.14) also that $\Delta v_{e-1} = k_2/\Delta m_{e-1}$ in Eq. (6.9). As a result, in Eq. (6.7) there could also be

$$\Delta(p_u)_{e-1} = \frac{3}{2}\pi^2 \cdot \Delta\lambda^4 = \epsilon_u \cdot [\Delta(v/n)_{e-1}^4]$$
$$= \epsilon_u \cdot \Delta(\Phi_u/n)_{e-1}^4 = \epsilon_u \cdot k_2^4/\Delta(n \cdot \Phi)_{e-1}^4 \quad (6.10)$$

and consequently, the Eq. (6.9) could further appear as

$$[\Delta p_{e-1} = \epsilon \cdot \Delta(n \cdot m_{e-1})^4 = \epsilon \cdot \Delta(n \cdot \Phi)_{e-1}^4]$$
$$= k/\{\Delta(p_u)_{e-1} = \epsilon_u \cdot \Delta(v/n)_{e-1}^4\}$$
$$= \epsilon_u \cdot k_2^4/\Delta(n \cdot \Phi)_{e-1}^4. \quad (6.11)$$

Therefore, the Eq. (6.11) actually has revealed an equality inbetween all possible known & unknown Supersymmetric gauge fields with corresponding scales of curved spacetime as gravitation and all possible unknown Supersymmetric antigauge fields with curved antispacetime as antigravitation in every scale of GBs or GSBs as PSs in Eq. (2.28).

2. A Unified Equation for whole Quantum-Reality

The Eq. (6.11) is not only a non-inertial unified definition for all known & unknown possible forces within category of M_6 (also as a range of all non-zero & non-infinity discrete values of all those 5+5 inverse 10-CIPs) but has also shown all those left-handed and right-handed fields of forces in same category of M_6 for all scales of PSs for every infinitesimal duration of discrete time as GBs or GSBs from micro-most to macro-most scales (that includes also the whole BBCOU).

Therefore, the Eq. (6.11) would be a unified non-inertial definition including all left-handed and right-handed fundamental natural forces. It has included not only those four known fundamental forces but could also predict about the possibilities of additional unknown left-handed gauge-fields of forces along with right-handed antigravitational and antigauge-fields of forces within the category of M_6.

All those presently known left-handed material-forces are three conventional Supersymmetric gauge-fields of (electromagnetic, strong and weak nuclear) forces. Those known forces are involved as the scale-specific sum of all homogeneous smallest-bound quantized material contents of respective scales of PSs in Eq. (2.28) in any specific scale of GBs or GSBs which have equated with curved spacetime of same scale-specific GBs or GSBs for one infinitesimal discrete moment of inertial time.

Then, both of the Eqs. (2.28) & (6.11) would be universally applicable in all such scales of PSs or GBs or GSBs including the whole universe or BBCOU as macro-most scale in category of M_6.

Therefore, the Eq. (6.1) could further write from Eq. (6.11) as

$$\Delta M_6 = [\Delta p_{e-1} = \in \cdot \Delta(n \cdot m_{e-1})^4 = \in \cdot \Delta(n \cdot \Phi)^4_{e-1}]$$

$$= k/\{\Delta(p_u)_{e-1} = \in_u \cdot \Delta(v/n)^4_{e-1}\}$$

$$= \in_u \cdot k_2^4 / \Delta(n \cdot \Phi)^4_{e-1}. \qquad (6.12)$$

The ΔM_6 in non-inertial unified Eq. (6.12) has comprised all the quantized non-zero & non-infinity valued for all 5+5 CIPs i.e. real and discrete type of PSs or GBs or GSBs irrespective of scales from micro-most to macro-most, and also all those comprising same PSs or GBs or GSBs are only exchangeable by any similar real and discrete type of signals to ours like real and discrete type of observers. Therefore same Eq. (6.12) would be ultimately the non-inertial unified equation for whole real and discrete values within category of M_6 or discrete-reality or quantum-reality. Subsequently, that category of M_6 with quantum-reality seems to have 'quantized-limitations' for all its comprising GBs or GSBs as observances, exchangeable quantized signals for observations and ours like observers in physical nature.

Such a quantized-limitation has already described in Section-3 of Chapter-1 and again in Sections 2 & 3 of Chapter-2 but when every quantum-reality have assumed under inertial isolations in absence of natural forces (i.e. in absence of any quantized-signals exchanges with same). But any such inertial-state for those PSs have now revealed remain an inertial-state during every infinitesimal discrete duration of quantized inertial-time of same PSs during every non-inertial discrete rate of changes as any GBs or GSBs.

Consequently, that inertial quantized-limitation of Eq. (6.12) could also be assumed as the non-inertial quantized-limitation in every scale of quantum-real GBs or GSBs. Because, those GBs or GSBs are fundamentally integrations of all integer micro scales of inertial quantum-realities or inertial-PSs having such similar quantized-limitations during every discrete values of inertial-time Δt for infinitesimal discrete rate of changes while in quantized acceleration or non-inertial state.

As a result, that quantized-limitation of quantum-reality would be finally the limitation not merely in one inertial-state but also in non-inertial state beyond which an observer like us with such limitation could not 'see' anything. Moreover, such quantized-limitations within every observances, signals and observers like us would create the maximum possible range for observations that seems have imposed by the physical nature itself.

That quantum-limitation, in broader sense, could be imagined to exist up to the geodesic of entire spacetime as gravitation and simultaneously up to the geodesic of entire antispacetime as antigravitation of category of M_6. Such quantum-limitation through such effect of gravitation would be like all homogeneous smallest-bound-mass-energy quantum-real-signals as one ultimate Supersymmetric gauge-field ($= \Delta m_{e-1}$) but simultaneously through the effect of antigravitation as homogeneous highest-bound-motion quantum-real-signals of one Supersymmetric antigauge-field ($= \Delta v_{e-1}$) within same category of M_6.

Therefore, the whole of our quantum-real cognizable limit in physical nature, that has finally defined in Eq. (6.12), would not only unify all possible known & unknown non-inertial quantum-real basic forces but also have included everything as quantum-real within the category of M_6.

That unification of quantum-reality in Eq. (6.12) would be up to that maximum quantum-limitation till the discrete values of all 5+5 inverse 10-CIPs would have non-zero & finite magnitudes for and PSs as GBs or GSBs, and remain would be an integrated part of the category of M_6.

But if there any other thing in same physical nature, say that will have any zero & infinity magnitudes for same 5+5 inverse 10-CIPs must not be any integrated part of same M_6. That entity must exist beyond our quantum-real limitations to be exchanged by any quantum-real limitations of signals (even beyond the effects of gravitation and antigravitation) with all non-zero & non-infinity discrete magnitudes of corresponding 5+5 inverse 10-CIPs. Any such entities beyond that quantum-real limitations in physical nature with all zero & infinity values might have no quantum-reality instead the virtuality.

That virtuality with all zero & infinity values could be one ridiculous vacuum that extended up to infinity in same physical nature beyond the category of M_6. The same could be also a kind of non-quantized continuum for same 5+5 inverse 10-CIPs as 10-dimensions but with all non-quantized zero values for non-existent matters, space & time as well as non-quantized infinity values for infinite-existent motions, anti-space & anti-time beyond all ours quantum-real exchange limits of signals.

For convenience, such a non-quantized continuum of all zero & infinity of 5+5 inverse 10-CIPs as 10-dimensions beyond the quantum-signals exchange limits of M_6 in same physical nature can be termed as 'vacuum-infinity virtuality' or simply as 'vacuum-virtuality'.

The same ideas of vacuum-virtuality have assumed earlier within the inertial-state in Chapters 1 & 2. But through the Eq. (6.12) in non-inertial conditions also similar vacuum-virtuality have appeared.

3. A Mirror-Image Duality in Quantum-Reality

The unified equation for everything as quantum-real that has defined in Eq. (6.12) also has included all possible basic forces those are involving in all scales of GBs or GSBs. Where the Eq. (4.47) has unified all known gauge-fields of forces with gravitation in inertial-state, and Eq. (2.28) has unified everything quantum-real in inertial-state as infinitesimal discrete rate of changes in acceleration (for any non-inertial state).

In Chapter-1, the Eq. (2.28) also has described that inertial quantum-reality as the intrinsic scale-specific quantized magnitudes having the mandatory non-zero and non-infinity values for all 5+5 inverse 10-CIPs which are involved with any PSs irrespective of scales. In Eqs. (4.47) and (6.11), the non-inertial GBs or GSBs are primarily also involved with all same 5+5 inverse 10-CIPs having any non-zero & non-infinity scale-specific instantaneous quantized inertial values for any infinitesimal moment of discrete time for every discrete rate of changes in accelerations also the similar quantum-real in type. Therefore, the Eqs. (4.47), (6.11) & (6.12) are unifiedly also the quantum-real in type as like as it was quantum-real in type in the unified inertial Eq. (2.28).

The same Eq. (2.28) has also appeared as LE-C in Eq. (2.29) following the Eq. (1.25), and also there was a simultaneous RE-

C in Eq. (2.30) by following its previous Eq. (1.26). Both Eqs. (2.29) & (2.30) are then appeared as mutual mirror-images to each other in Eq. (2.31) as one unified inertial quantum-reality.

Subsequently, any such infinitesimal inertial mutual mirror-imaged quantum-reality in Eqs. (1.25) & (1.26) for PSs there have similar scale-specific non-inertial left-handed Eq. (4.47) as LE-C would be also corresponding non-inertial right-handed Eq. (4.48) as RE-C.

Therefore, the Eq. (6.12) actually defines such left-handed direction of LE-C for ΔM_6, and that intrinsic direction, where an observer like us is integrated with, onward expansion cycle of BBCOU

$$\Delta M_6 \equiv [\Delta p_{e-1}] = k/[\Delta(p_u)_{e-1}] \approx \xrightarrow{LE-C} \qquad (6.13)$$

where obviously $[\Delta p_{e-1} = \epsilon \cdot \Delta(n \cdot \Phi)^4_{e-1}]$ and $[\Delta(p_u)_{e-1} = \epsilon_u \cdot k_2^4/\Delta(n \cdot \Phi)^4_{e-1}]$ would be in Eq. (6.13), and in Eq. (2.29) those were considered in inertial condition. Conversely, as was in Eq. (2.30) the same ΔM_6 in Eq. (6.12) simultaneously there would be appeared for RE-C as

$$\xleftarrow{RE-C} \approx [\Delta(p_u)_{e-1}] = k/[\Delta p_{e-1}] \equiv \Delta M_6 \qquad (6.14)$$

for those mirror-imaged observers unlike us who are onward the simultaneous collapse cycle RE-C of BBCOU.

Finally, there would have a grand unified such mutual mirror-imaged quantum-reality for all scale-specific gravitational, antigravitational, gauge and antigauge fields forces in Eq. (6.11) along with left-handed expansion cycle LE-C and right-handed collapse cycle of BBCOU under category of ΔM_6

$$\begin{cases} \Delta M_6 \equiv [\Delta p_{e-1}] = k/[\Delta(p_u)_{e-1}] \approx \xrightarrow{LE-C} \\ \xleftarrow{RE-C} \approx [\Delta(p_u)_{e-1}] = k/[\Delta p_{e-1}] \equiv \Delta M_6 \end{cases}. \qquad (6.15)$$

Then, Eq. (6.15) is an ultimate mirror-imaged duality of the grand unified non-inertial quantum-reality that includes everything in the cognizable quantum-real range ΔM_6 of physical nature which includes all scales of GBs or GSBs and all known & unknown basic forces.

Therefore, anything quantum-real, those are defined in above Section-2 as well as in preceding Eq. (6.12) as unified and integrated parts of quantum-real cognizable realm of whole physical nature, irrespective of scales from micro-most to macro-most, are also appeared to have simultaneous LE-C & RE-C mutual-mirror-images. Those two mutual-mirror-images are intrinsic too in every quantum-real scale of GBs or GSBs as were in any inertial-PSs as definite cyclic oscillating parts of the whole BBCOU as defined in Eqs. (1.24) & (1.27) or in non-inertial manner in Eq. (6.15). Consequently, the corresponding directions of the both LE-C as well as RE-C for each quantum-real GBs or GSBs or inertial-PSs as any quantum-reality also have correspondingly intrinsic LE-C and RE-C those already assumed in Section-2 of Chapter-1.

That is, the unified quantum-reality has appeared as any scale-specific GBs or GSBs in Eq. (6.13) as like as inertial-PSs in Eq. (1.25) in the Eq. (6.12) which have ultimately the intrinsic mirror-imaged duality in same unified quantum-reality of Eq. (6.15). Then, our cognizable non-virtual i.e. quantum-real part of the physical nature within the category of ΔM_6 would have not merely

(i) quantum-reality with non-zero & non-infinite values of all 5+5 inverse 10-CIPs,
(ii) intrinsic discretenesses,
(iii) intrinsic scale-specificnesses,
(iv) intrinsic quantized-limitations for exchangeable signals, and
(v) non-inertial unifications,

but also, the simultaneous mirror-imaged 'duality' in all intrinsic directions of LE-C & RE-C.

4. Consequences:

The Eq. (6.15) would be the most basic expression for everything as quantum-real. It has simultaneously both individual as well as wholistic definitions for any single or all quantum-reality/ies. Therefore, it may resolve many present inconsistencies in relevance of observational comprehensions about physical nature. Particularly, two of those present inconsistencies in physical nature are stated in Sub-sections 4.1. & 4.2. below:

4.1. Symmetry in Quantity of Matter & Antimatter

In Eqs. (1.25), (2.29) and (6.13), the observers like us along with all observables and signals are not only intrinsically quantized in scale-specific manners but are also conceptually incrementing with expansion cycle onward the direction of LE-C in BBCOU. Hence, any such LE-C observers, signals and observables as integrated parts of such expanding phase of cyclic oscillating universe or LE-C of BBCOU, never could turn onward the simultaneous opposite direction of RE-C of same BBCOU. Even, if the direction from LE-C to RE-C or vice versa could forcefully (with input of energy) occur to turn the direction of same in opposite way, the corresponding observer or observable or signal cannot manage to withstand forever or steadily exist forever with that respective changed direction of LE-C or RE-C.

Because, from current observations it has comprehended that in such a situation of any forceful change in direction of LE-C (or RE-C) for any particle or system-of-particles either will be either 'annihilated' to other different scale of particle/s with corresponding direction of RE-C (or LE-C) or instantaneously 'turned back' to its previous direction of LE-C (or RE-C), and vice versa. Because, direction of one LE-C observer or observable or signal could not manage to withstand or steady forever along the direction of RE-C or vice versa as an integrated part of the direction of LE-C of BBCOU. Whatever could be the force could be applied on the same to change in its intrinsic direction in BBCOU.

In practical observations, this actually observes presently in all of quantum-real observations, as the same observers like us, our signals and observances have intrinsic associations with the direction onward LE-C of BBCOU in Eqs. (1.25), (2.29) & (6.13). Where all those are appeared to withstand or steady forever in direction of same LE-C.

All the cognizable quantum-real observables as well as exchangeable signals onward direction of expansion cycle LE-C of BBCOU within the category of M_6 those are only appearing steadily co-exist with us because we have also the similar association with that LE-C direction. Since, we are being quantum-real observers are also integrated onward expansion direction of LE-C in the BBCOU within that category of ΔM_6 in physical nature.

Hence, this would be true for all quantum-realities of not only for the scales of any quantum-real particles or systems-of-particles but also equally for all GBs or GSBs onward direction of LE-C of same BBCOU.

But the anti-particles or systems-of-antiparticles, (or say anti-GBs or anti-GSBs), what are observed in different observations of today, are appearing to exist for very infinitesimal moments of time. That is, those anti-particles or systems-of-antiparticles are appearing to have no steady existences like conventional particles or systems-of-particles onward direction of LE-C of BBCOU.

If it can assume that one such anti-particle or system-of-antiparticles is intrinsically integrated to the simultaneous direction of RE-C of same BBCOU in Eq. (6.14), would have obviously such an intrinsic opposite direction unlike ours like LE-C directional observers, observables and signals. Subsequently, any such anti-particle or system-of-antiparticles never could be existed steadily forever along our direction of LE-C instead of all such instantaneous existences onward our direction of LE-C for conventional time (Δt). Although, onward the simultaneous direction of RE-C of same BBCOU that anti-particle or system-of-antiparticles might have steady existence onward direction of RE-C for anti-time (Δt_u).

That is, unlike all quantum-real LE-C directional observables in Eq. (6.13), those quantum-real RE-C directional (anti-particles or systems-of-antiparticles) observables in Eq. (6.14), as anti-PSs or anti-GBs or anti-GSBs, cannot withstand on the LE-C direction unlike ours PSs or GBs or GSBs. Instead, those would withstand on the RE-C direction. Then, ultimately, any such RE-C directional quantum-real antiparticles or systems-of-antiparticles cannot be appeared to us onward ours LE-C direction as any steadily or forever existing observables or signals but can be defined by the same Eq. (6.14).

Therefore, as in Eqs. (1.27), (2.31) & (6.15), in spite of all equal and simultaneous co-existences of both LE-C quantum-real particles or systems-of-particles as well as RE-C quantum-real anti-particles or systems-of-antiparticles irrespective of scales in the same BBCOU, all those LE-C particles or systems-of-particles as quantum-real observables only appear to steadily withstand only onward direction of any LE-C observers like us.

That is, it can infer, why we can only see today onward ours LE-C directional quantum-reality only those similar intrinsic LR-C directional PSs or GBs or GSBs to steadily withstand but not intrinsic RE-C directional anti-PSs or anti-GBs or anti-GSBs in BBCOU. Although, being the mirror-imaged counterpart there would be equal numbers of RE-C directional anti-PSs or anti-GBs or anti-GBSs for all those LE-C directional PSs or GBs or GSBs with BBCOU.

Even in cases of very micro scales, it requires huge interventions of energies, to turn the direction of RE-C of any tiny or micro scale of such quantum-real antiparticles onward LE-C direction in particle colliders. But even that could occur for an instantaneous or almost for some negligible fraction of a second or infinitesimal duration of LE-C directional time. After that, such quantum-real antiparticles have turned onward LE-C direction either be annihilated or be turned back into its earlier steady intrinsic RE-C direction as defined in the Eq. (6.14). But in cases of macro scales, to see the same systems-of-antiparticles onward direction of LE-C, there would be proportionately required huge interventions of energies that seems now beyond of our present technological capabilities. Equally, that LE-C directional existences for such macro scales of quantum-real RE-C system-of-antiparticles would be appeared in more infinitesimal duration of LE-C time. As a result, in practical observations, only the RE-C antiparticles of very micro scales are now observed for durations of very infinitesimal moments of LE-C time proportionately with intervention of huge energies in todays particle accelerators.

Conversely, for one RE-C observer with Eqs. (1.26)), (2.30) & (6.14), those same scales of RE-C anti-particles or systems-of-antiparticles including ΔM_6 would appear remain simultaneously steady in existences. But all our LE-C intrinsic directional quantum-real PSs or GBs or GSBs irrespective of scales would appear to that RE-C directional observer to exist very instantaneous duration of anti-time.

As a result, to a quantum-real LE-C observer like us, who is onward direction of cyclic expansion or LE-C of ΔM_6 are seeing only the all LE-C quantum-real observables like PSs or GBs or GSBs to withstand in his all around but do not find any quantum-real RE-C directional anti-PSs or anti-GBs or anti-GSBs though those would have equal existences in BBCOU.

As a result, an asymmetry in existence of LE-C directional quantum-real particles / systems-of-particles over RE-C directional quantum-real anti-particles / systems-of-antiparticles within BBCOU in our present observations, in reality would not be there. That half of the RE-C directional quantum-real antiparticles / systems-of-antiparticles are simultaneously existing but just beyond our intrinsic direction of LE-C observations in same BBCOU. The Eqs. (1.27), (2.31) & (6.15) are suggesting that there always a symmetry rather than asymmetry for co-existences of those equal quantities of both LE-C & RE-C in BBCOU.

But conversely from the right-hand side, one RE-C observer would find similar apparent asymmetry in existence of RE-C quantum-real anti-particles or systems-of-antiparticles over LE-C quantum-real particles, although that observer, through same Eqs. (1.27), (2.31) & (6.15), would have a symmetry of both RE-C & LE-C in BBCOU.

Hence, from the Eqs. (1.27), (2.31) and (6.15) there would be no more asymmetries in-between total quantity of quantum-real LE-C PSs and quantum-real RE-C anti-PSs in BBCOU. The similar asymmetry in total quantities of matters and antimatters, those are correspondingly the sums of all LE-C particles or systems-of-particles or GBs or GSBs and all RE-C antiparticles or antisystem-of-particles or anti-GBs or anti-GSBs has observed [13] in current phase of LE-C for BBCOU. The current LE-C phase of BBCOU is appearing as if full of all matters but only with some negligible quantity of instantaneously exiting antimatters that can appear in estimations of particle colliders or astrophysical observations. But in reality, there would be no such asymmetry between total matters and antimatters in universe. Because, as discussed in above paragraphs, the matters, irrespective of scales and structures, are quantum-real LE-C directional, similar to ours LE-C direction of observation. While the antimatters, irrespective of scale and structures, are simultaneously existing as quantum-real RE-C direction in same universe that is opposite to our direction.

4.2. Resolving the E-P-R Paradox

The quantum-reality, in above Eq. (6.13), has unified in non-inertial state irrespective of scales for GBs or GSBs, can also be imagined as quantum-real fold equivalent to entire category of ΔM_6 in the physical nature. That fold has also an exchangeable quantum-real signal-limit for a similar quantum-real observer like us to observe any such scales of quantum-real observances like GBs or GSBs and all of those being quantum-real would have non-zero & non-infinity discrete values for 5+5 inverse 10-CIPs. Similarly, such a non-inertial quantum-real unification for all GBs or GSBs in Eqs. (6.13) would have also the inertial quantum-real unification for the all corresponding PSs in the Eqs. (1.25) & (2.29) as its every infinitesimal discrete rate of change in acceleration for every discrete duration of time.

Moreover, all those quantum-real inertial-PSs in Eqs. (1.27) & (2.31) as well as non-inertial quantum-real GBs or GSBs in Eqs. (6.15) also possess mutual mirror-images. Then conceptually, for any of such two mutual-mirror-images in the pair, left-handed or right-handed, cannot be separated from each other.

Such an intrinsic inseparability in-between a pair of mutual mirror-images for every scale of quantum-real PSs or GBs or GSBs would be a key to resolve the E-P-R paradox like events for two separated entangled particles of pairs.

However, in same Eqs. (1.27), (2.31) & (6.15), each of those mutual mirror-imaged entangled paired inertial or non-inertial quantum-real PSs or GBs or GSBs internally also having a mesh of total 5+5 mutual mirror-imaged inverse 10-CIPs with all scale-specific intrinsic quantized magnitudes.

That is, the mirror-imaged pair quantum-reality for those PSs or GBs or GSBs in Eqs. (1.27), (2.31) & (6.15) would be nothing but the intrinsic mirror-imaged pair quantum-realities out of those internal parameters like 5+5 inverse 10-CIPs.

Above Section-4 has revealed one mirror-imaged unification of the entire quantum-reality in Eq. (6.15). One part of such unification has intrinsic LE-C direction for every quantum-reality which have defined as particles or systems-of-particles in Eqs. (1.25), (2.29) & (6.13) including the observers like us, exchangeable signals of us & any observances for us. Internally, all of those are configured by total 5+5 mirror-imaged inverse 10-CIPs are integrated with one macro-most scale like BBCOU under category of ΔM_6.

Simultaneously, the Section-4 also has revealed in same Eq. (6.15) for that mirror-imaged unification for entire quantum-reality, but those have intrinsic opposite RE-C direction would be any antiparticles or systems-of-antiparticles in Eqs. (1.26), (2.30) & (6.14) including any anti-observer of us, anti-signals of us & anti-observances for us. Although each of those would have internally same 5+5 mirror-imaged inverse 10-CIPs as integrated parts of the same BBCOU under category of ΔM_6.

Then, during occurrence of any so-called E-P-R Paradox [14] 'event', that event must be associated with any of those mirror-imaged entangled (particle & antiparticle) pair PSs or GBs or GSBs under category of ΔM_6 in Eq. (2.31) & (6.15). If those two entangled particle & antiparticle in a pair become separated in space and time and if one of those two also as entangled mutual mirror-images in that pair has any change in magnitude of its one CIP, then due to the Eqs. (2.31) & (6.15) there would occur spontaneous and instantaneous changes in all corresponding magnitudes of other CIPs in both of those entangled particle & antiparticle pair. No matter how far those are separated in space and time.

Therefore, due to the Eqs. (2.31) & (6.15), if there any changes in magnitude of any one CIPs, either onward direction of LE-C or RE-C, whether such change occurs in separated spacetime, there must be always instantaneous changes in all other co-related inverse and mirror-imaged quantized magnitudes of all others corresponding CIPs. But without any need for signal exchanges in-between those two entangled mirror-images in the pair or particle-antiparticle in the pair due to Eqs. (2.31) & (6.15).

Then E-P-R Paradox would be very one consistent and common universal phenomenon for every occurring 'event' if that would link with any PSs or GBs or GSBs as entangled mirror-imaged pair quantum-reality with all non-zero & non-infinity discrete magnitudes of 5+5 inverse 10-CIPs under category of ΔM_6 in quantum-real fold of physical nature due to the Eqs. (2.31) & (6.15).

Summary

The Eqs. (6.12) & (6.13) are actually the non-inertial unified definition for quantum-real everything onward LE-C of BBCOU as the unified inertial definition for same quantum-real everything in Eqs. (2.28) & (2.29) for the same. In Chapter-4, the non-inertial (i.e. accelerated) state which has defined as an instantaneous and infinitesimal discrete rate of change. That is, unified non-inertial state of quantum-reality would be instantaneously an infinitesimal discrete form of inertial state of quantum-reality.

Then, the unified non-inertial quantum-reality in Eq. (6.13) having one mirror-image duality in Eq. (6.15) as an extension of the unified inertial quantum-reality in Eq. (2.31) for same. That is, entire quantum-reality in physical nature, where the BBCOU is macro-most scale and all other smaller scales are as ingredient of it, is possessing the simultaneous LE-C & RE-C mirror-imaged duality in both inertial and non-inertial states. Therefore, the Chapter-6 has derived the same quantum-real duality in everything under influences of basic natural forces or in non-inertial-state with all non-zero & non-infinity quantized values for all 5+5 unfolded ten inverse dimensions or CIPs.

Most importantly, that quantum-real duality in every scale of particles and systems-of-particles reveals some new inferences for current physics. Those are including

(i) presence of simultaneous Anti-gauge and Anti-gravitational forces with every quantum-real duality of particles or systems-of-particles,

(ii) a solution for current asymmetry in observing matters & anti-matters in BBCOU where anti-matter particles or systems-of-particles have appeared as the simultaneous RE-C directional ingredients of same BBCOU (and the matter particles or systems-of-particles have LE-C direction in same) which cannot steadily appear to sustain onward the LE-C direction of our observation, and

(iii) an explanation of entanglement between separated two components in one particle-antiparticle pair as any quantum-real duality in Eqs. (2.31) & (6.15).

If any *one* common-internal-parameter would change in Eqs. (2.31) & (6.15), in terms of both magnitude & direction, then instantaneously and automatically all other *nine* common-internal-parameters as CIPs or dimensions would be changed in magnitudes & directions onward respective LE-C & RE-C. Because all those 5+5 unfolded 10 dimensions are inversely corelated through respective seven universally invariant inverse constants in Chapter-2. Even, all the mutual mirror-images to each other under such simultaneous duality would have no need for any kind of signal-communications in between those if two are become separated the components in space and time.

Moreover, such universal quantum-real duality in Eq. (6.15) is interlinked by all universally invariant inverse constants $k_1, k_2, k_3, k_4, k_5, k_6$ & k, and consequently any event that will occur in such 5+5 inversely co-related 10-dimensions with all non-zero & non-infinity values should be determined if and only if at least the value of any one among all those ten common-internal-parameter could measure through direct observations. As a result, the entire unified quantum-real duality in physical nature would appear as "deterministic" through all those inverse universal constants $k_1, k_2, k_3, k_4, k_5, k_6$ & k. That will be a topic of next Chapter-7.

CHAPTER-7:

CAUSALITY IN EVERY QUANTUM-REALITY
[For Seven Universally Invariant Inverse Relations]

"The ability to perceive or think differently is more important than the knowledge gained."   - David Bohm

1. Plank's Constant h With Scale-Specific Values	272
2. Causality Involving with Every Quantum-Real-Event	274
3. Quantum-Real Determinism Due to Inverse-Relations	280
Summary	282

Present Chapter would be the fifth set of inferences in the series of total eight from those postulates have taken in Chapters 1 & 2 about physical nature on the basis of current understandings in physics.

In previous Chapter-6, there was ultimately a unification for quantum-reality including all particles or systems-of-particles and basic forces having non-zero and non-infinity values. Moreover, each of those non-inertial quantum-real entities also have shown a simultaneous left-handed and right-handed mirror-imaged duality in Eq. (6.15) alike to the inertial mirror-imaged quantum-reality for same in Eq. (2.31) in Chapter-2.

Those mirror-imaged dualities, irrespective of inertial and non-inertial realms for any quantum-real entity, also depict that there would be always a determinism in defining the other part if one part of the mirror-imaged pair is known through measurements/experiments/observations.

Furthermore, each of those quantum-real mirror-imaged part, in inertial state for Eq. (2.29) or Eq. (2.30) as well as in non-inertial state for Eq. (6.13) or Eq. (6.14), has 5-left-handed CIPs or dimensions those are inversely co-related to 5-right-handed-CIPs or dimensions. Each of those inverse relations has corresponding universally invariant inverse constants $k_1, k_2, k_3, k_4, k_5, k_6$ & k in Eqs. (2.1), (2.14), (2.17), (2.19), (2.21), (2.23) & (2.28) or (2.29) or (2.30). Therefore, if the magnitude for any one of those 5+5 inversely related 10-CIPs or 10-dimnsions can similarly measure through direct experiments/observations then all other nine could be known deterministically.

Hence, it seems that within the range of non-zero & non-infinity discrete values for entire quantum-reality there every quantum-real entity or event, irrespective of inertial and non-inertial in types, those are involving in any scales of particles or systems-of-particles or GBs or GSBs having the determinisms.

However, todays convention of 'indeterminism', that is involving in basic forms of the Uncertainty Principles in Quantum Mechanics, particularly in measurements of simultaneous 'location-of-occurrence' vs 'amount-of-mass', and also the simultaneous measurements of 'time-of-occurrence' vs 'amount-of-energy-release' in an 'event'. That 'event' has to be associated with any 'particle' or 'system-of-particles'. However, such indeterminism becomes ultimately determinism respective to the universally invariant inverse relation or de Broglie's wave-corpuscular (inverse) relationship. That inverse relationship has derived one universally invariant inverse constant k_1 in Eq. (2.1).

However, such convention of determinism now can redefine in some newer perspectives. Where it may count all non-zero & finite quantized-magnitudes for all mentioned 5+5 *inversely co-related* 10-CIPs in places of merely two of such inversely co-related CIPs like quantized magnitudes of mass & wavelength for all same particles or systems-of-particles in Eq. (2.28). Moreover, all those particles or systems-of-particles irrespective of scales are integrated parts of the macro-most scale like

270

BBCOU as well as Unified Quantum-Reality through same Eq. (2.28) with universally invariant inverse constant k in physical nature. As a consequence, in Chapter-2, there are all other corresponding universally invariant inverse constants k_2, k_3, k_4, k_5 & k_6 in corresponding Eqs. (2.14), (2.17), (2.19), (2.21) & (2.23) beside that k_1 in Eq. (2.1) in inertial state of all scales of particles or systems-of-particles. Therefore, all the 'events' which can occur in inertial state as mentioned in the Chapter-2, and those are involving in any scales of inertial particles or systems-of-particles must be ultimately deterministic through all such universally invariant inverse relations as well as constants. That is any such event within unified quantum-reality in inertial state through all those Eqs. (2.1), (2.14), (2.17), (2.19), (2.21), (2.23) & (2.28) and all corresponding universal invariant constants $k_1, k_2, k_3, k_4, k_5, k_6$ & k must be deterministic.

The Chapter-7 would reveal same determinism in all quantum-real events those are occurring or existing in non-inertial-state of unified quantum-reality in Eqs. (6.12) & (6.15) beside some other relevant inferences.

The below Section-1 would derive the Plank's constant h as one Scale Specific Universal Constant (i.e. SSUC) that will appear dependent on scale-specific magnitudes of all scales of particles or systems-of-particles. Therefore, the h would be no more any Universal Constant (UC) unlike those inverse constants like $k_1, k_2, k_3, k_4, k_5, k_6$ & k irrespective of scales of the particles or systems-of-particles. The Section-2 would deduce all those conventional indeterministic quantum-real phenomena, involve with position & mass and time & energy, as any integrated part of that whole BBCOU or Unified Quantum-Reality inclusive scope of the Eq. (6.15) as deterministic through all/any of those inverse Eqs. (2.1), (2.14), (2.17), (2.19), (2.21), (2.23) & (2.28) and corresponding Universal Constants $k_1, k_2, k_3, k_4, k_5, k_6$ & k. On the basis of the two Sections 1 & 2, the Section-3 would infer that the whole

Unified Quantum-Real-duality is in Eq. (6.15) ultimately as a deterministic or 'causal' where all its 'integrated' observers like us, signals & observables having all corresponding non-zero & finite quantized values for all 5+5 inverse 10-CIPs.

1. Plank's Constant h With Scale-Specific Values

The inertial-motion, as one of 5+5 inverse related 10-CIPs in all particles or systems-of-particles, has already considered as intrinsic quantized magnitudes i.e. Δv irrespective of the scales similar to all other intrinsic quantized magnitudes of other 9-CIPs e.g. [Δm, $\Delta \lambda = \Delta r_u$, Δr, $\Delta s(\Delta x, \Delta y, \Delta z)$, Δt, $\Delta s_u(\Delta x_u, \Delta y_u, \Delta z_u)$ & Δt_u] in Chapters 1 & 2. As a consequence, intrinsic quantized magnitudes for any such scale of Δv including all other 9-CIPs are as SSUCs. But never as UCs (like k_1, k_2, k_3, k_4, k_5, k_6 & k) in Eq. (2.28).

The inertial constancy in speed of light $c = 2.99792 \times 10^{10} sec. cm^{-1}$, as considered in foundation of SRT, would be also one of such intrinsic quantized speeds for the one specific scale of photons in EMS. Where that $c \equiv \Delta v_c$ as one of those SSUCs intrinsic quantized magnitudes for the scale-specific CIP like Δv. That has revealed, the same $c \equiv \Delta v_c = \Delta v$ is an obvious universal invariant value to every observer in corresponding IFRs everywhere in universe. But if the specific scale of photon in EMS with inverse quantized mass-energies would change, then instantaneously its SSUC-magnitude of such Δv_c would change to another SSUC-magnitude correspond for another inverse quantized magnitude of mass-energies as described in Eq. (2.14).

Subsequently, all the different magnitudes for $c = \Delta v_c = \Delta v$ are obviously any universal constants irrespective of IFRs spread all over the space and time in universe but those must be any SSUCs in type as assumed in earlier Chapters 1 & 2.

But, as was derived in Chapter-2, there are all inverse constants k_1, k_2, k_3, k_4, k_5, k_6 & k in corresponding Eqs. (2.1),

(2.14), (2.17), (2.19), (2.21), (2.23) & (2.28) are universal constants UCs unlike those SSUCs. Although, both of those types (SSUCs & UCs) of universal constant-values are internally associated with every scale of PSs, but the UCs would never change in values universally along the changes in scales of the same PSs.

However, the same $c = \Delta v_c = \Delta v$ being one CIP in each scale of PSs is inversely related to another CIP inertial quantized mass-energy $\Delta m_c = \Delta m$ in Eq. (2.14). Then, the same $\Delta m_c = \Delta m$ in Eq. (2.1) has also inverse relationship with another intrinsic quantized CIP de Broglie wavelength as $\Delta \lambda_c = \Delta \lambda$. In Eq. (2.5), there has such relationship

$$\Delta m_c \times \Delta \lambda_c \times \Delta v_c = h \qquad (7.1)$$

where only one parameter, that is h as Plank's Constant, is not appearing with similar scale-specific or SSUCs quantized magnitudes alike those 3-CIPs $\Delta m_c, \Delta \lambda_c, \& \Delta v_c$ in respective photon-particle. But, if those SSUCs quantized magnitudes of same inversely co-related 3-CIPs would change in Eq. (7.1) automatically the corresponding value of same h would also be changed. Therefore, Plank's Constant would have one of similar SSUCs-magnitudes in Eq. (7.1) say $h = \Delta h_c$ for one particular scale of photon-PSs in EMS with all corresponding scale-specific or SSUCs discrete values of $\Delta m_c, \Delta \lambda_c, \& \Delta v_c$ as

$$\Delta m_c \times \Delta \lambda_c \times \Delta v_c = \Delta h_c . \qquad (7.2)$$

Since, all those three corresponding CIPs $\Delta m_c, \Delta \lambda_c, \& \Delta v_c$ are SSUCs discrete magnitudes for one particular scale of photons and for every scale of PSs there would have corresponding $\Delta m_c = \Delta m, \Delta \lambda_c = \Delta \lambda \& \Delta v_c = \Delta v$, there the $h = \Delta h_c$ as Plank's Constant would have also the all similar scale-specific or SSUCs quantized magnitudes $h = \Delta h_c = \Delta h$. Therefore, for every scale of PSs there would be the SSUCs quantized magnitudes for same Plank's Constant in the Eq. (7.2) as

$$\Delta m \times \Delta \lambda \times \Delta v = \Delta h. \quad (7.3)$$

Therefore, the Plank's Constant, which has also appeared as the universal constant $h = \Delta h_c = \Delta h$ in Eqs. (7.2) & (7.3) but ultimately have all scale-specific or SSUCs as *local* discrete values respect to every scale of PSs, and would be similar to another local universal constant (also as SSUC) in SRT $c = \Delta v_c = \Delta v$ unlike 7 number of UCs k_1, k_2, k_3, k_4, k_5, k_6 & k.

From Eq. (2.10), where $(\Delta m \times \Delta \lambda) = k_1$, Eq. (7.3) would be

$$k_1 \times \Delta v = \Delta h \quad (7.4)$$

and the Eq. (7.4) reveals one direct proportional relationship in-between two such SSUCs discrete magnitudes of Δv & Δh in every scale of PSs respect to the UC k_1. Then, by proceeding gradually onward micro scales of particles or systems-of-particles with scale-specific gradual higher quantized magnitudes of inertial-motions there would be expectedly all corresponding gradually incrementing scale-specific higher magnitudes for such Plank's Constant.

2. Causality Involving with Every Quantum-Real-Event

The term 'quantum-reality' or rather entire 'unified quantum-reality', that has described for inertial state in Eqs. (2.28) & (2.31) and also for non-inertial state in Eqs. (6.11) & (6.15), basically emerges as quantized entity meshed up with all 5+5 inversely corelated 10-CIPs [Δm, Δv, $\Delta s(\Delta x, \Delta y, \Delta z)$, $\Delta s_u(\Delta x_u, \Delta y_u, \Delta z_u)$, Δt & Δt_u] where each would have:

(i) any SSUCs intrinsic quantized magnitude, and
(ii) each of such magnitudes must be 'real' in a sense must have any non-zero and non-infinity SSUCs quantized magnitudes.

However, in other hand, the general convention involves in the Uncertainty Principles of Quantum Mechanics is based on the principles of indeterminism in relevance of

(i) simultaneous detections of 'location in space and measurement of mass' (obviously any quantum-real in type) or 'time & energy' (also quantum-real in type) of a particle (i.e. also quantum-real in type) if and only if those could be exchanged through any signal (also quantum-real) transmissible to observer like us (also ultimately quantum-real in type as well) [1], and

(ii) also, the detection (in creation/destruction) of any such similar quantum-real particles out of the virtual Non-quantized energies background [2].

In first situation of Uncertainty Principles, the relevant everything like observances (i.e. space & mass as well as time & energy of any particle), signals of us and observers like us all are not only associated with any quantum-reality but are also associated with the unified quantum-real background as mentioned in first paragraph of this Section.

But in second situation of Uncertainty Principles, the relevant everything like an observance as creating /destructing particle that likely to form as a quantum-real particle would whether be detected through exchange of any quantum-real signal with one quantum-real observer like us.

But the background from / to where such a quantum-real particle as observance could emerge / destroy would not necessarily be always any such quantum-real (with all non-zero & finite values of 5+5 inverse CIPs) in type. Such a background could be also a Non-quantized virtual in type of zero & infinity non-quantized magnitudes for same 5+5 inverse 10-CIPs.

As a consequence, the background could be either a 'Non-quantized and Virtual' or 'quantum and real' in type.

---

1. having non-zero & non-infinity quantized values of all 5+5 inverse 10-CIPs.
2. having zero & infinity magnitudes of all same 5+5 inverse 10-CIPs.

Therefore, it would be either 'a so-called uncertainty of any quantum-real particle regarding creation from / destruction into any quantum-real background' or 'an uncertainty of same quantum-real particle regarding creation from / destruction into any non-quantized virtual background'. Then, for convenience, the first and second situations uncertainty, those are involving with Uncertainty Principles of today can state correspondingly as

(i) Uncertainty within Quantum-Reality; and
(ii) Uncertainty from Non-quantum-Virtuality.

The Section-2 will reveal remain the certainty or determinism or causality in spite of such 'uncertainty within quantum-reality' situation of the Uncertainty Principle due to those Eqs. (2.1), (2.14), (2.17), (2.19), (2.21), (2.23) & (2.28). That determinism would appear within every quantum-reality of observance, signal & observer (like us) those are integrated with entire unified quantum-real background under category of ΔM_6 in Eqs. (6.12), (6.13) & (6.15) in physical nature.

However, in contrary, the occurrence of any uncertain or indeterministic event as 'uncertainty from Non-quantum-virtuality' background in Uncertainty Principle, with all zero & infinity non-quantized magnitudes of same 5+5 inverse 10-CIPs will appear remain Uncertain or Indeterministic or Non-causal. Such Non-causal occurrence of events out of that all zero & infinity non-quantized magnitudes for 5+5 inverse 10-CIPs would not be any integrated parts of the same unified quantum-reality under category of ΔM_6 in Eqs. (6.12), (6.13), (6.14) & (6.15) within same physical nature. However, that type of 'uncertainty from Non-quantum-virtuality' of background would be issues in proceeding Chapters 9 & 10.

2.1. Defining Quantum-Real Certainty Within Uncertainty of Quantum Mechanics:

That 'uncertainty within quantum-reality', which has now perceived through the Uncertainty Principles of Quantum Mechanics, while trying to measure simultaneous location in space & amount of mass or location in time & amount of energy for a PSs, due there presence of only de Broglie wave-corpuscular universal inverse relationship between two CIPs like Δm & $\Delta \lambda$ in Eq. (1.1) with only one UC like k_1 in Eq. (2.1). Because, respect to only such available universal inverse relationship in Quantum Mechanics that has not included other CIPs like inertial-motion Δv or space Δs or time Δt CIPs within same unified quantum-real background. Then corresponding wave-equation for PSs in Quantum Mechanics cannot link such Δv or Δs or Δt with the Δm & $\Delta \lambda$ through Eq. (2.1) with certainty and deterministic manner. But with the other corresponding universal inverse Eqs. (2.14), (2.17), (2.19), (2.21), (2.23) & (2.28) such 'uncertainty within unified quantum-reality' could overcome.

However, the Eq. (2.1) is not only one extension of the de Broglie's equation in Eq. (1.1) but there is also a set of another five new inverse Eqs. (2.14), (2.17), (2.19), (2.21), (2.23) & (2.28). All those inverse equations have the universal capacities to link any uncertain mass with its inertial-motion with all other simultaneous space & mass or time & energy to show all certainties within all those known uncertainties of PSs out of mere k_1 in Eq. (2.1). Hence, there would be no more such uncertainties in respect of all those newer universal inverse relationships in a set of Eqs. (2.1), (2.14), (2.17), (2.19), (2.23) & (2.28).

Particularly, the conventional Uncertainty Principle, in Quantum Mechanics as any indeterministic Quantum-Real-Events must be appeared as deterministic, if there would be those two Eqs. (2.1) & (2.14).

Through both of those Eqs. (2.14) & (2.1), all 'observers', in respective inertial frames of reference in physical nature, can measure accurately and almost simultaneously all other intrinsic-quantized magnitudes of all CIPs ($\Delta m, \Delta \lambda, \Delta v$ &Δh) in every scale of particles or systems-of-particles, if can measure directly the scale-specific intrinsic-quantized magnitudes of at least any one of those parameters or CIPs. For example, the simultaneous 'position' and 'inertial mass-energy' for a particle or system-of-particles can be known through k_1 in Eq. (2.10) from Eq. (2.1) and k_2 in Eq. (2.14), if we can define or measure any one of its CIPs like: (Δm or Δv or $\Delta \lambda$ or Δh) through any direct observations or measurements. Suppose, if Δm is known, then its simultaneous intrinsic-quantized magnitudes of ($\Delta v = k_2/\Delta m$) in Eq. (2.14) and ($\Delta \lambda = k_1/\Delta m$) in Eq. (2.1) & ($\Delta h = k_2/\Delta \lambda$) in Eq. (2.14) can be instantaneously defined.

Again, from the known intrinsic SSUC quantized magnitude of Δv, the 'position' of that particular particle or system-of-particles with Δm at any given moment of 'time' can also be calculated. Hence there will be also no 'uncertainty' or 'indeterminism' in measurements of simultaneous 'position' and 'mass-energy' for all 'discrete' particles or system-of-particles.

If 'time' is $t = (c = \Delta v_c/\Delta \lambda)$ and 'energy' is $E = mc^2 = \Delta h \cdot c/\Delta \lambda$, and if magnitude of any one CIPs out of the ($\Delta m, \Delta v_c = c, \Delta h$ s $\Delta \lambda$) is known, then all the simultaneous uncertain magnitudes for both (t s E) of the same PSs can be obtained. For example, if ($\Delta v_c = c$) is known for a PS, then from Eq. (2.14) we have $\Delta m = k_2/(\Delta v_c = c)$, and from Eq. (2.1) we have $\Delta \lambda = k_1/\Delta m$, and obviously $E = m \cdot (c = \Delta v_c)^2 = \Delta h \cdot (c = \Delta v_c)/\Delta \lambda$, as well as $t = (c = \Delta v_c)/\Delta \lambda$; and from both k_1 in Eq. (2.1) and k_2 in Eq. (2.14) we can define all other non-definable CIPs simultaneously in the same particle or system-of-particles. So, there will be no 'uncertainty' or 'indeterminism' in simultaneous measurements of 'time' and 'energy' in all discrete, or intrinsically-quantized, scales of particles or system-of-particles in the physical nature.

That is, one "quantum-real-event" that occurs in the unified quantum-real background, once appears as uncertain through the only available universal inverse relationship of wave-corpuscular inverse relation in Eq. (1.1), would be thereafter no more uncertain due to the emergences of those two universal inverse relations in Eqs. (2.1) & (2.14). Consequently, there are all certainties in every quantum-real-event when all same particles or systems-of-particles which are involving with total 5+5 inverse 10-CIPs through a series of Eqs. (2.1), (2.14), (2.17), (2.19), (2.21) & (2.23) in a unified quantum-real background of Eq. (2.28).

Furthermore, any quantum-real-event, as a web of 5+5 inverse 10-CIPs, must be always certain or deterministic if the magnitude of any one CIP out of total 10-CIPs can directly measure or known. Say its CIP like quantized mass-energy Δm is known, then its other 9-CIPs from the respective inverse Eqs. (2.1), (2.14), (2.17), (2.19), (2.21) & (2.23) can be calculated as

$$\left\{ \begin{array}{c} \Delta\lambda = k_1/\Delta m; \\ \Delta v = k_2/\Delta m, \ \Delta v = k_6/\Delta r; \\ \Delta r = k_3/\Delta\lambda, \ \Delta r = k_6/\Delta v; \\ \Delta s = \frac{3}{4}\pi \cdot \Delta r^3, \ \Delta s = k_4/\Delta s_u; \\ \Delta s_u = \frac{3}{4}\pi \cdot \Delta\lambda^3, \ \Delta s_u = k_4/\Delta s; \\ \Delta t = 2\pi \cdot \Delta r, \ \Delta t = k_5/\Delta t_u; \\ \Delta t_u = 2\pi \cdot \Delta\lambda, \Delta t_u = k_5/\Delta t. \end{array} \right\}. \quad (7.5)$$

The same thing can also occur through directly knowing or accurate measurements of the magnitude for any other CIPs and calculation of the rest of other 9-CIPs those are associated with the same quantum-real-event linked with any particles or systems-of-particles.

Then, the Eq. (2.28), for irrespective of occurrence of all quantum-real-events, would mandatorily link to any or all scales of particles or systems-of-particles and nothing but the ultimate quantum-real background of any such quantum-real-events.

Because, any so-called quantum-real-event having the uncertainty must have the certainty through all those inverse Eqs. (2.1), (2.14), (2.17), (2.19), (2.21) & (2.23) in the unified quantum-real background of Eq. (2.28).

3. Quantum-Real Determinism Due to Inverse-Relations

The inertial common definition for all particles or systems-of-particles in Eq. (2.28), as described in above Section-2, is not only the unified quantum-real background for occurrence of every quantum-real-event in inertial state, but such inertial state has also defined as any instantaneous discrete non-inertial state for an infinitesimal discrete moment of time. Then, the same unified quantum-reality in an inertial-state would subsist in every non-inertial state of any quantum-real-event as described in Eq. (4.47). The gravitational forces in Eq. (6.12) along with all other known & unknown Supersymmetric gauge-fields of forces in such unified non-inertial quantum-real background under category of ΔM_6 has also defined in the Eq. (6.13).

Then, inertial certainty, as one inertial certainty of quantum-real-events in Eq. (7.5), that could occur or create from /destroy into that unified inertial quantum-real background in Eq. (2.28), could also be imagined as if equivalent but different infinitesimal discrete *slices* of the non-inertial certainty of any non-inertial quantum-real-event in Eq. (6.12) that can occur in entire unified non-inertial quantum-real background as ΔM_6 of Eq. (6.13).

For example, if the magnitude of Δm_{e-1} is known, then in any scale-specific GBs or GSBs through Eq. (6.13), either all its instantaneous inertial quantized magnitudes for the rest of its 9-CIPs could be known deterministically through the Eq. (7.5) within same (GBs or GSBs) or the whole non-inertial quantum-real-event like GBs or GSBs could be known deterministically for similar instant through the respective Eqs. (4.41), (4.47) and (6.12).

Then, if the Δm_{e-1} becomes known, then scale-specific non-inertial magnitude for whole quantized magnitude of entire mass-energy as in Einsteinian Field Equations of GRT as $(T_{\mu\nu})_{e-1}$ in Eq. (4.40) and also the magnitude of whole Δq_{e-1}^4 in Eq. (4.41) for the same scale-specific GBs or GSBs could be deterministically known. Subsequently, from that direct proportionality relationship in same Eq. (4.41) in between

$$\{[(G_{\mu\nu} + g_{\mu\nu}\Lambda)_{e-1}]\} \propto \{[(T_{\mu\nu})_{e-1} \cdot (\Delta m_{e-1})^4]\} \quad (7.6)$$

or in the same Eq. (4.41) from earlier Eq. (4.19) there would be the direct proportionality

$$\left[\Delta p_{e-1} = \tfrac{3}{2}\pi^2 \cdot k_6^4/\Delta V^4 = (G_{\mu\nu} + g_{\mu\nu}\Lambda)_{e-1}\right] \propto \Delta q_{e-1}^4 \quad (7.7)$$

where ϵ would be one *proportionality constant* equal to $(\tfrac{3}{2}\pi^2 \cdot k_6^4/k_2^4)$ in that Eq. (4.41). From those known deterministic or certain values of all non-inertial parameters in Eq. (7.7) like $(\Delta p_{e-1} = \epsilon \cdot \Delta q_{e-1}^4)$ there must be the non-inertial mirror-imaged deterministic or certain magnitudes where only the gravitation has involved not any Supersymmetric gauge-fields

$$[(\Delta p_u)_{e-1} = \epsilon_u \cdot (\Delta q_u)_{e-1}^4] = k/(\Delta p_{e-1} = \epsilon \cdot \Delta q_{e-1}^4). \quad (7.8)$$

But, when all the known and unknown gravitation & antigravitation as well as gauge & antigauge force would be considered within those same GBs or GSBs under category of ΔM_6 inclusive of all scales of non-inertial quantum-real-events in Eq. (6.13) and its mutual mirror-imaged in Eq. (6.14), for the known values of $[\Delta p_{e-1} = \epsilon \cdot \Delta(n \cdot \Phi)_{e-1}^4]$ involves gravitation & all gauge forces, there must be the all similar deterministic or certain values for the antigravitation and all antigauge forces

$$[\Delta(p_u)_{e-1} = \epsilon_u \cdot k_2^4/\Delta(n \cdot \Phi)_{e-1}^4]$$

$$= k/[\Delta p_{e-1} = \epsilon \cdot \Delta(n \cdot \Phi)^4_{e-1}]. \tag{7.9}$$

The Eq. (7.9) as well as Eqs. (6.13) & (6.14) ultimately defines the non-inertial certainty or determinism in every quantum-real-event those are involving with every known and unknown gravitational & gauge-fields of forces as well as antigravitation & antigauge-fields of forces in the background of non-inertial unified quantum-reality in Eq. (6.15) with mutual mirror-imaged 'duality'. As a result, whole non-inertial quantum-reality under category of ΔM_6 that has appeared unified as discrete part of physical nature in Eq. (6.15) should be the non-inertial unified-quantum-real background of certainty or determinism.

That is, the whole physical nature, conceptually, which is also comprised by all such scales of discrete and deterministic particles or systems-of-particles, whether are in infinitesimally discrete 'inertial-slices' of the non-inertial state of entire category ΔM_6 in different scales of GBs or GSBs in Eq. (6.15), ultimately would appear not only as *discrete* with *duality* but also simultaneously as a *deterministic* with *causality* (having all causal sequences of causes and effects). Consequently, such entire quantum-real part of the physical nature, where observers like us are also integrated with it as similar quantum-real, would have not only the simultaneous left-handed & right-handed duality due to the Eq. (6.15) due to ΔM_6 in the background as Unified Quantum-Reality, but the same also through the Eq. (7.5) ultimately be a Deterministic or Certain or Causal.

Summary

The entire unified quantum-reality in Chapter-5, not only was deduced to possess an intrinsic left-handed and right-handed duality in Chapter-6 but same has revealed further to have the determinism (instead of indeterminism) in above Section-3 of the Chapter-3.

The Chapter-7 has actually derived universal determinism within entire unified quantum-real duality in Eq. (7.9) having all non-zero & non-infinity quantized magnitudes.

A quantum-real-event cannot occur without its link to any quantized particles or systems-of-particles within the unified quantum-real background of physical nature. Any such quantized particle or system-of-particles should have all non-zero & non-infinity values for the 5+5 inverse 10-CIPs to become exchangeable to any similar type of quantum-real observers like us through any similar quantum-real signals. Therefore, to observe the occurrence of any such quantum-real-event, it is unavoidable in any quantum-real observers like us to exchange with anything through any of the quantum-real type signals, within the quantum-real range of occurrence of the event within all non-zero & non-infinity discrete magnitudes of all those 5+5 inverse 10-CIPs. Because, a quantum-real observer like us, having the only quantum-real signal exchange limitation, where we can only communicate merely to any similar kind of quantum-real events or observables, as an integrated part of the entire unified background of quantum-reality.

Therefore, if there any other 'event', which can occur in values of zero & infinity for same 5+5 inverse 10-CIPs, that 'event' never can be communicated or exchangeable beyond such non-zero & non-infinity (quantum-real) values for same 5+5 inverse 10-CIPs of quantum-real signals for the observers like us. Then, occurrence of any such event that occurs *in* the zero & infinity values of same 5+5 inverse 10-CIPs must be beyond the communicable limit of the same quantum-real signals and observers like us are involving with all non-zero & non-infinity values of same 5+5 inverse 10-CIPS. Consequently, that event with all zero & infinity values should appear as any indeterministic event for ours like deterministic (non-zero & non-infinity valued) observers in physical nature.

Therefore, such an indeterministic event should remain be beyond of our quantum-real-limits as ours like observers' as well as beyond the limits of quantum-real-signals in physical nature. The entire unified quantum-reality, which has all corresponding non-zero & finite discrete magnitudes of 5+5 inverse 10-CIPs should be deterministic to ours like similar type quantum-real observers as well as signals.

Moreover, same quantum-real determinism not merely would occur not only in the arena of inertial states but also could be in the non-inertial states as well in presence of all basic forces. Because, any such inertial quantum-real determinism would ultimately be any infinitesimal discrete non-inertial quantum-real determinism.

Moreover, Planck's Constant h has derived as Δh i.e. one of SSUCs with discrete magnitudes like $c = \Delta v_c$ for specific scale of photon-particles in electromagnetic spectrum.

However, finally, the entire unified quantum-reality now seems deterministic. Since observers like us having only the Eqs. (7.5) & (6.15) along with the other Eqs. (2.1), (2.14), (2.17), (2.19), (2.21), (2.23), (2.28) & (2.31), that never could find any event to occur/destroy within such a realm of unified quantum-reality to have any broken sequence in causes & effects. That is to have any indeterministic occurrences in any event definable through same equations.

Since the unified quantum-reality, in those Eqs. (7.5) & (7.8), is ultimately a sequence of all causes & effects, that governs by the laws is also must be any *Causal Logics* or *Causal Laws*. Subsequently, any event that is occurring through such a sequence of all causes & effects, in deterministic manner, must also be an independent to any interference of its detection or signal-exchange, i.e. an observer-independent or an *objective* type. Then such objectivity, in entire unified quantum-real-duality in Eqs. (7.5) & (7.8), that governs by the Causal Laws, would be main topic of next Chapter-8.

CHAPTER-8:

OBJECTIVITY IN EVERY QUANTUM-REALITY
[Observation Independent Quantum-Real Events]

"The true laboratory is the mind, where behind illusions we uncover the laws of truth."   - Jagdish Chandra Bose

1. Defining the Quantum-Real Objectivism	288
2. Quantum-Real-Objectivity of Exchangeable Signals	294
3. Quantum-Real-Objectivity of Observables	295
4. Quantum-Real-Objectivity in Our Cognition	297
5. In Search of Subjectivity in Quantum-Reality	308
Summary	310

The word 'objectivity', in ordinary senses or in basic conventions of Classical Mechanics, for creation or destruction of an 'event' [1] is a process that follows the sequence of cause and effect or causal logic [2]. That trails through all causal step(s) in sequence from the creation/occurrence, without any break, up to the end/destruction. Consequently, that causal sequence does not care about anything from external (or internal) to influence it's happening. That is, one objective event is ultimately an external influence independent or in other way an observer's intervention independent phenomenon.

Later this fundamental presumption related to one objective-event in Classical Mechanics is doubted through Uncertainty Principle of Quantum Mechanics. That has emerged out with ideas of non-causality [3]. That is conventions of broken-

---
1. Should associate with any particle or system-of-particles of scale-specific quantized magnitude of 5+5 inverse 10-CIPs.
2. If once the magnitude is known all the future events could be defined.
3. First ever in the core mechanical issues of current physics.

causalities in sequence of any such causal quantum-real events in creations and destructions. That has raised as one of supportive universal hypotheses of 'Subjectivity' with broken-causality incorporates into the core issues of Mechanical descriptions of events in current physics.

Therefore, any subjective-event, in contrast of objective, seems not only to have one broken-causal sequence in its logic (if any), but also there another concern to observe it by any observers like us. That is, whether we can able to exchange with such broken-causality through any of those quantum-real-signals in causal processes 'before' the creation of one causal quantum-real event. In other words, whether the occurrence any of such so-called non-causal or subjective events out of broken-causality would be another type of logical sequence in same physical nature. Whether such a different type of logic actually 'differs in its type of sequence' compare to our conventional type of causal sequence of logic beyond any quantum-real exchange-limits of signals as well as observers like us.

Then, to comprehend whether an event, that occurs or destroys in physical nature, is basically an objective or a subjective in type also be one debatable topic in current physics. Therefore, before entering into the details of such arguments in relevance of objective or subjective type of any event in physical nature, it would also be indispensable to reveal whether the cognition along with the observers like us (who can 'see' such events through available exchangeable quantum-real signal limits) could actually be an objective or a subjective in type.

Moreover, in Chapters 2 & 6, that entire unified quantum-real-causality, as defined in Eqs. (2.31) & (6.15), has appeared intrinsically to have all scale-specific non-zero & non-infinity quantum-real values. That unified quantum-reality is also a maximum quantum-real cognizable range or *limit* in observation of any quantum-real observers like us who has limitations to exchange only within the range of any quantum-real signals to see anything within that limit. However, every ingredient particles or systems-of-particles of BBCOU within that unified quantum-real range would have all individual intrinsic quantum-reality. That means, all those quantum-reality, that is involving with any scales of particles or systems-of-particles are basically a web of all non-zero & non-infinity scale-specific quantized values of 5+5 inverse 10-CIPs in Eq. (2.28) and each of those values of CIPs are deterministic due to a set of inverse Universal Constants ($k_1, k_2, k_3, k_4, k_5, k_6$ & k).

Moreover, the observers like us, who is also basically a sum of similar configuration of all specific quantum-real particles or systems-of-particles, are also the integrated parts of that unified quantum-reality in Eqs. (2.31) & (6.15). Furthermore, all the possible exchangeable signals, available to us, also have the similar kind of quantum-reality. Consequently, any such specific scales of signal particles or systems-of-particles would be also the integrated parts of the same unified quantum-reality. Therefore, all those integrated parts of unified quantum-reality, as any observable-event or exchangeable-signals or observers-like-us, having all similar types of non-zero & non-infinity discrete values for all 5+5 inverse 10-CIPs along with quantum-reality, determinism and causality, would be objective in type.

Therefore, entire unified quantum-reality has defined in Eqs. (2.28), (6.12), (2.31) & (6.15), there would have no indeterminism (i.e. non-causality) due to any of its integrated quantum-real particles or systems-of-particles involving as any events, signals and observers (like us) due to all corresponding universally invariant inverse relations $k_1, k_2, k_3, k_4, k_5, k_6$ & k.

Then, any so called non-causal event earlier in Principles of Uncertainty in Quantum Mechanics, which seems now as integrated part in the same unified quantum-reality, can be a certain through all those universally invariant inverse relations.

Consequently, within the causal & deterministic range of entire unified quantum-reality in Eqs. (2.31) & (6.15), inclusive of all the observers like us, exchangeable quantum-real signals

& events, everything those are exchangeable to us must be ultimately objective in type. That is, any such 'quantum-real objectivity' must be within the range of all scale-specific discrete non-zero & non-infinity values of 5+5 inverse 10-CIPs for all respective communicable signals, events and observers. As a result, conversely, any such objective observers & signals within scale-specific discrete non-zero & non-infinity valued limits for all 5+5 inverse 10-CIPs never can find any exchangeable 'event' if that has zero & infinity magnitudes for the same 5+5 inverse 10-CIPs.

The current Chapter-8, would define such universal objectivity within the quantum-real-causality of all observables, signals and cognition of observers like us. The Chapter would reveal the entire unified quantum-reality, in Eqs. (2.28), (2.31), (6.12), (6.15) & (7.15), ultimately as the realm of all unified quantum-real causal-laws or physical-laws. It would also be the sixth set of consequences as inferences in a series from the basic postulates of Chapters 1 & 2.

The Section-1 would define that 'objectivity' in universal quantum-reality. The Sections 2 to 4 would depict such quantum-real objectivity in every integrated quantum-real observables, signals and cognition of observers (like us) of the unified quantum-reality in physical nature. Subsequently, this would unfold our sense of 'life' in different fashion, i.e. in quantum-real manner where our mind ultimately appears as a result of such quantum-real-objectivity. The Section-5 would reveal no exchangeable subjectivity in the forms of subjective-event or subjective-observer as any integrated part of such unified quantum-reality in Eqs. (6.12), (6.13) & (6.15) through any quantum-real signals and observers.

1. Defining the Quantum-Real Objectivism

The Quantum-Reality, in its ultimate form of mutual mirror-image duality, has revealed in unified manner in the Chapter-6.

Every such quantum-reality also has appeared as deterministic or causal sequence of causes and effects having all observer independent SSUCs discrete (non-zero & non-infinity) magnitudes for 5+5 inversely co-related 10-CIPs in Chapter-7. Then Section-1 would derive every quantum-real event as Objective (i.e. independent to its observations and observers). Subsequently, it would argue to define such objectivism in any quantum-real event in three aspects:

(i) whether any quantum-real event that occurs in unified quantum-real background would have always the 'common' non-zero & non-infinity SSUCs-discrete-magnitudes for 5+5 inverse 10-CIPs;

(ii) whether any quantum-real event that occurs in unified quantum-real background would have always the common 'determinism' due to inverse co-relations of all those same non-zero & non-infinity SSUCs-discrete-magnitudes for 5+5 inverse 10-CIPs; and

(iii) whether any quantum-real event that occurs in unified quantum-real background would have always common 'objectivism' due to intrinsic objectivism in all those same non-zero & non-infinity SSUCs-discrete-magnitudes for 5+5 inverse 10-CIPs.

If the observer-independent 5+5 CIPs have (non-zero & non-infinity) discrete-magnitudes as SSUCs would involve with any event, then such an event must be quantum-real. The same 5+5 CIPs are always inversely corelated through total seven inverse relationships and 7-UCs. So, the quantum-real event would be deterministic through all those 7-UCs. One such deterministic quantum-real event with all such observer-independent 5+5 CIPs as SSUCs in discrete magnitudes as well as 7-UCs would also be always observer-independent. That is, any such quantum-real event with all non-zero & non-infinity discrete magnitudes as observer-independent must be one objective in physical laws.

Moreover, one such individual quantum-real event, must be within category of ΔM_6 in Eqs. (6.12), (6.13) & (6.15), would be a part of:

1.1. A Unified Quantum-Reality of Events: Because there

(i) A Quantum-Real-Event or Quantum-Reality is always, linked to any quantum-real inertial particles or systems-of-particles in Eq. (2.31) which is also as any infinitesimal discrete rate of changes during accelerations of the non-inertial GBs or GSBs those have defined unifiedly in Eq. (6.15). The same also have the mutual mirror-imaged duality in same Eq. (6.15). Then, one quantum-real-event cannot differ from any quantum-real particle or system-of-particles. Hence, any such quantum-real-Event that 'occurs' as integrated part of the unified quantum-reality under category of ΔM_6 must have any scale-specific non-zero & non-infinity quantized magnitudes for all its 5+5 inverse 10-CIPs. Therefore, any quantum-real-event that associates with any scale-specific particles or systems-of-particles in Eqs. (2.31) & (6.15) say $\Delta\Gamma$ must possess any quantized magnitudes within the range of

$$\Delta\Gamma = \begin{bmatrix} \Delta m \neq 0 \text{ or } \infty, & \Delta s(\Delta x, \Delta y, \Delta z) \neq 0 \text{ or } \infty, & \Delta t \neq 0 \text{ or } \infty \\ \Delta v \neq \infty \text{ or } 0, & \Delta s_u(\Delta x_u, \Delta y_u, \Delta z_u) \neq \infty \text{ or } 0, & \Delta t_u \neq \infty \text{ or } 0 \end{bmatrix}$$

(8.1)

and all those non-zero & non-infinity discrete values must be inversely co-related in 5+5 CIPs through corresponding 7-UCs. All the scales of particles or systems-of-particles or GBs or GSBs which can instantaneously definable through the Eqs. (2.31) & (6.15) must be also assumed as any such quantum-real-event $\Delta\Gamma$ as in Eq. (8.1).

(ii) The occurrence of any such quantum-real-event could have a start and an end in the sequence through specific changes in all those 5+5 inversely corelated 10-CIPs. Consequently, occurrence of any such quantum-real-event would be in such fashion would be a 'progression' from one such specific start to an end. After that end there can be again a fresh start and also another end, or can be a cycle of all start and end sequences as a part of the whole unified quantum-reality.

(iii) As a result, ultimately, that progression of any such quantum-real-event in the cycle would be nothing but a progression through 'causality'. Since, every quantum-real particle or system-of-particles irrespective of scales having intrinsic (i.e. also as one invariant) quantization in respective SSUCs magnitudes for all 5+5 inverse 10-CIPs, therefore the effects of such causality through same progression of any corresponding quantum-real-event must be also universally invariant as well as observer-independent to all IFRs as parts of the ΔM_6 in Eq. (6.15).

1.2. A Unified Quantum-Reality of Determinism: Because

(i) Any such quantum-real-event $\Delta \Gamma$ in Eq. (8.1), which is occurring through one causal progression and ultimately through changes in SSUCs in intrinsic quantized magnitudes of all 5+5 inverse 10-CIPs under non-inertial states in Eq. (6.15), can also be a series of all 'deterministic' infinitesimal discrete rate of changes in same quantum-real-event of Eq. (8.1) and those can only definable through all inertial Eqs. (2.1), (2.14), (2.17), (2.19), (2.21), (2.23), (2.28) and (2.31).

(ii) Since, any quantum-real-event which is universally invariant to all IFRs through the Eqs. (3.42) - (3.45), (4.41) & (6.11) within the range of unified quantum-real-events $\Delta \Gamma$, would be also the universally invariant "deterministic" as any integrated parts of that range of unified quantum-real determinism.

(iii) Therefore, our quantum-real cognizable and exchangeable part of the physical nature must be a Unified Quantum-Real Deterministic.

1.3. A Unified Quantum-Reality of Objectivism: Because

(i) Since a quantum-real-event which has the progression that follows all causal quantized or discrete steps and all those causal discrete steps are universally invariant, i.e. observer-independent to all IFRs, therefore every quantum-real-event would be an Objective in type. Because, any such quantum-real-event does not depend on the intervention of any signals from observers throughout the causality sequence. Alternately, the same causality cannot 'break' or cut off through any such signal exchangeable interventions of the interactive quantum-real forces. Therefore, any quantum-real-event must be associated not only with the determinism but simultaneously with the Objectivism.

(ii) It has revealed that every quantum-real-event has an integration with the unified quantum-reality in Eq. (6.15). Therefore, such Objectivism, that involves with quantum-real-event would be also an integrated part of that unified quantum-real-event. Subsequently, that unified quantum-real-event would have also a Unified Quantum-Real Objectivism.

(iii) Since, recognizing any such quantum-real-event needs three quantum-real components: an exchangeable signal, an observance as any particles or systems-of-particles, and an observer like us where each of those would have all intrinsic non-zero & non-infinity SSUCs in quantized magnitudes for each of 5+5 inverse 10-CIPs, the same quantum-real-event in unified quantum-reality under category of ΔM_6 in Eq. (6.15) must be also the objective in type. As a consequence, causal sequences in any or all of those three parameters like one quantum-real 'event', one quantum-real 'signal' and one quantum-real 'observer' cannot be 'broken' without presence of any zero & infinity non-quantized magnitudes of the 5+5 inverse 10-CIPs throughout the causal progression of the same quantum-real-event from its start to end or in cycle. Conversely, to recognize one such quantum-real-event, with all intrinsic non-zero & non-infinity quantized magnitudes for 5+5 inverse 10-CIPs, as one 'Subjective' (i.e. observer and/or observer-dependent) in type, any or all those three quantum-real components have stated above must need to be broken anywhere through interventions of any or all zero & infinity magnitudes of same 5+5 inverse 10-CIPs. That zero or infinity magnitude would discontinue or stop in the same causal sequence progression anywhere in between so-called start and end. That is, if the three components like: signals, observances & observers like us are basically as well as intrinsically quantum-real with all non-zero & non-infinity values for 5+5 inverse 10-CIPs, then it would be impossible for any of those three components to break such causality sequence within the progression of respective quantum-real-events to have any zero & infinity values for any or all those 5+5 inverse 10-CIPs to appear as any Subjective type event. Oppositely, if any or all those three components (observances, signals & observers) would have any or all zero & infinity magnitudes for all those same 5+5 inverse 10-CIPs, that/those zero & infinity magnitudes would create all corresponding discontinuities within causal sequence of all non-zero & non-infinity 5+5 inverse 10-CIPs transforming same into one non-causal or subjective event in type with all zeros & infinites in same 5+5 inversely related 10-CIPs. As a result, the quantum-real causality would be broken, and ultimately the quantum-reality would change to one Subjective-Event in its occurrence anywhere within the cycle of progression from start to end through intervention of any or all zero & infinity values of same 5+5 inversely 10-CIPs.

(iv) In proceeding Sections 2 to 4, any signal, any observance and any 'cognizable-mind' of an observer like us would deduce as quantum-real as well as objective in types due to all those non-zero & non-infinity 5+5 inverse 10-CIPs or all those same could appear as subjective in type if there found any zero & infinity magnitudes within all or any values of those 5+5 inverse 10-CIPs of any such signal, observance and 'cognizable-mind'.

2. Quantum-Real-Objectivity of Exchangeable Signals

The Unified Quantum-Reality, as described in above Section, and that has defined in Eqs. (2.28), (2.31), (6.12) & (6.15), comprises all Quantum-Real particles or systems-of-particles or GBs or GSBs irrespective of scales. Such a Quantum-Real particle or system-of-particles can either be a Quantum-Real-Event $\Delta\Gamma$ in Eq. (8.1) to occur or be a Quantum-Real observer like us to aware or acknowledge such an occurrence or be a signal to exchange in-between that event $\Delta\Gamma$ & the observer.

In practical sense, that exchangeable signal is ultimately the interaction of basic forces carrying Quantum-Real particle or system-of-particles, with all scales specific 5+5 inverse non-zero & non-infinity 10-CIPs in Eq. (8.1), that includes all possible gauge-fields of forces and gravitation. Hence, any signals in the Unified Quantum-Reality must be intrinsically quantized as any Quantum-Real-signals. Hence, any intrinsic Quantum-Real-signals must be an observer independent, i.e. Quantum-Real-Objective.

Moreover, any such Quantum-Real-Objective signals, with non-zero & non-infinity 5+5 inverse 10-CIPs, cannot exchange or interact in-between any observance (or event $\Delta\Gamma$) and observer which have non-quantized entity; i.e., with any or all zero and infinity magnitudes for those same 5+5 inverse 10-CIPs is/are associated with those event and observer. Then, any Quantum-Real-Objective signal as integrated part of whole Unified Quantum-Reality in Eqs. (2.31) & (6.15) irrespective of scales must have a *limit* to exchange or interact with non-quantized event or observer.

Then, all such exchangeable signals in Unified Quantum-Reality in Eq. (6.15) should have Quantum-Real-Limits with all non-zero & non-infinity quantized values for 5+5 inverse CIPs. Hence, all those signals should be Objective in type with Quantum-Real-Objective limitations.

Consequently, all those Quantum-Real-Objective type signals with all non-zero & non-infinity magnitudes of 5+5 inverse 10-CIPs never can able to exchange with any zero and infinity magnitudes of all same 5+5 for inverse 10-CIPs involve with any observer and event. As a result, such a Quantum-Real-Objective signal never could able to break the causality sequences in any Quantum-Real-Event (as we have described in earlier Section) with all causal progressions in corresponding non-zero & non-infinity 5+5 inverse 10-CIPs to turn the same it into any Subjective Event with zero & infinity magnitudes 5+5 inverse 10-CIPs. That is, any such Quantum-Real-Objective signal is limited with non-zero & non-infinity scale-specific quantized 5+5 inverse CIPs and never find any Subjective observers and events in same Unified Quantum-Reality in Eqs. (2.31) & (6.15) through such quantum-reality.

3. Quantum-Real-Objectivity of Observables

Those Quantum-Real particles or systems-of-particles irrespective of scales, as defined in Eqs. (2.28) & (6.15), are also integrated parts of the whole unified quantum-reality ΔM_6 as equivalent as any Quantum-Real-Events $\Delta \Gamma$ as defined in Eq. (8.1). Because each of those are possessing non-zero & non-infinity 5+5 inverse 10-CIPs. Then, any such quantum-real-event would be also an integrated part of the unified-quantum-reality ΔM_6 in Eq. (6.15) that cannot occur without any association or link with such quantum-real particles or systems-of-particles irrespective of scales and also without such 5+5 inverse 10-CIPs with intrinsic quantized (non-zero & non-infinity) magnitudes. Moreover, if any such intrinsic (non-zero

& non-infinity) quantized magnitudes of 10-CIPs would associate with any quantum-real-event then only such an event could define as a quantum-real Objective Event.

Moreover, any Quantum-Real-Signal, with similar quantum-real property in magnitudes & quantum-real Objectivity, due to intrinsic non-zero & non-infinity quantized magnitudes of all 5+5 inverse 10-CIPs, cannot interact/exchange with anything as observance or as event in the same unified quantum-reality in Eqs. (2.28), (2.31), (6.12) & (6.15) if such observance or event would possess other than non-zero & non-infinity quantized magnitudes for same 5+5 inversely corelated 10-CIPs with it. Because, due to the presence of any or all zero & infinity magnitudes in all those 5+5 inverse 10-CIPs within such observance or event, the same would be no more quantized to become quantum-real exchangeable for any similar quantum-real signals.

Therefore, anything, that is with 5+5 inverse 10-CIPs of all intrinsic quantized non-zero & non-infinity magnitudes, cannot exchange/interact/communicate with anything that is non-quantized with zero & infinity magnitudes for those 5+5 inverse 10-CIPs unlike $\Delta\Gamma$ in Eq. (8.1). Because as defined in Eq. (8.1), there $\Delta\Gamma$ has always corresponding ranges in values of all 5+5 inverse 10-CIPs in-between greater that zero and lesser than infinity.

Subsequently, any quantum-real-event in such unified quantum-reality in Eq. (8.1) also have any quantum-real scale-specific SSUCs-discrete magnitude limitations for same 5+5 inverse 10-CIPs to emit/radiate or exchange any non-quantized values of signals. As a result, in practical term, any such quantum-real-event could not emit/radiate/affect any exchangeable quantum-real signal that would possess any magnitude for any or all those 5+5 inverse 10-CIPs with zero and infinity magnitudes.

Moreover, in addition, if magnitude of any one CIP could be known for one quantum-real-event through exchange of a quantum-real-signal having non-zero & non-infinity magnitude for its 10-CIPs, then instantaneously the values for rest of the 9-CIPs could be known from all those Eqs. (2.1), (2.14), (2.17), (2.19), (2.21), (2.23), (2.28), (2.73)-(3.75), (4.41), (4.47), (6.12) & (6.15) as well as corresponding inverse UCs k_1, k_2, k_3, k_4, k_5, k_6 & k.

Therefore, every such quantum-real observances / events as integrated parts of that unified quantum-reality in Eq. (6.15) would appear ultimately as any

    (i) Quantum-Real-Objective,
    (ii) Quantum-Real-Limited, and
    (iii) Quantum-Real-Deterministic.

Consequently, all such quantum-real-observance / events, as any particles or systems-of-particles or GBs or GSBs in unified quantum-reality has defined in Eq. (6.15), would be an Objective in type with quantum-real objective scale-specific range of SSUCs in discrete magnitudes, and could not emit/radiate /affect any exchangeable signal that might have magnitudes for all same 5+5 inverse 10-CIPs with zero and infinity magnitudes. As a result, all those quantum-real-objective type of observances/events as $\Delta\Gamma$ in Eq. (8.1) would remain be in a causal sequence with all corresponding non-zero & non-infinity magnitudes for those 5+5 inverse 10-CIPs of same $\Delta\Gamma$ even after intervention of any such quantum-real signals due to not having any zero & infinity magnitudes in corresponding 5+5 inverse 10-CIPs. That quantum-real objective type event would not transform anymore into any non-quantized subjective type of event in such exchange.

4. Quantum-Real-Objectivity in Our Cognition

The category of ΔM_6 as entire Unified-Quantum-Reality, in Eq. (6.15) as well as its infinitesimal inertial state in Eq. (2.31), also includes all observers like us who have cognition or cognizable-mind or simply cognizance or say intelligence [1] to comprehend the same. Which seems sometime as Subjective in type. Although, all those Observers like us with cognition are also basically the specific measurable sums of all constituent quantum-real particles or systems-of-particles which are as quantum-real Objective in type. This has already appeared in previous Sections 1 to 3.

As a consequence, each of those quantum-real Observers would have intrinsic configurations within itself any quantized range in magnitudes for the same Objectivity as obvious non-zero & non-infinity SSUCs for all corresponding 5+5 inverse 10-CIPs similar to one Objective quantum-real Signal and Objective quantum-real Observance. But one Objective quantum-real Observer like us in addition seems to have such self-awareness or cognizance which sometime reflects as different or Subjective in type and does not linked to such universally unified quantum-real Objectivity of quantum-reality for everything as parts of the ΔM_6 in Eqs. (6.15) and (2.31).

Is that cognizance a separate entity within same Objective quantum-real Observers integrated with ΔM_6? Is that cognizance attached with the observers like us a part of something other kind of 'substance' having separate spread in unified Quantum-Reality of universe or in whole physical nature? Is it something having Subjectivity links beyond unified quantum-reality in physical nature?

Or, is it also a result of similar kind of causal sequences within the same quantum-reality of particles or systems-of-particles comprising bio-chemical basis of such intelligence actually unfolding from deeper level within the bio-physical body of quantum-real Observers like us where one such cognizable-mind attached?

So, this is now most important to comprehend whether such cognizable-mind in any quantum-real Observer would be an

---

1. Intelligence in somebody 'to take notice of and consider something, especially when judging' something in its surrounding.

integrated part of the entire Objective unified-quantum-reality in Eq. (6.15) being ultimately one Objective or Subjective in type. This would be also very important to know whether such a quantum-real Observer, having with such cognizable-mind acknowledges quantum-real Events, could be at all any integrated part of same unified-quantum-real Objectivity.

(i) Quantum-Reality in one Observer (like us): In Biological Sciences that has now already revealed that ours like quantum-real Observers, in most of its basic forms, are configured primarily with certain causal-arrangements of mostly carbon-based chains of RNA or DNA [1] molecules beside other atoms and molecules. Therefore, any such physical body of quantum-real Observers like us would be also basically any quantum-real Objective in type as like as any quantum-real Event or quantum-real Signal that has already defined in Eq. (8.1).

But in addition, those basic units like RNA or DNA molecules as any quantum-real in type are also possessed always one 'unique mechanism' beside following all its internal causal sequences out of intrinsic property of quantum-reality. Such 'unique mechanism' is actually as if encoding the 'life' within same quantum-real Observers like us.

However, such an encoding of 'life' through the same DNA or RNA molecules can also be assumed otherwise as if one unfolding of a quantum-real algorithm involves within the same quantum-real molecule from its beginning to the end.

That could finally be defined as if one quantum-real causal cycle as a part of entire cyclic causal sequence of the unified quantum-reality from start to end then again end to a fresh start that has already described in previous Section-1.

But alternately such an unfolding process of quantum-real algorithm in specific DNA or RNA could seems like an *autorun* by its own throughout in 'life' for one quantum-real Observer. That often can emerge as feeling of automatic decoding in itself.

---

1. Ribonucleic Acid (as RNA) and Deoxyribonucleic Acid (as DNA).

(ii) Quantum-Reality of Observer has always an Algorithm. However, a 'sense of living' would appear from such a decoding of quantum-real algorithm or a feeling of autorun of 'life' but that would always possess one so-called intelligence or 'cognizable mind' or cognition from the beginning to end.

Therefore, such sense of intelligence or cognizance or cognizable-mind or cognition in one of ours like quantum-real-observers then seems to appear somewhere from those scales of quantum-real molecules like RNA or DNA. Then obviously any of those RNA or DNA molecules as quantum-real systems-of-particles would be also basically fabricated by same 5+5 inverse 10-CIPs with definitely all non-zero & non-infinity SSUCs discrete magnitudes. Moreover, all those RNA or DNA could be equally definable by all the gauge and gravitational forces through Eqs. (6.12) & (6.15) in further deeper levels and obviously integrated parts of the unified-quantum-reality of Objectivity for ΔM_6.

In other way, in a bit of philosophical words, the same quantum-reality of RNA or DNA could also be imagined as if any individual black boards. On that black board, every time a new quantum-real-algorithm could write before a fresh start of one of so-called 'life' (along with its cognizance) for every corresponding new quantum-real Observer like us. Such a process of every new writing i.e. fresh encoding of every new quantum-real algorithm in the budding of every new 'life' would also merely one causal 'mathematics'.

(iii) Source of such Algorithm is Mathematical. That 'mathematics' is a permutation-combination of the all defragmented genetical codes of information or data which were involving with other similar quantum-real RNA or DNA (from parents) integrated parts of the unified quantum-reality. Such defragmented quantum-real-arrangements of genetical codes from different sources would start a fresh 'life' in one zygote. The beginning of another new 'life' with another new cognizable-mind.

Although, permutation-combination mathematics would have uncertainty in such process of combination, but that could be apparent. Through intervention of inverse laws along with such 7-UCs in Chapter-2, such permutation-combination process among all quantum-real particles or systems-of-particles.

But what that cognizable-mind or cognition would ideally mean for so-called 'sense of living' emerging through the decoding of a quantum-real-algorithm? Could it be further integrated with the unified quantum-reality ΔM_6 in Eq. (6.15)? Or, could be a separate entity parallelly existing with that ΔM_6 in Eq. (6.15)?

The cognizable-mind coexists with the existences of any quantum-real-observers. That cognizable-mind of such quantum-real-observers rather emerges from the 'sense of living' of same. Moreover, that 'sense of living' emerges from decoding through unfolding of quantum-real-algorithm associated with every quantum-real RNA or DNA of the quantum-real-observers.

(iv) 'Sense of Living' is a 'Sense of Dynamism' Through Unfolding of Algorithm. Such a 'sense of living', could further be elucidated as the decoding of one quantum-real-algorithm within every quantum-real Observer from its start to end for a 'life'. This 'sense of living' also have perceived as if a kind of 'dynamism' [1] that seems to release all along the duration of such decoding of corresponding quantum-real algorithms or 'life' that one so-called living-being would have unlike one non-living thing.

Then such dynamism within one 'life', through decoding of any such quantum-real-algorithm, is also as if differentiating

---

1. Something as 'non-static' or 'not dead', that apparently differentiates one so-called living being and one non-living being.

between one quantum-reality 'with life' and another quantum-reality 'without life' but as integrated parts of same unified quantum-reality of category ΔM_6 in Eq. (6.15).

However, with the decoding of such dynamism in its most fundamental level, the impression of so-called 'sense of living' within one 'with life' quantum-real-system with precise count of quantum-real particles or systems-of-particles, would initiate as quantum-real Observer.

Consequently, any such quantum-real algorithm (of quantum-real Observer) based 'sense of living' would have ultimately a causal sequence. Any such causal sequence has defined in previous Sections as quantum-real Objective in type. But that so-called 'dynamism' through such decoding of quantum-real algorithm needs to be clarified further.

(v) Varying Degrees of 'Sense of Living'. Because, that 'sense of living' in different classes of quantum-real-observers are observed with variable amplitudes or degrees proportionate to the variable degrees in evolutionary bio-physical complicacies in different modified quantum-real-algorithms.

In one higher order or species of quantum-real Observers (like us), through the process of evolutions, it becomes now so psycho-somatically [1] sophisticated that corresponding cognizance sometime reasons something other than any quantum-real Objective in type or as Subjective with 'consciousness' [2]. That seems to co-exist separately within the same bio-physical-body of quantum-real Observer.

However, in contrary, in one lower form or degree of such 'sense of living' there could also be noticed similar kind of quantum-real-algorithms. For example, in one quasi-living micro-organism like a unicellular virus existing on the edge of so-called living and non-living boundary. In that degree for

---

1. Rather psycho-neural process of cognizance through electro-chemical exchange of signals.
2. That can govern by its own wills to occur any event or with broken sequence of causality.

'sense of living', there would also have a process of decoding in corresponding quantum-real algorithm and would also release the corresponding dynamism for its 'life'. But that 'sense of living' as well as 'released dynamism' would be simple or less complicated compare to one sophisticated 'sense of living' and 'released dynamism' in one quantum-real observers like human. Subsequently, the cognizance based on higher degree 'sense of living' through such sophisticated 'release of dynamism' in one human observer sometimes appears difficult to describe same as if any quantum-real Objective in type.

(vi) 'Tasks of Living' for the 'Sense of Dynamism. Then, what would actually be the meaning for such term 'dynamism' used in above paragraphs could be also a key to describe further the 'sense of living' in deeper level. That quantum-real dynamism could be comprehended from some of its basic common features those have often noticed within all varying degrees of 'sense of livings' based on corresponding quantum-real-algorithms in respective all so-called living beings.

However, such dynamism through decoding of quantum-real algorithm in any living being can assume to execute as if some of the basic 'tasks' of life or as if for sum of specific 'tasks of living' from its beginning to end. Primarily, few of those 'tasks for living' in broader meanings can be like

(1) accumulations,
(2) interactions,
(3) adaptations,
(4) replications,
(5) diversifications,

and so on in its quantum-real surrounding that is integrated with the same unified-quantum-reality of ΔM_6 in Eq. (6.15).

The cognizance of one quantum-real Observer would be also a result of such dynamism, that dynamism arises through execution of such 'tasks of living', those 'tasks of living' occur through unfolding of respective quantum-real algorithm in

Observer (having quantum-reality), and finally, such quantum-real algorithms integrated with the unified quantum-reality that is Objective in type.

Therefore, the cognizance is appearing as integrated part of any quantum-real Observer as so-called living being which is also an integrated part of the unified quantum-reality in Eq. (6.15). So, the impression of such dynamism emerges in any so-called living-being in all along its course of executing the 'tasks of living' as the emerging 'sense of living' merely out of decoding of one's own individual quantum-real-algorithm.

Therefore, without presence of any such decoding of quantum-real-algorithm within the 'life' so-called living being from start to end there would be impossible to have any such 'dynamism' as well as 'cognizance' even in observers like us. Subsequently, one such cognizable mind out of a quantum-real algorithm in the course of execution of that sum of all 'tasks of living' would be ultimately anything quantum-real, and anything such quantum-real 'tasks of living' must have the quantum-real Objectivity (instead of consciousness or Subjectivity) as integrated part of the unified quantum-real Objectivity in Eq. (6.15).

Hence, cognizable minds of the quantum-real Observers like us is ultimately appearing as quantum-real Objective instead of anything separately co-existing Subjective within there.

However, from this comprehension of quantum-real Objectivity for any cognizable-mind, that is involved with quantum-real Observers in ours like psycho-somatic observers as integrated parts of the whole unified-quantum-reality in Eq. (6.15), such ideas of quantum-real cognizance can proceed further even beyond the boundaries of so-called livings.

(vii) 'Tasks of Living' in all Living & Non-living Quantum-Realities. Because, in most basic form, all such 'tasks of living' can also be apprehended even beyond the borderline of so-called living beings. In the alleged non-living quantum-real systems, as integrated parts of same unified-quantum-reality of Eq. (6.15), similar type of 'tasks of living' could perceive to exist.

However, each of those 'tasks of living' related to one quantum-reality, irrespective of any living or non-living systems, fundamentally, needs to be consistent with all existing primary conservational laws in physics. For example, mass-energies, momenta, or other CIPs those are involved in any quantum-reality must be conserved from its start to end and then end to a fresh start as part of the unified quantum-reality in Eq. (6.15).

As a result, in some broader aspects, similar 'tasks of living' following those universal conservation laws can be explained within a creation-destruction-creation cycle of one so-called non-living quantum-reality of particles or systems-of-particles or even GBs or GSBs in Eqs. (2.28), (4.47) and (6.15) alike to one creation-destruction-creation cycle of one living quantum-reality within unified quantum-reality.

It could observe frequently in 'life' span of a star or other astronomical objects under dominating gravitational fields. In the 'life' of one quantum-real-system like a star, being one scale-specific GSB, there would also be a kind of unfolding of such sense-of-dynamism through decoding of its inherited quantum-real-algorithm from initial accumulation of mass-energies from surroundings vis-à-vis curved spacetime under hydrostatic balance would have also one specific 'tasks of living' to execute from its beginning to end.

The 'tasks of living' of such a star would follow the sequence of *accumulations* (of nearby quantum-real-mass-energies or astronomical dusts or debris of other dead stars), *interactions* (with other quantum-real-mass-energies in surroundings), *adaptations* (to achieve its specific hydrostatic equilibriums), *replications* (into another new star from its own astronomical debris by the end), and *diversifications* (in varieties of stars or other astronomical objects of different scales in long run) and so on.

In same way, there would have also the similar kind unfolding of sense-of-dynamisms for 'tasks of living' can observe as if through decoding of respective quantum-real-algorithms within different atoms and molecules under electromagnetic fields or within sub-nuclear quantum-real systems of different fermions and bosons under gauge fields of forces.

Within an atom or a molecule (including a quantum-real-RNA or DNA molecule as a 'seed' to earlier ideas of 'sense of living' for so-called basis of quantum-real Objectivity in cognizance of sophisticated quantum-real Observers like us), would have also the scale-specific 'tasks of living'. Those would also follow the tasks of *accumulation* of specific sub-atomic quantum-real particles/systems to initiate, *interaction* (with quantum-real-surrounding), *adaptations* (through suitable chemical bonding or other structures formation), *replications* (into another atom or molecule from its debris at the end for a fresh start of another new quantum-real-algorithm), and *diversifications* (into different millions of varieties of molecules or more than a hundred types of scales for atoms in long run).

Then, similar 'sense of living' through executions of all such 'tasks of living' in most rudimentary features like man made artificial quantum-real-systems. Every such artificially created quantum-real-system like a cyber software-hardware systems, quantum-real machine and instrument etc. are ultimately unfolding through all respective quantum-real-algorithms from beginning to end and from end to a fresh beginning. Those are also ultimately the integrated parts of the unified-quantum-reality in same Eq. (6.15).

(vii) Objectivity in Cognizance of Every Quantum-Real Observers. Therefore, the cognizance linked to one quantum-real algorithm would be not merely any quantum-real Objective in type referring any cognizable-mind in restricted senses of only one psycho-somatic system of quantum-real-objective Observers like us. But that can also appear to associate with all quantum-real-Objective types of Observances and Signals irrespective of living and non-living quantum-real entities as integrated parts of the unified quantum-reality. Although such Objective type cognizance ultimately from the so-called 'sense of living' may have many degrees of variations. Since, all those same quantum-real Objective particles or systems-of-particles with individual degrees of quantum-reality have Objective-cognizance are integrated parts of the unified-quantum-reality in Eq. (6.15), therefore conceptually one such integrated sum of all quantum-real Objective-cognizance the entire unified quantum-real-Objectivity would have also a unified Objective-Cognizance through Eq. (6.15). Such a unified cognizance through one respective sense-of-dynamism seems to execute its unified 'tasks of living' to unfold the entire quantum-real-algorithm along the Bigbang/Bigcrunch Cyclic Oscillations for the macro-most scale of BBCOU.

Ultimately, being integrated with all such quantum-real Objective-Cognizance, irrespective of quantum-real Objective-Observers like us or any other quantum-real Objective particles or systems-of-particles, there could also have all similar quantum-real-Objective type exchange/interaction in-between any quantum-real Objective types of observable Events and Signals as integrated parts of same unified-quantum-reality.

Because, anything that is non-quantized and non-real, i.e. with zero & infinity magnitudes in any/all 5+5 inverse 10-CIPs, such observers cannot recognize it being one such quantum-real Objective-Observer. Even that Observer cannot 'draw' or imagine anything through his highest possible degrees of cognizable-mind anything about such non-real or non-quantized entity with all zero & infinity values for 10-CIPs of no space, on time, no matter until same is emerging out for that quantum-real Observer as something similar to anything non-zero & non-infinity values to 'draw' or 'imagine' for same 10-CIPs having space, having time & having matter as any quantum-real Objective-Event or $\Delta\Gamma$ in Eq. (8.1). That can be exchanged through similar quantum-real Objective-Signal to one such Observer with similar quantum-real Cognizable-Mind. So, there must have all the definite limits for all 5+5 inverse 10-CIPs in Eq. (8.1) for any cognizable imaginations for drawing or exchanging with any such not-real Event

$$\Delta \Gamma = \begin{bmatrix} 0 < \Delta m < \infty, & 0 < \Delta s < \infty, & 0 < \Delta t < \infty \\ \infty > \Delta v > 0, & \infty > \Delta s_u > 0, & \infty > \Delta t_u > 0 \end{bmatrix} \quad (8.2)$$

and the same not-real Event beyond that exchange limit would never follow any conventional quantum-real-algorithm of all non-zero & non-infinity vales. Also, that would not be any part of the unified quantum-real Objective-Cognizance as well as unified-quantum-reality in Eq. (6.15).

Therefore, all such cognizable-minds as any particles or systems-of-particles or GBs or GSBs in unified-quantum-reality defines in that Eq. (6.15) should have quantum-real-limits with all non-zero & non-infinity quantized magnitudes in 5+5 inverse 10-CIPs. Hence, all those cognizable-minds would be objective in type with quantum-real Objective-Limitations.

Finally, those Cognizable-Minds should never able to break the causality sequences within any of the Quantum-Real-Objective-Events have described in Eqs. (8.1) & (8.2) with all non-zero & non-infinity magnitudes for 5+5 inverse 10-CIPs through interventions of any zero & infinity magnitudes for same 5+5 inverse 10-CIPs to become one Subjective Event.

5. In Search of Subjectivity in Quantum-Reality

The Unified-Quantum-Reality of Eq. (6.15), which is inclusive of all intrinsic quantizations as Quantum-Realities, is comprised the whole Unified Quantum-Real-Objective Cognizable part as ΔM_6 in Eq. (6.15) of the physical nature and has defined in above Section-1 not only as Deterministic but also as Objective.

The Quantum-Real-Events $\Delta \Gamma$ as defined 'limited' in Eq. (8.1) are along with every exchangeable Quantum-Real-Signals having only the non-zero & non-infinity quantized magnitudes for all 5+5 inverse 10-CIPs; and also, the Quantum-Real-Observers with Quantum-Real-Algorithm with Cognizable-Minds are Quantum-Real Limited in Eq. (8.2) within all non-zero & non-infinity quantized magnitudes for all 5+5 inverse 10-CIPs.

Therefore, any Quantum-Real-Objective-Observer in Eq. (8.2) cannot exchange, even if there would have any signal that might possess any zero & infinity magnitudes in 5+5 inverse 10-CIPs, to that Quantum-Real-Events as $\Delta\Gamma$ in Eq. (8.1). Because, any such Quantum-Real-Observer in Eq. (8.2) only able to exchange through any Quantum-Real-Signals with all non-zero & non-infinity quantized magnitudes for all 5+5 inverse 10-CIPs and never find any Event through such exchange any zero & infinity magnitudes for same 5+5 inverse 10-CIPs. Any non-zero & non-infinity value cannot exchange with any zero & infinity value.

Consequently, in Unified-Quantum-Reality, since every Quantum-Real-Event as $\Delta\Gamma$ in Eq. (8.1) is Objective in type, there would not be any possibility to exchange through any non-quantized events like zero & infinity valued 5+5 inverse 10-CIPs signals and to break the sequences of Quantum-Real-Causalities through any such Quantum-Real-Signals by the Quantum-Real-Observers like us.

As a result, in same Unified-Quantum-Reality in Eq. (6.15), there should not be any non-quantized or Subjective events without having sequence of Quantum-Real-Causality within exchangeable limits of any Quantum-Real-Signals as well as Quantum-Real-Observers.

Then, in that Unified-Quantum-Reality there would not be any integrated Quantum-Real exchangeable Subjective type of Events with such zero & infinity values of 5+5 inverse 10-CIPs. Not only that, any similar type of non-quantized Signal would be also beyond the exchangeable limit of the Quantum-Real-Objective-Observers like us within Eq. (8.2). And finally, that

Quantum-Real-Objective-Observer who would be limited by only quantized values of non-zero & non-infinity 5+5 inverse 10-CIPs in Eq. (8.2) never could detect any such non-quantized observer or Subjective-Observer (unlike us) of no causal sequences in same Unified-Quantum-Reality of Eq. (6.15).

Hence, within the Quantum-Real-Limitations of all Quantum-Real-Observers in Unified-Quantum-Reality there would not have any exchangeable Subjective-Observers, Subjective-Signals & Subjective-Observances or Events (with no causal sequences).

Summary

Any quantum-real event, as discrete ingredient of the whole unified-quantum-reality in Chapter-6, which has deduced further as deterministic in Chapter-7 starting from the Eq. (2.28), is also appeared as objective in this Chapter. Not merely those objective events but all our communicable signals as well as cognizable observers like us are equally have revealed as quantum-real deterministic objective ingredients of same unified quantum-reality. The 'life' in observers like us has also ultimately appeared as if a 'dynamism' from the unfolding of the corresponding quantum-real algorithm that is involving in every quantum-real system. A cognizable-mind of any quantum-real observer like us are also assumed as respective quantum-real deterministic objective algorithms. But, such quantum-real deterministic objectivity of cognizable-mind in observers like us seems to be any quantum-real and no more indeterministic with subjectivity in Uncertainty Principles of Quantum Mechanics. It has found in above Sections that a cognizable-mind could not have any such quantum-real indeterminism and subjectivism. Because that could never subjectively create or destroy any such quantum-real-event through exchange of any quantum-real deterministic objective signals as integrated parts of the unified quantum-reality. Moreover, being one quantum-real observer, neither it can

310

break in similar way the sequence of causality due to Eqs. (2.1), (2.14), (2.17), (2.19), (2.21), (2.23) & (2.28) nor can exchange through any signals with the values of zero & infinity for the corresponding 5+5 inverse 10-CIPs. Consequently, no such quantum-real indeterministic subjective observer can be found as discrete ingredient parts of the same unified quantum-reality in Eqs. (2.31) & (6.15). Hence, within entire unified quantum-reality there would not be any such quantum-real but subjective type of observers. At least those cannot be exchanged within the quantum-real limit of the objective type of observers like us, who are integrated parts of the same unified quantum-reality. Consequently, in most fundamental levels, every observer, signal and event (or observable) which are as integrated parts of unified quantum-reality in Eqs. (2.31) & (6.15), with all scale-specific quantum-real non-zero & non-infinity magnitudes for 5+5 inverse 10-CIPs, are not merely deterministic through Eqs. (2.1), (2.14), (2.17), (2.19), (2.21), (2.23) & (2.28) but are also objective within quantum-real limit of observations for the same observers like us.

Therefore, the entire unified quantum-real-duality is deterministic and objective which is only governed by the Causal Laws where the magnitudes of all the 5+5 unfolded 10 inverse CIPs or dimensions have non-zero & non-infinity scale-specific discrete magnitudes. As a result, anything, if that possesses say magnitudes like zero & infinity for the same 5+5 inverse 10-CIPs or dimensions must be always stayed beyond that quantum-real limit of observations for the observers like us in same physical nature. Since such zero & infinity magnitudes never could be deterministically exchanged through any quantum-real discrete non-zero & non-infinity magnitude signals, that vents would be indeterministic beyond the limits of quantum-reality. That could be any event of non-real or say 'virtual'. Any such virtual-event must be indeterministic and then obviously seems to be subjective in type, and that would be the issue in next Chapter-9.

CHAPTER-9:

BEYOND THE EDGE OF QUANTUM-REALITY
[A Non-Quantized Continuum of Higher-Energy]

"There is no learning without having to pose a question. And a question requires doubt."   - Richard P. Feynman

1. Defining Subjectivism in Physical Nature:	315
1.1. One Event beyond that Quantum-Reality	316
1.2. A Realm of Indeterminism beyond Quantum-Reality	321
1.3. A Realm of Subjectivism beyond Quantum-Reality	326
2. Subjectivism Manifests in every Higher-Energy-State	327
3. Non-Quantum-Virtuality is one Higher-Energy-State	332
4. Subjectivism is only with Non-Quantum-Virtuality	334
Summary	335

The Chapters from 1 to 8 have actually depicted stepwise a unification for entire Quantum-Reality with all non-zero and non-infinity values for 5+5 inverse CIPs or dimensions for every 'event' or particle or system-of-particles in both inertial and non-inertial conditions. That unified quantum-reality has further shown the intrinsic properties of dualism, determinism and objectivism within it. The same unified quantum-reality also governs by all Causal Laws through a set of total seven universally invariant inverse constants $k_1, k_2, k_3, k_4, k_5, k_6$ & k. That unified quantum-reality is also a sum of all Quantum-Real-Events, Quantum-Real-Signals and Quantum-Real-Observers.

Present Chapter-9 would reveal something beyond that unified quantum-reality as seventh set of inferences in series from those basic postulates of Chapters 1 & 2.

That beyond something of unified quantum-reality would be in a range of all zero & infinity values for those same 5+5 inverse 10-CIPs or 10-dimensions in contrary to all non-zero &

non-infinity discrete magnitudes for any quantum-real-events, quantum-real-signals & quantum-real-observers in physical nature. Obviously, such a realm of all zero & infinity values for same 5+5 inverse 10-CIPs should not be equivalent to the non-zero & non-infinity quantized values of quantum-reality.

Such a *continuum* of all zero & infinity values for 5+5 inverse 10-CIPs in same physical nature has assumed as *Non-quantized-continuum*. That ridiculously appears as any non-quantized *vacuum* but seems simultaneously spread over the infinity values. As if a realm for happening of all non-quantized events or *virtuality*. Then, anything that is involving such zero & infinity values can be stated as *Non-quantized-virtual* in physical nature. A continuum with no values for space, time & matter.

But such a Non-quantum-virtuality, with all zero & infinity values of 5+5 inverse 10-CIPs or dimensions, should be always beyond the quantum-real exchange limits of any quantum-real-signals as well as quantum-real-observers like us (with all non-zero & non-infinity discrete values of same 10-CIPs or dimensions).

As a result, any event, that is 'going to occur' in such Non-quantum-virtuality, never could be exchanged by the quantum-real-signal of quantum-real-observers like us until the same finally 'evolves out' as any quantum-real-event within exchange limit for that quantum-real-signal of us.

That exchangeable limit is actually equivalent to the range of all non-zero & non-infinity quantized values for those 5+5 inverse 10-CIPs. Then until the event that is 'going to occur' within Non-quantum-virtuality is fitting into confinements of all non-zero & non-infinity values of same 5+5 CIPs for any equivalent similar non-zero & non-infinity valued quantum-real-signal of quantum-real-observers like cannot be exchanged. But if any same 5+5 inverse 10-CIPs would have at least one zero or infinity magnitude in any Quantum-Real-Event. As a result, that should not have any causal sequences due to any such zero or infinity value but with all broken causal sequences.

Therefore, 'where', 'when' and 'how' such a quantum-real 'event' would be emerged out or destroyed into that Non-quantum-virtuality, and would appear within the exchangeable range or limit of unified quantum-reality, that cannot be causally forecast in quantum-real range of Eqs. (2.31) & (6.15). But, if there any kind of logical sequence that may 'flow' in such a realm of Non-quantum-virtuality never could be comprehended in our quantum-real limit of cognizable mind. That would remain beyond the quantum-real exchangeable limit of all causal logical pattern of sequence out of all non-zero & non-infinity magnitudes of 5+5 inverse 10-CIPs in physical nature.

Then, any kind of intrinsic logical pattern, suppose if that is involving there in realm of Non-quantum-virtuality, should appear as a non-causal and with all broken causal sequences to the quantum-real observers like us.

Hence, that non-causal and Non-quantum-virtuality would obviously be appeared to us as 'indeterministic' due to the absence of deterministic quantum-realities as described in Chapter-7. Therefore, same Non-causality and Non-quantum-virtuality, which is also indeterministic, would to have Non-objectivity as defined in Chapter-8.

But how one Non-quantum-virtuality subjectively transforms itself into one Quantum-reality or vice versa cannot be known due to same broken sequences in any quantum-reality.

Another thing about such a Subjective Non-quantum-virtuality is related to 'why' and 'how' in physical term, apart from few such mere mathematical numbers (like zero & infinity) involving within all those 5+5 inverse 10-CIPs, actually *differs* from an Objective Quantum-reality (like non-zero & non-infinity) is also unknown. In that context, the physical nature seems ultimately to have two basic folds – (a) unified quantum-reality defines in Eqs. (2.31) & (6.15), and (b) another is Non-quantum-virtuality [14] beyond that Unified Quantum-Reality in a form of something like Non-quantized continuum of energy i.e. 'Non-quantum-energy'. Moreover, due to the presence of Subjectivity in creation/destruction of any event within it, the same Non-quantum-energy seems to follow all Non-Causal Laws like any Virtual Wills.

In below Section-1, such Non-quantum-virtuality would reveal in details. The Section-2 would describe subjectivity in physical nature as if the manifestations of higher energy states compare to observers are observing those states. In Section-3, the subjectivity of Non-quantum-virtuality would also describe as if such a manifestation of any higher energy state beyond the quantum-real objectivity of universal gravitation. Finally, in Section-4, any subjective observers and its corresponding subjective observations to create/destruct any quantum-real-event in such higher energy state would also be the parts of same higher-energy-state or Non-Quantum-Virtuality in physical nature.

1. Defining Subjectivism in Physical Nature:

The concept about Non-quantum-virtuality has initiated in the above paragraphs. The same could define in this Section more elaborately. But that could be a conjecture beyond the realm of unified quantum-reality under category of ΔM_6 in Eqs. (2.31) & (6.15) in physical nature.

But being one quantum-real observer like us, who has objective cognizable-mind, the same Non-quantum-virtuality can be only conceptualize by starting from those same unified quantum-real Eqs. (2.31) & (6.15). Because, that Non-quantum-virtuality with all magnitudes of zeros & infinities never could be exchanged by quantum-real magnitudes of all non-zeros & non-infinities although both are the two basic folds of the same physical nature. The following three Sub-Sections would conceptually deduce mainly three major characteristics those are seemingly associated with the realm of such Non-quantum-virtuality in physical nature:

1.1. One Event Beyond that Quantum-Reality

The unified quantum-reality in inverse Eq. (2.28) in inertial state that has appeared with a mirror-imaged duality in Eq. (2.31). Those have corresponding non-inertial extensions in Eqs. (6.12) & (6.15) which are also based on inverse relations of all 5+5 CIPs with all non-zero & non-infinity discrete values. Each of those in association with any event can define through such 10-CIPs as any inertial particles or systems-of-particles and non-inertial GBs or GSBs. Subsequently in Eq. (8.1), any such events involving with all those particles or systems-of-particles have derived as quantum-real, deterministic & objective in types.

The Eqs. (2.28) & (2.31) or (6.12) & (6.15) are as if looking like a boundary in between the two domains: for all non-zero & non-infinity scale-specific discrete magnitudes of 5+5 inverse 10-CIPs as integrated parts of unified quantum-reality under category of ΔM_6 in one side, and for all zero & infinity non-quantized magnitudes of same 5+5 inverse 10-CIPs outside of category ΔM_6 in other side for occurrence of any event in physical nature. That would also be a same boundary in-between the 'quantum-reality and Non-quantum-virtuality', the 'quantum-real-determinism and Non-quantum-virtual-indeterminism', and the 'quantum-real-objectivism and Non-quantum-virtual-subjectivism' for description of any event in that physical nature.

The paragraphs below would describe the main features of such an 'event' if that will occur within the realm of Non-quantum-virtuality for its any/all 5+5 inverse 10-CIPs might have zero and/or infinity magnitudes in physical nature. That is if that event would occur elsewhere outside of the unified quantum-reality ΔM_6:

(i) The 'Event' with Non-Quantum-Virtuality: If we start from same unified quantum-real Eq. (2.28) that has defined any 'event' as quantum-real if the same possesses all non-zero &

non-infinity discrete magnitudes for 5+5 inverse 10-CIPs. If any/all of those same 10-CIPs for same event would possess one or all zero & infinity magnitudes, then same inverse Eq. (2.28) would show that 'event' in different type. That 'event' would occur somewhere in all same 5+5 inverse 10-CIPs but with 'no space', 'no time' and 'no matter' though in 'infinity anti-space', 'infinity anti-time' and 'infinity inertial-motion', or vice versa.

The territory, which is also beyond all quantum-real signal exchange limit with all non-zero-space, non-zero-time and non-zero-matter but non-infinity anti-space, non-infinity anti-time and non-infinity inertial-motion or vice versa.

Such 'event', which might occur beyond the exchangeable limit of any quantum-reality must be one conceptual 'Non-quantized' type having infinitely spreads in 'anti-space' and 'anti-time' with 'infinite-motion' but puzzlingly within 'zero-space' and 'zero-time' with 'zero-matter' or vice versa. So, that Non-quantized values for quantum-real conventional space $\Delta s(\Delta x \cdot \Delta y \cdot \Delta z) = 0$, time $\Delta t = 0$ and mass-energy or matter $\Delta m = 0$, but inverse anti-space $\Delta s_u(\Delta x_u \cdot \Delta y_u \cdot \Delta z_u) = \infty$, anti-time $\Delta t_u = \infty$ and inertial motion $\Delta v = \infty$, or vice versa.

Moreover, if any one of those 5+5 inverse 10-CIPs in Eq. (2.28) reaches in the magnitude either zero or infinity, then all other 9-CIPs would be also instantaneously transformed either with zero or infinity magnitudes depending on particular CIPs.

The events, those might occur & destroy in that realm with all such zero & infinity magnitudes for same 5+5 inverse 10-CIPs, would never detectable within the quantum-real exchangeable range of signal for the observers like us. Because one such quantum-real signal (including gravitation) would have only non-zero & non-infinity quantized type values. Since anything having zero or infinity value would be discrete magnitude type exchangeable signals and observers like us.

As a result, those zero & infinity magnitudes for any event would be anything that must have a continuous & unlimited property or as a continuum in contrast of a quantized & limited property of unified quantum-reality under category of ΔM_6. That is one 'virtual continuum' that involves with such 'Non-quantized values'. That can be collectively termed as 'Non-quantum-virtuality' for an event for convenience in contrast of quantum-reality of other type events as integrated parts of the unified quantum-reality in physical nature.

Therefore, in contrast of any individual quantum-real-event or unified quantum-real-event $\Delta \Gamma$ in Eq. (8.1) as an event in that Non-quantum-virtuality within spreads of all zero and infinity magnitudes for all 5+5 inverse 10-CIPs say $\vdots {}_\infty^0\Gamma \vdots$, and that can be defined as

$$\vdots {}_\infty^0\Gamma \vdots = \begin{bmatrix} \Delta m = 0 \text{ or } \infty, & \Delta s(\Delta x, \Delta y, \Delta z) = 0 \text{ or } \infty, & \Delta t = 0 \text{ or } \infty \\ \Delta v = \infty \text{ or } 0, & \Delta s_u(\Delta x_u, \Delta y_u, \Delta z_u) = \infty \text{ or } 0, & \Delta t_u = \infty \text{ or } 0 \end{bmatrix}$$

(9.1)

which reveals one different realm for one or all event/s that can occur in the same physical nature starting from the same Eq. (2.28), and never could be exchanged or found anywhere within the confinement of the unified quantum-reality in Eqs. (2.31), (6.15) & (8.1) & (8.2).

(ii) The 'Event' having No Causal Start & End: The occurrence of any such Non-Quantized-Virtual-Event in Eq. (9.1) obviously should have no start and no end. It conceptually occurs in the zero co-ordinates as well as in inverse infinity co-ordinates. Anything that is with zero and/or infinity values would not have any separation (in the senses of geometrical as well as physical) in-between its start and finish. Consequently, one Non-Quantized-Virtual-Event in Eq. (9.1) should not have any backward and forward progresses in terms of quantum-real space, time & matter.

But the motion would have infinity value. That means, there would be always the zero differences in between two 'ticks' of any sense of quantum-real time before and after the occurrence of any such Non-quantized-virtual event.

The quantum-real matter would have also zero value. That could mean matter does not require any spatial value to 'connect' with any quantum-real 'event' that has not yet quantum-really occurred or emerged out to become quantum-really exchangeable through similar kind of quantum-real-signal for detection of similar kind quantum-real-observer like us.

Consequently, all those Non-quantized-virtual-events might have always occurrences but would remain be non-exchangeable through quantum-real signals until emergences of quantum-real matters with quantum-real space & time. As a result, any such Non-quantized-virtual-event should have no progress in quantum-real time & space to have any start for any end, or would have 'no start no end' occurrence unlike those as quantum-real events in Eq. (8.1).

(iii) The 'Event' would have Broken Causality: That non-progression, in occurrence of any such Non-quantized-virtual-event, in the realm of Non-quantized-virtuality or any non-causality, would be also nothing but one that occurs beyond our conventional logical processes of causality or cause and effect sequences of occurrence/destruction. Since, such a Non-quantized-virtual-event also has mandatorily any intrinsic zero & infinity magnitudes of its 5+5 inverse 10-co-ordinates, it should have ever broken causal sequence with all non-zero & non-infinity magnitudes for all those same 5+5 inverse 10-co-ordinates.

From the realm of our conventional type quantum-real causality one never could know 'where', 'when' and 'what type' of above-mentioned quantum-real event could be occurred or emerged out that could be exchanged through one quantum-real signal. Since one quantum-real event could occur or emerge with broken causality out of Non-quantized-virtuality could exchange either instantly soon after its occurrence or could have to be waited for infinity period of quantum-real time for its occurrence. It could be similarly observed instantly at adjacent or nowhere or in infinity apart in quantum-real space. It could have new instantly emerged exchangeable quantum-real matter or no quantum-real matter for exchange or could have infinity quantity of matter.

In other way, this would appear until those broken causal sequences, with all zero and infinity values for 5+5 inverse 10-co-ordinates of any such Non-quantized-virtual-event in Eq. (9.1), would become anyhow turning into one quantum-real exchangeable non-zero & non-infinity valued range for causal sequences of any quantum-real-event, any quantum-real-observer like us never could observe it. But 'where', 'when' & 'what type' of that quantum-real-event would emerge and exchange by one quantum-real-observer like us could be completely an indeterministic due to all broken causality or Subjectivity within the Non-quantized-virtuality of continuum in Eq. (9.1).

Alternately, one Non-quantized-virtual subjective-event in same Eq. (9.1) should remain be one non-causal and indeterministic as a part of the Subjectivity of that Non-quantized-virtual continuum until it would be exchanged by one quantum-real signal as causal quantum-real-event in Eq. (8.1) with all non-zero & non-infinity discrete magnitudes of same 5+5 inverse 10-CIPs. Then after the same quantum-real event would be causal and integrated part of the unified quantum-reality ΔM_6 in Eq. (6.15) exchangeable by any quantum-real-signal for observation of one quantum-real-observer.

Therefore, finally, any Non-quantized-virtual-event in Eq. (9.1) would be always non-causal or broken sequence due to the presence of zero & infinity in its non-discrete magnitudes of all 5+5 inverse 10-CIPs. But within the range of non-zero & non-infinity magnitudes of same 5+5 inverse 10-CIPs in Eqs. (8.1)

& (8.2) for entire unified quantum-reality ΔM_6 in Eq. (6.15) there would be no such zero & infinity magnitudes. So, there would be no non-causality or broken sequences in any quantum-real events as the parts of same unified quantum-reality under category of ΔM_6 in Eq. (6.15) in physical nature.

1.2. A Realm of Indeterminism beyond Quantum-Reality

(i) Continuing from the previous Sub-section-1.1, any deterministic event exchangeable to any quantum-real-observers like us, that has defined in Eq. (8.1) from the Eqs. (2.28), (6.12) & (6.15), would appear deterministic if and only if it occurs within the range or limit of any non-zero & non-infinity 5+5 inverse 10-CIPs. Alternately, the same would also be an occurrence of any event in quantum-real discrete valued non-zero & non-infinite 10-co-ordinates system having a causal sequence from start to end or conversely from end to start for such event. But conversely, within the zero & infinity valued 5+5 inverse 10-CIPs, the Non-quantized-virtual 10-co-ordinates system has defined in Eq. (9.1) and derived from the same Eq. (2.28), the occurrence of any event might be two types of origins.

In first type, an event that occurs with all zero & infinity magnitudes 5+5 inverse 10-co-ordinates as one conceptual 'Non-quantum-virtual category' within (all zero & infinity magnitudes of corresponding 5+5 inverse 10-CIPs) Non-quantized-virtual realm non-causality has defined in Eq. (9.1). So, that will never exchangeable through any quantum-real signals by the quantum-real-observers like us. As a result, we would have no alternative way to observe the occurrence of any event within such Non-quantized realm of non-causality in physical nature. But conceptually, any observer who would be an integrated part of that Non-quantum-virtuality unlike us could see the same Non-quantized-virtual category event who would have all similar zero & infinity magnitudes for same 5+5 inverse CIPs. So, first type of such 'Non-quantum-virtual category' event would always be a part of the same Non-quantized-virtual realm.

But in second type, that event could occur conversely as 'quantum-real category' with all non-zero & non-infinity discrete magnitudes of same 5+5 inverse 10-co-ordinates in Eq. (8.1) but out of the same Non-quantized-virtual realm in Eq. (9.1). Conversely, instead of such occurrences, one such quantum-real-event can also destroy in similar way in such Non-quantum-virtual realm.

However, in both of these processes of occurrence & destruction situations there would seem to have all transformations in respective magnitudes from one zero to one non-zero as well as from one infinity to one non-infinity, and vice versa.

But, how those zero and infinity magnitudes of one Non-quantized-virtual event, beyond the quantum-real exchange limit, could transform into non-zero and non-infinity discrete magnitudes of another quantum-real event under the quantum-real range of signal exchanges? That cannot be defined through neither Eqs. (8.1) & (9.1) nor Eqs. (6.12), (6.15), (2.31) & (2.28) due to beyond exchangeable capacity to learn directly by our conventional quantum-real signals. But in proceeding Sections 2 & 3 would try to define in other way through the conventions of energy level differences between observers and observances for occurrence of one such second type of Non-quantized-virtual event.

However, the occurrence of one such second type of event out of the Non-quantum-virtuality would appear indeterministic respect to one lower energy level observer, but would be deterministic to the observers having equal or higher energy levels.

But whatever might be the indeterminism or determinism respect to the observers' energy level specific observations, one such second-type of quantum-real event would have always the 'fresh sequence of causality' starting from its very moment of occurrence. Subsequently, that would have all proceeding causes and effects sequences from start to end and the end to start in cycle through all those Eqs. (8.1), (6.15), (6.12), (2.31) and (2.28).

Similarly, in case of destruction for any quantum-real event into that Non-quantum-virtuality would be also one breaking of such proceeding sequence of causes and effects. Then one such event like destruction of causality into the Non-quantum-virtuality would also be assumed as indeterministic in type through similar fashion. Because, there would actually be a transformation of earlier 'non-zero' discrete magnitudes into -zero' and 'non-infinity' discrete magnitudes into 'infinity' of that event. Through such transformation of event from quantum-reality to Non-quantum-virtuality it seems that 'causality' with all 5+5 inverse 10-CIPs of non-zero & non-infinity corresponding discrete magnitudes are breaking into 'non-causality' of same 5+5 inverse 10-CIPs with all zero & infinity magnitudes. That is, when one quantum-real-event in Eq. (8.1) would destroy within continuum of Non-quantum-virtuality in Eq. (9.1), all causal progressions of causes and effects or progressions as the cycle of causality could involve with same quantum-real event must be instantaneously ended up. One cannot deterministically predict or no more able to exchange with same quantum-real event further through any causal process within unified quantum-reality by Eqs. (8.1), (6.15), (6.12), (2.31), (2.28) and so on.

But, one conceptual Non-quantum-virtual-observer [1], who is imagined to integrate with that Non-quantized-virtual realm of continuum in Eq. (9.1), can 'see' or exchange with that Non-quantized-virtual-event of all 5+5 inverse 10-CIPs with zeros & infinities. Because, that Non-quantized-virtual type observer unlike us would have also the similar or equal zero & infinity magnitudes of 5+5 inverse 10-CIPs to exchange with such

---

1. Basically one subjective in type.

Non-quantized-virtual event. Moreover, the same Non-quantized-virtual observer even could conceptually exchange one quantum-real event. Because, such a Non-quantized-virtual observer would have within itself 'infinity values' > 'non-infinity values' of such one quantum-real event.

So, a Non-quantized-virtual observer could have exchange capabilities with both Non-quantized-virtual event in Eq. (9.1) as well as quantum-real event in Eqs. (8.1) & (8.2) in physical nature.

However, in other hand, one quantum-real-observer like us, as an integrated part of the unified quantum-reality of ΔM_6 in Eqs. (2.31), (6.15), (8.1) & (8.2), could be only limited to see or exchange with any similar type quantum-real observances through identical type of quantum-real signals. Because, each of those three as observer, observance & signals have equivalent type of non-zero & non-infinity magnitudes for all 5+5 inverse 10-CIPs within same unified-quantum-reality ΔM_6 in Eqs. (2.31), (6.15), (8.1) & (8.2). Furthermore, the same quantum-real-observer as well as its quantum-real signals would have all 'non-infinity values' < 'infinity values' of one Non-quantized-virtual observance, signal and observer. So, any Non-quantized-virtual observance or event as a part of the Non-quantized-virtual continuum or any quantum-real event that have destroyed into that Non-quantized-virtual continuum in Eq. (9.1) cannot be exchanged by that 'limited' quantum-real observer integrated with ΔM_6 in Eqs. (2.31), (6.15), (8.1) & (8.2).

However, any creation/destruction of same quantum-real-event after evolving from or before destroying into the Non-quantized-virtuality (in above second type of creation/destruction) of Eq. (9.1) never could be deterministically exchanged by any similar quantum-real observer like us. Subsequently, all those quantum-real observations or exchanges of signals before the creation or after the destruction of any similar quantum-real event (as above second-type) in Non-quantized-virtual continuum would be always non-causal or indeterministic to any quantum-real-observers like us within ΔM_6.

Therefore, one such quantum-real-observer like us, who is an integrated part of the unified quantum-reality and can see creation/destruction of that quantum-real-event $\Delta\Gamma$ in Eq. (8.1) deterministically within ΔM_6 would see the creation/destruction of the same quantum-real-event $\Delta\Gamma$ out of the Non-quantized-virtuality $:{}_\infty^0\Gamma:$, only through the non-causal manner

$$\Delta\Gamma \rightleftharpoons\ :{}_\infty^0\Gamma:\ . \tag{9.2}$$

But after such a non-causal creation of causal event from or a non-causal destruction of causal event into Non-quantum-virtuality, the same quantum-real event would continue through a cycle of causal sequences objectively or independent of observations of any quantum-real-observers like us until that cycle non-causally destructed or broken at any point.

Then, such a non-causal origin of quantum-real-event $\Delta\Gamma$ out of a Non-quantum-virtuality of $:{}_\infty^0\Gamma:$ in Eq. (9.2) would be independent to all quantum-real-observer like us in ΔM_6 in Eqs. (2.31), (6.15), (8.1) & (8.2). But that same $\Delta\Gamma$ must be dependent to one Non-quantum-virtual-observer unlike us conceptually as an integrated part of that non-causal Non-quantum-virtuality with all same zero & infinity co-ordinates in Eq. (9.1). Because, a signal with zero & infinity magnitudes (for such Non-quantum-virtual observer) can exchange with both quantum-real-event as well as Non-quantum-virtual-event unlike a signal with non-zero & non-infinity magnitudes.

(ii) In Eq. (9.2), any quantum-real-event $\Delta\Gamma$ that creates / destroys in non-causal ways out of the Non-quantum-virtuality $:{}_\infty^0\Gamma:$ must be an indeterministic quantum-real-event.

(iii) Since, each of those quantum-real-event $\Delta\Gamma$ in Eq. (9.2) is integrated part of the whole unified quantum-reality ΔM_6 in Eq. (6.15), subsequently all the creations/destructions of same unified quantum-reality ΔM_6 out of the Non-quantum-virtuality : $_\infty^0\Gamma$: must also be a non-causal or indeterministic event to any quantum-real-observer like us.

1.3. A Realm of Subjectivism beyond Quantum-Reality

(i) In above paragraphs, particularly through Eq. (9.2), it has revealed that the indeterministic creations/destructions of any quantum-real-event $\Delta\Gamma$ in Eq. (8.1) out of that Non-quantum-virtuality : $_\infty^0\Gamma$: must be a quantum-real-observer independent phenomenon after its such non-causal creation or before its non-causal destruction. That is, every such quantum-real-event $\Delta\Gamma$ out of Non-quantum-virtuality : $_\infty^0\Gamma$: must be any Objective in type for all similar quantum-real-observers like us after its indeterministic creations or up to its indeterministic destructions. That is why no such quantum-real but subjective in type of observers unlike us not yet has observed to integrate with the same unified quantum-reality ΔM_6 in Eq. (6.15) as well as in Section-5 of previous Chapter-8.

But in Eq. (9.2), it has also revealed that the Non-quantum-virtuality : $_\infty^0\Gamma$: , with all zero & infinity magnitudes for 5+5 inverse 10-CIPs, can non-causally create out of itself or destroy into itself any quantum-real-event $\Delta\Gamma$ in Eq. (8.2) with all non-zero & non-infinity quantized values for all same 5+5 inverse 10-CIPs.

(ii) Therefore, the same Non-quantum-virtuality : $_\infty^0\Gamma$: with all zero & infinity magnitudes for 5+5 inverse 10-CIPs in Eq. (9.2) is also showing an essential subjectivism in both of such non-causal creation and destruction of any quantum-real-events out of or into itself as : $_\infty^0\Gamma$: . That can call as Non-quantized-virtual-subjectivism.

(iii) Consequently, we never know, due to our inability to exchange any quantum-real-signal with any such Non-quantized-virtual-observer be an integrated part of that entire Non-quantized-virtual-subjectivism : $^0_\infty\Gamma$: in Eq. (9.2) and also whether the same Non-quantized-virtual-observer would be a subjective in type within same : $^0_\infty\Gamma$:. But if there really any such subjective type Non-quantized-virtual-observer exists and might integrate with entire Non-quantized-virtual-subjectivism : $^0_\infty\Gamma$: in Eq. (9.2), that must have also similar type of Non-quantized-virtual-subjectivism as like as entire Non-quantized-virtual-subjectivism : $^0_\infty\Gamma$: unlike any quantum-real-observers like us in $\Delta\Gamma$. Subsequently, everything, in such conceptual Non-quantized-virtual-subjectivism : $^0_\infty\Gamma$:, including all those integrated subjective type observers, signals and events would have always zero & infinity magnitudes for all 5+5 inverse 10-CIPs. Then, in that beyond our quantum-real exchangeable range of state for Non-quantized-virtual-subjectivism, everything would be equivalent to zero & infinity magnitudes for all 5+5 inverse 10-CIPs.

That is all those 5+5 inverse 10-CIPs as 10-dimensions for Non-quantized-virtual-subjectivism as a continuum would be a state that appears to has no-space, no-time, no-matter but infinity-anti-space, infinity-anti-time and infinity-inertial-motion beyond exchangeable range or limit of all quantum-real-objective type observers like us.

2. Subjectivity Manifests in Every Higher Energy State

In above paragraphs of Sub-section-1.3, the whole physical nature for Eq. (2.28) has actually appeared to us (as quantum-real-observers) in two folds: one is as the unified quantum-real-objective which could be defined through the Eqs. (6.15) & (8.1) and another is as the Non-quantized-virtual-subjective might be defined by the Eqs. (9.1) & (9.2).

That unified quantum-real-objectivity could create from or destroy into that Non-quantum-virtual-subjectivity. Such two corresponding folds can also be imagined as if the two different levels or states of energies. If the former state can say as if the unified quantum-real-states-of-energies (UQRSE) then later would be as the Non-quantized-virtual-states-of-energies (VVSE). But both as a web of all same 5+5 inverse 10-CIPs. The UQRSE is comprised only by all SSUCs discrete non-zero & non-infinity magnitudes and the VVSE is with all zero & infinity magnitudes for same 5+5 inverse 10-CIPs.

Since, conceptually, the VVSE has all zero & infinity magnitudes for those 5+5 inverse 10-CIPs, there value of spacetime would have also the same zero & infinity magnitudes as described earlier. Consequently, in such an absence of any non-zero & non-infinity magnitudes for spacetime within VVSE, even there would have no effects of gravitational force (as quantum-real in type) to exchange with the same VVSE. That is, VVSE would have a separate energy state beyond even the exchange limits of gravitation.

Suppose, there are two observers who are integrated with corresponding UQRSE and VVSE levels of energies. The quantum-real-observer, who is an integrated with the UQRSE can only exchange with those which are quantum-real. That is, such an exchange could be conducted only in-between one Real vs. Real type of observer and observance through exchange of quantum-real type signals only. But one quantum-real-observer vs. one Non-quantized-virtual-observance type of exchange of signals could not be possible due to the limits of quantum-reality within every such quantum-real-signals.

Conversely, Non-quantized-virtual-observer who is conceptually integrated with the VVSE can able to exchange with not only one Non-quantized-virtual event as observance within that VVSE fold but being one subjective observer also can able to exchange with any subjectively creating/destructing quantum-real event as observance too within the UQRSE fold. Because one such Non-quantized-virtual-observer vs. one similar Non-quantized-virtual observance and one Non-quantized-virtual observer vs. one quantum-real observance type exchanges there would the signals those would be definitely beyond the quantum-real limits in Eq. (8.2) or say as virtual-signals for that virtual-observer within Eq. (9.1).

Hence, any event that could occur in physical nature, whether would be an objective or subjective in type, may appear to dependent [1] on the corresponding type of observers who is observing it. But those observers would further dependent to its specific states of energies where the respective observers are integrated. Therefore, occurrence of any 'subjective type of event' could also assume as if an event of one higher state of energy compare to the energy state of its respective observer. Then creation / destruction of any quantum-real-event, that appears as subjective from the VVSE respect to one quantum-real observer who is integrated with UQRSE could also be assumed as any event of higher energy state (VVSE) in physical nature. Because, those two states of energies like UQRSE and VVSE cannot be the same. One must have higher energy state than the other.

Therefore, the whole unified quantum-reality $\Delta\Gamma$ in Eq. (8.1) can subjectively create/destroy out of the entire Non-quantum-virtuality as $: {}^{0}_{\infty}\Gamma :$ in Eq. (9.2), then VVSE as Non-quantum-virtuality $: {}^{0}_{\infty}\Gamma :$ would have higher state of energy than the UQRSE (for BBCOU) as unified quantum-reality $\Delta\Gamma$.

However, those assumptions of higher and lower states of energies, as considered in above paragraphs, in occurrence of a subjective type event can also be comprehended through any two different states of energies even within the realm of UQRSE if respective observer would be integrated with lower energy state.

---

1. If it is one quantum-real type of observer, the event would be a quantum-real objective type but who never notice any Non-quantized-virtual subjective type event due to its limited exchange capacity through quantum-real-signals. If it is one Non-quantized-virtual category observer, the events could be both quantum-real-subjective as well as Non-quantized-virtual-subjective types due to its exchange capacity through conceptual Non-quantized-virtual-subjective signals.

For example, we can imagine one such lower-energy-state quantum-real observer within the realm of UQRSE. Who is say exclusively made by ice (i.e. in one lower thermal energy state) and is trying to exchange with one event that is occurring within a quantity of liquid water (i.e. in comparatively higher thermal energy state compare to that ice) through one similar ice-made signal or an 'ice-ball' (i.e. also in identical lower thermal energy state alike to that icy-observer than the liquid water). With his ice-signal, the icy-observer never could able to make any to and from exchange with that event occurring within the liquid water. Because, every time that ice-ball signal would reach to the higher-energy-state liquid water be melted and ultimately never could come back to that icy-observer to make any effective exchange. Consequently, any finite quantity of (quantum-real) liquid water in comparative higher energy state, where such event has occurred, would ultimately appear to that icy-observer either as something as 'nothing' (i.e. all zeros) and simultaneously as something of infinite spread and magnitudes (i.e. all infinities) [1]. Moreover, due to any uneven causal process of distribution for the hydro-thermal energy, if one part of that liquid water become frozen earlier than its other parts, that would become exchangeable to that icy-observer through his ice-ball, but could assume such a formation of ice-formation-event within the same liquid water as one subjective creation of a quantum-real-event. But to one observer who is involving with the higher energy states equal to that liquid water or even much higher than it would observe the same event still as objective in type.

The similar circumstances could be repeatedly observed in-between all the relevant circumstances of other differences in energy-states of liquid-water vs water-vapor, water-vapor vs plasmas, plasmas vs electromagnetic energies, or might be in

---

1. Although that chunk of liquid water we know being one quantum-real quantity and would have all non-zero & non-infinity discrete magnitudes.

between visible-matters vs dark-matters, dark-matters vs dark-energies. That could be continued even up to the UQRSE vs VVSE.

Bu, if all those observers in corresponding lower energy-states, being massive entities, are smart enough to exchange through gravitational-signals would find there no such subjectivity in occurrence of events in all those corresponding non-zero & non-infinity magnitudes of all so-called higher energy-states. Because, one such conceptual gravitational-signal compare to one ice-ball-signal would have universal exchange capacity within unified quantum-reality of ΔM_6 in Eq, (6.15). Then, instead of all those earlier subjectivities in occurrence of events there would be all objectivities within all corresponding states of higher energies in unified quantum-reality.

But such a quantum-real objectivity would have really a limit even through possible exchange of any so-called universally effective gravitational-signals in physical nature. Because, those (quantum-real) observers, whatever may have their extreme degree of smartness through such quantum-real gravitational-signals, never could communicate with any Non-quantized-virtual event as objective in the circumstance two different energy-states UQRSE vs VVSE.

Because, there still would occur all Non-quantized-virtual-subjective creation/destruction of all events out of the VVSE beyond the exchange limits for even a gravitation-signal or gravitational-effect. Therefore, the state of VVSE could assume really as an absolute Non-quantized-virtual-subjective state in physical nature respect to the state-of-energies of UQRSE. Where all the observers like us with relevant quantum-real objective cognition are limited with all non-zero & non-infinity magnitudes of 5+5 inverse 10-CIPs.

However, any such subjective event in physical nature, irrespective of its UQRSE and VVSE folds, would be ultimately the manifestations of all corresponding higher energy states compare to the respective lower energy state where the observers are integrated.

That VVSE might be the highest state of energies for any subjective event to occur/destroy beyond exchange limits of gravitation-signals and all observers like us who are integrated with lower state of energies within UQRSE of physical nature.

3. Non-Quantum-Virtuality is one Higher-Energy-State

The Section-2 has presented the entire unified quantum-reality of ΔM_6 in Eqs. (6.12)-(6.15) as a unified quantum-real-objective as well as a UQRSE including all basic forces including gravitation.

Moreover, the Eq. (6.11) may suggest a possible state for same unified quantum-reality ΔM_6 where its specific phase of highest inward gravitational collapse force no one could exist to counter same collapse through outward counter force still to hold any value for its the radius. That is, the curvature of quantum-real spacetime of everything as quantum-real at that phase could be collapsed up to the radius of same ΔM_6 as the $\Delta r = 0$ to erode the everything quantum-real into the nothing as quantum-real.

As a result, the quantized magnitudes of space and time that could also be transformed into the corresponding values for $\Delta s = 0$ & $\Delta t = 0$ in Eqs. (1.17) & (1.21) through same Eq. (6.11). Then, simultaneously, from Eq. (2.17) there would appear $\Delta \lambda = \infty$. Therefore, in Eq. (2.1), the total mass-energies of UQRSE (also as the BBCOU) would also be (as if dissolved into the) $\Delta m = 0$. Conversely from all those corresponding Eqs. (2.19), (2.21) & (2.14) there would be the simultaneous values for its other CIPs like $\Delta s_u = \infty$, $\Delta t_u = \infty$ & $\Delta v = \infty$.

Therefore, conceptually, the whole unified quantum-reality in Eq. (6.11) can transform into a state of Non-quantized-virtuality or VVSE when all 5+5 CIPs in Eq. (2.28) would be with all zero & infinity magnitudes. Since, that state is nothing but a zero magnitude mass-energies ($\Delta m = 0$) with zero or so-called Non-quantized values of spacetime ($\Delta s = 0$ & $\Delta t = 0$) with simultaneous infinity value of inertial-motion ($\Delta v = \infty$) of such VVSE within infinity spreading anti-spacetime ($\Delta s_u = \infty$ & $\Delta t_u = \infty$). That is, a Non-quantized 'continuum' of VVSE [15] beyond the quantum-real-exchange limit of us as a 'something' out of that 'nothing'.

Furthermore, in the absence of unified quantum-reality of spacetime ($\Delta s = 0$ & $\Delta t = 0$) there would be no gravitation as curvature of such spacetime. Also, there would be no gauge-fields of forces in absence of the all unified quantum-reality-mass-energies ($\Delta m = 0$) in Eq. (6.11).

Consequently, that VVSE would also beyond all our unified quantum-reality-signal exchange limits due to same zero values or no quantum realities for the signals with ($\Delta m = 0, \Delta s = 0$ & $\Delta t = 0$) to exit anymore to carry any 'effects of force' or zero exchange capacity for the universal gravitation & other basic forces.

As a result, that Non-quantum-virtuality, out of that continuum of Non-quantized VVSE, can also be defined by the same 5+5 inverse CIPs in Eq. (2.14) but with all zero & infinity magnitudes ($\Delta m = 0$, $\Delta s = 0$, $\Delta t = 0$, $\Delta v = \infty$, $\Delta s_u = \infty$ & $\Delta t_u = \infty$) unlike UQRSE of us with all non-zero & non-infinity magnitudes ($\Delta m > 0$, $\Delta s > 0$, $\Delta t > 0$, $\Delta v < \infty, \Delta s_u < \infty, \Delta t_u < \infty$). Subsequently, that Non-quantized-virtuality, or VVSE would be always beyond all of our quantum-reality of signals exchange limits to perceive it.

Then, it now appears that such a conceptual VVSE with highest energy level is existing conceptually beyond all our quantum-real-signals exchanging limits even the furthest and smallest possible reaching of gravitational effects.

4. Subjectivism only within Non-Quantum-Virtuality

In above Sub-Section-1.3, the creation/destruction of a unified quantum-real event in Eq. (8.1) has appeared Subjective out of the Non-quantum-virtuality in Eqs. (9.1) & (9.2). That is, any quantum-real-objective-event is appearing to any quantum-real-objective-observers like us would be ultimately subjectively dependent for its creation / destruction out of that Subjectivity of Non-quantized-virtual continuum VVSE. Where one conceptual Non-quantized subjective-observer unlike us could be a part.

In Eqs. (2.14), (2.28) & (6.11), it also appears, if the objective magnitudes of any one CIP out of those total 5+5 inverse 10-CIPs becomes in anyway becomes zero or infinity magnitude within any quantum-real-event or quantum-real particles or systems-of-particles, then rest of its inversely corelated 9-CIPs would instantaneously transform into all zero and infinity magnitudes. As described in previous Sections, any such zero & infinity magnitudes could break the cause and effect sequence of a causal event. Subsequently, through such breaking of causal sequences due to presence of zeros or infinities in those transformed 10-CIPs of quantum-real event, any such transformed quantum-real event or entity would also transform from the UQRSE to the VVSE, and ultimately would be the Subjective.

Conversely, one similar quantum-real entity could possible to create from that VVSE as any quantum-real-event or as any quantum-real particle or system-of-particles to become a part within UQRSE. But that formation of such objective quantum-real particle or system-of-particles could also occur through subjectivity of the VVSE or that can imagine as if the creation of those particles or systems-of-particles from the subjectivity of those Non-quantized-virtual observers as part of that VVSE-continuum.

Therefore, the observer within lower energy state of UQRSE, who is able to see any event that is supposedly creating/destroying through out of higher energy state of VVSE in indeterministic manner, could see the same creation/destruction of event as Non-quantum-virtual-subjective in type. That is, the event is creating / destroying out of all the zero and infinity magnitudes of 5+5 inverse 10-CIPs of VVSE-continuum and the moment of creation/destruction of the event would be always beyond the exchange limit of that quantum-real-objective-observer who is integrated with the lower energy state UQRSE in physical nature.

Summary

The 5+5 inverse 10-CIPs, e.g. Δm & Δv, $\Delta s\ (\Delta x, \Delta y, \Delta z)$ & $\Delta s_u (\Delta x_u, \Delta y_u, \Delta z_u)$, Δt & Δt_u, Δr & $\Delta r_u \equiv \Delta \lambda$ in Chapter-1 through the all corresponding inverse relationships in Eqs. (2.2), (2.4), (2.15), (2.16), (2.18), (2.20) & (2.22) & (2.27) in Chapter-2 have described in current Chapter with two distinct sets of values for whole physical nature. In one set, there are all discrete non-zero & non-infinity values and have referred to all quantum-realities, and in another set, there are all continuity of zero & infinity values and have linked to Non-quantum-virtuality. Due to this, ultimately, the physical nature is now appearing in two folds – (a) the entire unified quantum-reality as UQRSE for BBCOU in Eqs. (2.31) & (6.15) and (b) the Non-quantum-virtuality in Eq. (9.1) beyond that UQRSE or BBCOU in same 5+5 dimensions.

That Non-quantum-virtuality, unlike quantum-reality, is beyond exchange limit of observers like us even outside the effects of gravitation or other natural forces in Eq. (6.15). A non-causal (indeterministic) form non-quantized continuum beyond all of our quantum-real-comprehensible limits of logic, and is obviously a subjective compare to our objective way of observations. The same Non-quantum-virtuality, in Eq. (9.1) through the Eqs. (2.31) & (6.15), can also assume as a realm, beyond unified quantum-reality, with 'no space', 'no time' and 'no matter' measurable by any finest or smartest quantum-real conceptual instruments. But with infinity values for 'anti-space', 'anti-time' and 'inertial-motion'.

That subjectivity in observation can also perceive in unified-quantum-reality as two different manifestations of energy states. Sometimes, if an event occurs in higher-energy-state (can appear subjective) compare to its observer in lower-energy-state. But such an apparent subjectivity in occurrence/destruction of event in one higher energy state can further become objective if the same can see from one further higher states of energy.

The subjectivity is appearing to involve in one quantum-real event in Non-quantum-virtuality in Eq. (9.1) and an integrated observer in unified-quantum-reality in Eq. (8.2) also seems to be two different states of higher and lower energies but there is no choice for the observers like us to the same subjective-event from further higher state of energies if there in physical nature. Because, all those higher states of energies are even beyond the effects of universal gravitation. Therefore, Non-quantum-virtuality possesses a higher energy state compare to any quantum-reality or entire unified quantum-reality in physical nature.

However, any such zero magnitude for all those 5+5 inverse 10-CIPs means not actually any zero or 'no entity' in physical sense. But that zero in physical sense would be anything that is a non-quantized continuum. As a consequence, physically a Non-quantum-virtuality is one non-quantized continuum in higher-state-of-energy compare to us that may appear us to spread to infinity.

Can those two states of energies be linked together or have any equivalence? But if so, that could be one conjecture for a quantum-real observer like us in the physical nature.

However, that subjectivity within Non-quantum-virtuality as Non-causality of the same, beyond the reach of gravitation, also appears like any Non-Causal-Wills with all broken sequences unlike the all Causal-Laws of sequences within objectivity of Unified-quantum-reality in physical nature.

Moreover, what could be the basic form of any such quantum-realities if those are really originate subjectively out of that Non-quantum-virtuality? These would be also an essential topic in next and final Chapter-10.

CHAPTER-10:

BASIC FORM OF QUANTUM-REALITY
[As if a Whirl in Non-Quantized Continuum]

"Everything we call real is made of things that cannot be regarded as real."   - Niels Bohr

1. Virtuality within Non-Quantized Energy	339
2. Creations/Destructions of Quantum-Reality	341
3. Subjective Creation of Quantum-Real 'Whirls'	344
4. Ten Spatial Dimensions in a Quantum-Real-Whirl	347
5. Quantum-Reality from the Virtuality	352
6. Magnitudes of Quantum-Real Limits	355
7. Ten Spatial Dimensional Duality of Physical Nature	357
Summary	359

Non-quantum-virtuality, as defined as a higher state of energy in previous Chapter-9, has conceptually emerged out through the inverse relationships of all 5+5 inverse ten numbers of common-internal-parameters involving with all ingredient quantum-real particles or systems-of-particles in unified quantum-reality of BBCOU. That Non-quantum-virtuality also found unexchangeable by any such quantum-real signals for the observers like us. Subsequently, the Non-quantum-virtuality would be a conjecture in physics in a sense of its direct scope of observations. But the previous Chapter-9 also has revealed some of basic features in Non-quantum-virtuality through which that Non-quantum-virtuality could be comprehended. It appears with all zero & infinity values for all same 5+5 inverse ten common-internal-parameters or 10-CIPs in same physical nature as an origin of all non-causal form of logical processes which are beyond all of our conventional causal logical sequences involving with the quantum-real-observers like us. Even universal gravitation would have no effect on such realm of Non-quantum-virtuality due to no space, no time, & no matter situation.

In current understanding of physics, the perception of 'Non-quantized-energy' or 'virtual-energy' [15] seems equivalent to such Non-quantum-virtuality. Because, basic properties of two are conceptually appearing the same.

It is also not yet clear, what would be the most basic form of such quantum-reality for micro-most scale of a quantum-real particle that could subsequently originate from such Non-quantum-virtuality. But, how such a particle, with non-zero & non-infinity discrete values for all 5+5 inverse 10-CIPs, can transmute from that zero & infinity continuity of subjective Non-quantum-virtuality? Is there any real meaning for such quantum-real inertial-matter $\Delta m \neq 0$ involving in such subjective creation/destruction of that particle?

The current Chapter is actually the eighth and final set of consequences as inferences from the basic postulates in Chapters 1 & 2.

In below Sections 1 & 2, such subjective Non-quantum-virtuality has assumed as Non-quantized-energy [15] and Sections–3, 4 & 5 defined the quantum-reality ultimately as 5+5 inverse spatial & anti-spatial dimensional correspond to each of those 10-CIPs. As a result, a smallest scale of discrete spatial magnitude for any quantum-reality will deduce in Section-6; and a grand unified 10-dimensional spatial definition for everything in physical nature inclusive of all quantum-reality & Non-quantum-virtuality will be deduced in Section-7.

1. Virtuality within Non-Quantized Energy

In current understandings of Quantum Physics [16] also supposes, there might be something in most fundamental state of physical nature spreads everywhere in 'vacuum' space or something Non-quantized space [1] beyond exchangeable discrete limits of any so-called quantized signals that fills with Non-quantized-energies or vacuum-energies [15].

Although such Non-quantized-energy cannot be directly exchanged but it has supported through some current comprehensions [17] in Physics that quantum-real particles can non-causally create/destroy out of such vacuum-energies or Non-quantum-virtual continuum. But such non-causal occurrence of particles would remain be virtual until it becomes exchanged by any quantum-real-signal to appear as one quantum-real particle or event for a similar quantum-real observer like us. That is all such creation/destruction of quantum-real particles or events out of the Non-quantum-virtual-continuum or vacuum-energies appears also non-causal i.e. Subjective in observations of any quantum-real-objective-observers [2].

The same has also inferred starting from different presumptions through Eqs. (2.31), (6.15), (8.1) & (9.1) in Sections 3 & 4 of Chapter-9.

Therefore, such 'vacuum-energy' beyond the reach of any quantum-real-signals in current understandings of Quantum Physics can be assumed equivalent to that Non-quantized virtual continuum of Eq. (9.1) starting from Eqs. (2.31) & (6.15) in same physical nature. The Non-quantized-virtual continuum in Eq. (9.1) has defined with all zero & infinity non-quantized magnitudes for all 5+5 inverse 10-CIPs, therefore its equivalent vacuum-energy would assume to have similar zero & infinity magnitudes of 5+5 inverse 10-CIPs.

Also, there in previous Chapters, anything with such zero & infinity magnitudes in 5+5 inverse 10-CIPs should be a virtual in type rather than real respect to any quantum-real-observers-

---

1. That is with no real space, no real time and no real mass-energies or matter but instead fills with infinity of anti-space, infinity of anti-time and infinity of inertial-motion or Non-quantized-energies of non-quantized higher energy state.
2. Which is integrated part of the unified quantum-reality with objective cognizance.

like us. Moreover, anything that is virtual must not be a part of the unified quantum-reality ΔM_6 in Eqs. (6.15) that has defined in Eq. (8.1), but an integrated part of entire Non-quantum-virtuality in Eqs. (9.1) & (9.2).

Then any quantum-real-event that can emerge from or destroy into such non-quantized vacuum-energy through non-causal process must be a Subjective within such Non-quantum continuum of energy as higher-state-of-energy of VVSE from the lower-state-of-energy like UQRSE of unified objectivity of quantum-reality for both quantum-real-observers like us as well as relevant quantum-real-signals.

As a result, such non-quantized vacuum-energy that is equivalent to Non-quantum-virtuality could be defined through same Eqs. (9.1) & (9.2). From the same Eq. (9.2) Non-quantum-energy appears subjective Non-quantized-virtual to one observer like us who is an integrated part of the unified quantum-reality in Eqs. (6.15) & (8.1).

2. Creations/Destructions of Quantum-Reality

Since, Non-quantized-energy is Non-quantized-virtual, and can definable by the Eqs. (9.1) & (9.2), therefore from the Eq. (9.2) anything that is quantum-real and objective in Eq. (8.1) could create / destroy from such subjective Non-quantized-energy. Likewise, since such creation or destruction is non-casual, therefore any such quantum-real-event should appear as subjective event respect to any objective quantum-real-observers like us.

Consequently, after any such non-causal creation or up to any such non-casual destruction in Non-quantized-energy, every such quantum-real-event must follow a causal or observer-independent sequence to any objective quantum-real-observers like us.

Therefore, any such subjective to objective or non-causal to causal or Non-quantized-energy to quantum-real-energy or

Non-quantized-virtuality to quantum-reality transformation would be basically a conversion in all corresponding values of same 5+5 inverse 10-CIPs from zero & infinity to non-zero & non-infinity or vice versa.

Subsequently, the creation/destruction of any quantum-real objective particles or systems-of-particles irrespective of micro-most and macro-most (i.e. BBCOU) scales would have also the similar subjectivity from that Non-quantum-virtuality. The macro-most scale like BBCOU or ΔM_6 in Eq. (6.15), as the integration of all micro scales including micro-most scale, in that fashion is appearing to have similar kind of subjective creation/destruction before and after Bigbang / Bigcrunch like events correspondingly along with all other sequential formations of different scales those are involved with it all along the structure formations.

Then, the conventional moment of occurrence for Bigbang, in the Bigbang-Bigcrunch model of universe, would have some different background. The occurrence (i.e. creation) of such event like Bigbang out of that background of zero & infinity magnitudes with Non-quantum-virtual continuum would be also one subjective type event for subjectivity of that Non-quantum-virtual continuum. Conversely, the event like Bigcrunch (i.e. destruction) of the BBCOU would be also another subjective in type into the subjectivity of the same Non-quantum-virtual continuum.

As a result, the moments of creation / destruction of the BBCOU (ultimately as one quantum-real entity) could not be defined causally if it would occur out of the background of all same zero & infinity magnitudes of 5+5 inverse 10-CIPs of Non-quantum-virtual continuum.

However, after transmutation of magnitudes anyway from such zero & infinity to non-zero & non-infinity of the same 5+5 inverse 10-CIPs i.e. subjective creation of that quantum-real-BBCOU initiating through that quantum-real-event like Bigbang all that causal sequence of metamorphoses of universe would proceed up to the end as Bigcrunch. But at that Bigcrunch (or end) either it could start back to the Bigbang (for another fresh start) in a cycle causal sequences or that could be destroyed in Non-causal (or broken causal) manner into the Non-quantum-virtual-continuum.

However, such a breaking of causality within the causal cycle of BBCOU could subjectively occur anywhere.

That is, creation/destruction of the same BBCOU out of the Non-quantum-virtuality would be also the non-causal and subjective to the same Non-quantum-virtual continuum, if once any of those magnitudes of 5+5 inverse 10-CIPs would reach anyway to zero and/or infinity.

This could happen, if in Eqs. (5.16) & (5.17), the magnitude becomes for $\Delta m_{e-1} = 0$ due to no more outward material-forces would be there in ΔM_6 in Eq. (6.15) to resist or balance the non-stoppable collapsing pressure from the curved spacetime (gravitation) of the Big-Crunching BBCOU. That could also imagine as if a last quantum-real outward pressure or force to resist gravitation subjectively before its either 'melt down' into the Non-quantum-virtual continuum or restore the BBCOU against such collapse for next Big-Bang.

But, such a destruction of BBCOU through such Big-Crunching if reaches to magnitude like Non-quantized-virtual $\Delta m_{e-1} = 0$ (i.e. all zero & infinity values for all other 9-CIPs) would break the causal sequence of same BBCOU and subsequently there would appear a subjectivity in creation of another (next) Bigbang out of that Non-quantum-virtual continuum. If there, through such Big-Crunching it reaches to any quantum-real value for same $\Delta m_{e-1} > 0$ (i.e. all other 9-CIPs would have corresponding non-zero & non-infinity values) there would be no breaking in cyclic causal sequence of BBCOU. The BBCOU would reach to the next Big-Bang, then for another Big-crunch.

3. Subjective Creation of Quantum-Real 'Whirls'

The Non-quantum-virtual continuum in Eq. (9.1) has assumed as equivalent as the Non-quantized vacuum-energy with all zero & infinity magnitudes for its 5+5 inverse 10-CIPs. But it is still needed to comprehend how a quantum-reality could create (or destroy) in such a Non-quantum-virtual continuum and also what could such a creating (or destroying) quantum-reality be looked like in most of its basic form out of that subjective Non-quantum-virtual continuum.

Because, prior to such subjective creation of or after subjective destruction of one basic form of (objective and causally-sequence) quantum-reality, out of that (subjective and non-causal) all zero and infinity inverse 10-CIPs Non-quantum-virtual continuum, cannot be exchanged through any similar quantum-real signals with all non-zero & non-infinity magnitudes of same 5+5 inverse 10-CIPs in lower energy-state. That higher energy-sate of Non-quantum-virtual continuum if assumes to have a 'continuity' due to such higher energy, compare to lower energy-state of our conventional quantum-real discretization of UQRSE due to its lower states of energies, can be also imagine as if another kind of apparent 'continuity'. Similar kind of apparent 'continuity' in higher energy-state has revealed earlier in cases of corresponding thermal energy state of liquid-water compare to the lower thermal energy-state of one observer, it's signals & the observances when all have imagined to made by solid ice cubes (a kind of discrete materials as like as our discretization in UQRSE).

The solidification of that apparently 'continuous' liquid-water occurs not in the whole stuff at a time. Due to scattered losses of thermal energies in one discrete pocket of it would form all instantaneous freezing of same liquid-water through crystals of ices. That 'loss of energy' in higher energy-state seems creating lower-energy-states within the apparent 'continuity' of liquid-water, and as if, it facilitates somehow 'discretization' in lower energy-state through formation of ice-crystals in comparative lower thermal energy-state.

Moreover, in Eq. (2.28), any (such discretized) quantum-reality would have one simultaneous intrinsic left-handed and right-handed 5+5 inverse 10-CIPs beside one simultaneous mirror-imaged duality in Eq. (2.31). Therefore, every such discretized ice-crystal within higher thermal energy-state of liquid-water could imagine in similar fashion with simultaneous lefthanded and righthanded duality comprising all 5+5 inverse 10-CIPs.

Consequently, formation of such scattered (discrete) ice-crystals with lower thermal energy-states out of the higher thermal energy-state of apparent continuity of liquid-water can also appear as 'subjective' formation or creation of such discrete ice-crystals of lower energy-state. But such a 'subjective' type of formation or creation of same ice-crystals would appear only to that observer who has formed by similar kind of ice-crystals (with same energy-state) and also able to exchange within the limits of similar ice-crystals.

Conversely, due to uneven distribution of higher state-energies on the surface of ice-crystals, when and where a ice-crystal would dissolve or destroy into the continuity of liquid-water of higher thermal energy-state would also become subjective to one similar observer of ice-crystals through all his communicable signals of ice-crystals.

This manifestation of subjectivity due to differences in thermal energies in-between two energy-states of ice-crystals and liquid-water would have similar inferences in all corresponding differences in energy-states within unified quantum-real-energy as UQRSE in Eqs. (6.15) & (8.1).

Moreover, same UQRSE in Eqs. (6.15) & (8.1) has lower energy-state and Non-quantized-virtuality in Eq. (9.1) as VVSE has conceptually higher energy-state. As a result, any subjective creation / destruction of quantum-reality within such higher energy-state continuity of VVSE in Eq. (9.2) have already described as simultaneous lefthanded and righthanded 5+5 inverse 10-CIPs with mirror-imaged duality in Eqs. (2.31) & (6.15).

Similarly, as like as one icy-observer as an integrated part of the corresponding lower energy-state, one quantum-real-observer would also be an integrated part of that unified quantum-real state of lower-energies and would have no scopes to communicate / exchange with such apparent continuity of higher energy-state of Non-quantized-virtual continuum. Even beyond the scopes of universally effective gravitational communicable / exchangeable limits.

Then, ultimately, any Objective quantum-real-event which is involving with any particles or systems-of-particles those are integrated to unified quantum-real state of lower-energies and exchangeable to ours like similar objective quantum-real-observers, could only definable by those Eqs. (2.28), (2.31), (6.15) & (6.15) must have any micro-most scale. Such micro-most scale of particles would be also the most fundamental scale of such discrete units of UQRSE as like as an ice-crystal of ices. That most fundamental scale or micro-most scale could non-causally and subjectively create / destroy with the non-zero & non-infinity discrete magnitudes of all same 5+5 inverse 10-CIPs out of that Non-quantized-virtual continuum state of highest-energy (VVSE) in Eq. (9.2) in physical nature as like as higher thermal energy-state of liquid-water.

'Whirls' of Lower Energy-state. We can further imagine, such a subjective formation of lower energy-state ice-crystal as discrete 'whirl' of such lower energy within the continuity of so-called higher thermal energy-state of liquid-water. Such a whirl would have lower energy-state within that continuity of higher energy-state. Each of such whirls would have also the slower speed of 'flow' compare to the speed of flowing of the concerned higher energy-state of continuity. A flow of one whirl in water having slower speeds compare to the speed of stream in river. Subsequently, that whirl of lower energy would contain the higher 'energy' of continuity in stream but that 'energy' of continuity in stream never could comprise by those whirls. That is, such a whirl would be the continuity but the continuity would not be a whirl.

Since, the UQRSE has appeared to have similar lower energy-state compare to the higher energy-state continuum of VVSE and all the observers & signals are as parts of such UQRSE have also quantum-real limitations in observations any subjective creation / destruction of a micro-most scale of particle out of that VVSE continuum, the same micro-most scale of particle can also imagine as one smallest scale of whirl of lower-energy compare to that highest energy-state of VVSE. So, there could be different scales of such energy-state of subjective whirls out of the Non-quantum-virtual continuum of VVSE (correspond to different micro to macro scales of particles or systems of particles) similar to different lower thermal energy-states of ice-crystals out of the continuity of higher-energy liquid-water.

4. 10-Spatial Dimensions in a Quantum-Real Whirl

The destruction of any quantum-reality within Non-quantized-virtual-state-of-energy, as quantum-real-state-of-energy which is as an integrated part of whole unified quantum-real-state-of-energies, occurs when the magnitudes becomes $\Delta r = 0$ and $\Delta \lambda = \infty$ in respective Eqs. (2.17) & (2.28) through the Eq. (6.11). Conversely, a creation of any quantum-real-event as an integrated part of same whole unified quantum-real-state-of-energies out of the same Non-quantized-virtual-state-of-energy can be possible if the magnitudes of same becomes $\Delta r \neq 0$ and $\Delta \lambda \neq \infty$. One cannot exchange it as quantum-real objective until that transmute to $\Delta r \neq 0$ or ∞ within the exchange limit of one quantum-real-objective-signals (i.e. at least within the effects of gravitational effects) for the corresponding quantum-real space $\Delta s = \frac{3}{4}\pi \cdot \Delta r^3 \neq 0$ or ∞ in Eq. (1.17) as well as time $\Delta t = 2\pi \cdot \Delta r \neq 0$ or ∞ in Eq. (1.21). As a result, that quantum-real-event from Eq. (2.17) will appear ultimately with $\Delta r \neq 0$ or ∞, and subsequently $\Delta \lambda \neq \infty$ or 0.

But, what that actually means for a creation like all those ($\Delta m \neq 0$ or $\infty, \Delta s \neq 0$ or $\infty, \Delta t \neq 0$ or ∞) and also ($\Delta v \neq \infty$ or $0, \Delta s_u \neq \infty$ or $0, \Delta t_u \neq \infty$ or 0) as all subsequent quantum-real-objective 5+5 inverse 10-CIPs in that most fundamental level of quantum-real-creations, and also how that quantum-real-creation could become separated or transmuted from that Non-quantum-virtuality ($\Delta m = 0$ or $\infty, \Delta s = 0$ or $\infty, \Delta t = 0$ or $\infty, \Delta v = \infty$ or $0, \Delta s_u = \infty$ or 0 & $\Delta t_u = \infty$ or 0) is most important.

However, in previous Section-3, such separation has described as formation of lower energy-state whirls. But, from Non-quantum-virtuality of all zero & infinity magnitudes of all 5+5 inverse 10-CIPs, as continuity of higher energy state with non-causal subjectivity in such Non-quantum-virtual continuum, when transmutes to an objective quantum-reality for one quantum-real-observer like us where both are integrated to unified quantum-real-state-of-energies UQRSE, would appear as all discrete low-energy scattered whirls of non-zero & non-infinity appear.

Such primary separation between a quantum-real-objective particles or systems-of-particles and the Non-quantized-virtual state of higher-energy can only be described in our conventional objective cognizance, in that most fundamental or macro-most scale of $\Delta r \neq 0$ or ∞ as well as $\Delta \lambda \neq \infty$ or 0, if and only if the same quantum-real-objective particles or systems-of-particles can be geometrically imagine as such any subjective whirl like configurations in that continuum of that Non-quantized-virtual state of highest-energy.

Such a whirl out of the Non-quantized-virtual state of higher-energy (with all zero & infinity magnitudes of 5+5 CIPs) will possess all non-zero & non-infinity 5+5 magnitudes of same inverse 10-CIPs in Eq. (2.28) as equivalent geometric parameters for one quantum-real-objective particles or systems-of-particles e.g. $\Delta s, \Delta s_u, \Delta t$ & Δt_u in all corresponding Eqs. (1.17), (1.18), (1.21) & (1.23). Such an exchangeable quantum-real-objective whirl to one quantum-real-observer out of the subjectivity of same Non-quantized-virtual state of highest-energy would be equivalent to one micro-most scale of any quantum-real particle (or could be any other scales of particles or systems-of-particles with similar configurations of corresponding $\Delta s, \Delta s_u, \Delta t$ & Δt_u in all corresponding Eqs. (1.17), (1.18), (1.21) & (1.23) if its corresponding quantum-real CIPs like radius $\Delta r \neq 0$ or ∞ and inverse anti-radius $\Delta \lambda \neq \infty$ or 0 for the same (quantum-real whirl). As a result, the quantum-real space and anti-space for that whirl will be in corresponding Eqs. (1.17) & (1.18) from Eqs. (8.1) & (9.2) as usual

$$\Delta s = \tfrac{3}{4}\pi(\Delta r^3 \neq 0 \; or \; \infty), \quad (10.1)$$

$$\Delta s_u = \tfrac{3}{4}\pi(\Delta \lambda^3 \neq \infty \; or \; 0), \quad (10.2)$$

and similarly, the quantized time and anti-time as equivalent to respective left and right-handed circumferential rotations of the same quantum-real whirl in Eqs. (1.21) & (1.23) would be correspondingly as

$$\Delta t = 2\pi(\Delta r \neq 0 \; or \; \infty), \quad (10.3)$$

$$\Delta t_u = 2\pi(\Delta \lambda \neq \infty \; or \; 0). \quad (10.4)$$

Now, the left-handed circumferential area of the respective left-handed volume of that quantum-real whirl of the Non-quantized-virtual state of highest-energy, that comprises Non-quantized-energy continuum, for the per unit rotation of (LAR) circumferential distance (as 1 unit of scale-specific quantum-real time Δt) would be equivalent to corresponding left-handed continuum of the quantum-real whirl when $\Delta m \neq 0$ or ∞ as

$$\Delta m = \pi \, \Delta r^2 / 2\pi \, \Delta r = \tfrac{1}{2}(\Delta r \neq 0 \text{ or } \infty), \qquad (10.5)$$

where, the Δm have no conventional unit like gm that was in earlier Eq. (1.1). But instead of that, due to the unit of Δr in cm, the same gm would have also a non-conventional unit in cm in Eq. (10.5) as like as quantum-real time Δt.

Conversely, the right-handed circumferential area (of RAR) of the corresponding right-handed volume of the same quantum-real-whirl out of that Non-quantized-virtual state of highest-energy state, logically, that would also comprise similar or any right-handed form of same Non-quantized-virtual state of highest-energy, for the per unit rotation of right-handed circumferential distance (as 1 unit of scale-specific quantum-real anti-time Δt_u) as equivalent to corresponding right-handed-inertia of the same quantum-real-whirl when $\Delta v \neq \infty$ or 0 as

$$\Delta v = \pi \, \Delta \lambda^2 / 2\pi \, \Delta \lambda = \tfrac{1}{2} (\Delta \lambda \neq \infty \text{ or } 0) \qquad (10.6)$$

where Δv as quantized parameter would not possess its conventional unit in $cm.sec^{-1}$ in Eqs. (1.14) & (2.14). Instead, it would have unit in cm due to the unit of $\Delta \lambda$ in cm in same Eq. (10.6).

But the units of all other CIPs like $\Delta s, \Delta s_u, \Delta t$ & Δt_u in above corresponding Eqs. (10.1), (10.2), (10.3) & (10.4) would remain be unchanged compare to Eqs. (1.17), (1.18), (1.21) & (1.23). As a result, the UCs like k_1, k_2, k_6 & k those have such newer units of Δm & Δv in all corresponding earlier Eqs. (2.1), (2.14), (2.23) & (2.28) need to be re-write accordingly. The k_1 would have unit in cm^2 instead of earlier $gm.cm$. The k_2 would have also unit cm^2 instead of earlier $gm.cm.sec^{-1}$. The k_6 would have its unit in cm^2 instead of earlier $cm^2.sec^{-1}$, and finally the k would have unit cm^{10} instead of earlier $gm.cm^9.sec^{-1}$.

However, the Eq. (10.5) also defines the intrinsic quantization in Non-quantum-virtual 'continuum' as mass-energies or as earlier mentioned as intrinsic-quantized inertial mass-energies in Eqs. (2.1) & (2.14) ultimately as left-handed circumferential

(LAR) area of that confined continuum of Non-quantized-energy or Non-quantized-virtual state of highest-energy within the LAR of the specific quantum-real-whirl. As described in last Chapter, that intrinsic-quantized inertial mass-energies would have causality as well as objectivity to any similar quantum-real-objective-observers like us after its formation or creation. But never could be exchanged before that formation or creation due to Non-causality and subjectivity within Non-quantized-energy or Non-quantum-virtuality of continuum.

Conversely, the Eq. (10.6) describes the intrinsic quantized inertia of motions or intrinsic quantized inertial motions in Eqs. (2.14) & (2.23) basically as right-handed circumferential area of the RAR confinements of the Non-quantized-energy within the simultaneous RAR of the same quantum-real-whirl.

Hence in Eqs. (10.1) – (10.6) for all 5+5 inverse 10-CIPs of such a quantum-real-whirl of subjective VVSE become equivalent to 5+5 inverse 10-CIPs for every quantum-real particles or systems-of-particles in Eq. (2.28). The same Eq. (2.28) for all scales of quantum-real particles or systems-of-particles is appearing equivalent to one quantum-real-whirl through all corresponding inverse relations in Eqs. (2.1), (2.14), (2.17, (2.19), (2.21) and (2.23) among 5+5 inverse 10-CIPs. Subsequently, those could finally deduce in most primary forms into one ultimate configuration for quantum-real everything but with total 10-quantum-real spatial dimensions i.e. with all 10-spatial units of dimensions as cm^{10}

$$\Delta r^5 \cdot \Delta \lambda^5 = k/\left(\tfrac{3}{4}\pi^2\right)^2 = k/\psi, \quad (10.7)$$

where $\Delta r \neq (0 \text{ or } \infty)$ as well as $\Delta \lambda \neq (\infty \text{ or } 0)$, and the constant $\psi = (\tfrac{3}{4}\pi^2)^2$ having no unit. Then, in Eq. (10.7), any quantum-real particles or systems-of-particles as quantum-real-whirls of objective unified quantum-real-state-of-energies (UQRSE) but would could have subjective creations from or destructions into that Non-quantized-virtual state of highest-energy continuum has defined in Eqs. (9.1) & (9.2).

As described in earlier Chapters 5 & 6, the unified Eqs. (2.28) & (2.31) are described as infinitesimally inertial due to all discrete rate of changes in accelerations. So, there have all discrete inertial expressions for the non-inertial unified Eqs. (6.11), (6.12) & (6.15) foe every scale of quantum-real particles or systems-of-particles for such UQRSE in Eqs. (8.1) & (8.2). Therefore, the ultimate 10-dimensional spatial definition for unified Eqs. (2.28) can also define through same as in Eq. (10.7). Therefore, the Eq. (2.31) can ultimately appear in such 10-dimensional cm^{10} mirror-imaged form

$$\begin{bmatrix} (0 < \Delta r < \infty)^5 \cdot (\infty > \Delta \lambda > 0)^5 = k/\left(\tfrac{3}{4}\pi^2\right)^2 = k/\psi \\ (\infty > \Delta \lambda > 0)^5 \cdot (0 < \Delta r < \infty)^5 = k/\left(\tfrac{3}{4}\pi^2\right)^2 = k/\psi \end{bmatrix} = \Delta \Gamma$$

(10.8)

for duality in every quantum-reality in Eq. (8.1). In non-inertial unified Eq. (6.15) with such mirror-imaged quantum-reality can also be replaced those two fundamental 2-CIPs like $\Delta r \neq (0\ \&\ \infty)$ and $\Delta \lambda \neq (\infty\ \&\ 0)$ through all corresponding inverse Eqs. (2.1), (2.14), (2.17), (2.19), (2.21) & (2.23) whenever needed to get non-inertial unified mirror-imaged quantum-reality through such 10-Dimensional spatial co-ordinates for the inertial in Eq. (10.8).

5. Quantum-Reality from the Virtuality

In above Sections 3 & 4, each of those quantum-realities, as any particles or systems-of-particles irrespective of scales, have imagined in most basic forms as 'whirls' of confined Non-quantized-virtual highest-energy continuum VVSE. But subjective creation / destruction of any such whirl out of that VVSE never could be causally predicted from UQRSE. Because, that creation / destruction occurs with zero & infinity magnitudes of 5+5 inverse 10-CIPs of VVSE which would be always non-zero & non-infinity magnitudes of same 5+5 inverse 10-CIPs of signals within UQRSE to exchange with such event of creation /destruction. But after creation out of or before destruction within such event in VVSE the same event would be one causal or causal sequence within exchange limit of quantum-real signals within UQRSE.

However, one such quantum-real whirl that confines the Non-quantum continuum of highest energy-state would have also any specific scale of particle or system-of-particle with mirror-imaged duality in Eq. (10.8).

Although, after subjective creation of one such whirl, the same would follow a (quantum-real) causal sequence (objectively) from its start to the end. But prior to that creation as well as destruction at any point of same causal sequence seems to be a non-causal or subjective process. That seems dependent on subjectivity of the Non-quantum continuum of that virtual-state-of-energy state or VVSE beyond the communicable limit of any quantum-real signal (even the gravitational exchange limit) in UQRSE.

However, the same creation and destruction of such quantum-real whirls has defined in Eq. (9.2). But, in the Eqs. (10.5), and also conversely in Eq. (10.6), it has revealed that a quantum-real whirl actually is a confinement of 'something', what we term as Non-quantized-energy having a non-quantized continuity (beyond all quantum-real signals exchange limits of the quantum-real observers like us).

That micro-most scale of whirl can assume to fill its left-handed & right-handed 'volumes' Δs & Δs_u with Non-quantized highest flow of continuum energy as VVSE with left-handed & right-handed duality.

The left-handed energy confinement in that whirl, which had initially $\Delta s = 0$ volume for that Non-quantized continuum of highest energy-state, could induce such emerging volume $\Delta s > 0$ to start rotating in left-handed direction. Simultaneously, the right-handed confinement of energy within same whirl, which had initially $\Delta s_u = \infty$ volume could induce to diminish in anti-volume to $\Delta s_u < \infty$ to start rotating with quantized value bit slower than the infinity value of right-handed direction.

Subsequently, that simultaneous discrete confinement of continuum-energy of VVSE as quantum-real-energy from Eq. (10.8) would form remaining 4-spatial dimensions or as the time Δt, antitime Δt_u, matter Δm and motion Δv like 4-CIPs. However, the magnitudes of all those quantum-real 5+5 inverse 10-CIPs would be compounded with the increments' onward micro to macro scales of such whirls or PSs.

But the Eq. (10.5) and conversely the Eq. (10.6) for the Eq. (10.7) or (10.8) as any such quantum-real whirl also has revealed that our so-called realization about quantum-reality seems to have no such *real significance* [1] in its most fundamental state. Such a quantum-reality or even a unified quantum-real-state-of-energies is rather confinements of virtuality or continuum-energy within some whirls of same. That is, everything as quantum-real are actually something out of the nothing or as if like the illusions of nothing. The quantum-reality is appearing as a delusion of something continuous we termed as Non-quantized-energy as Non-quantized-virtual-state-of-energy or Non-quantum-virtuality with no real sense of any finite quantized space, time, matter in infinite anti-space, anti-time & motion beyond any perceptions of quantum-real cognizable-mind of observers like us.

Oppositely, from the virtual end of the Eqs. (9.2), (10.5) & (10.6) anything like that quantum-real type of whirl could remain be any Non-quantized-virtual type including the quantum-real observers like us. An observer, as a part of that VVSE and similar Non-quantum-virtual in type with such highest continuum-energy state, would perceive all those whirls comprising the UQRSE and confining such Non-quantized continuum-energy as similar Non-quantum-virtual in type. As

---

1. a 'whirl' that would originate in Non-quantized-energies and filling with Non-quantum-virtuality made by something that can't be known beyond the limits of all our exchangeable signals.

one observer, for example made out of the liquid-water or H_2O molecules under thermal state of higher-energy, can perceive one or all ice-crystals are made of similar molecules but under lower thermal energy state.

6. Magnitudes of Quantum-Real Limits

The quantum-reality has appeared obviously as not continuous in all quantized magnitudes for the 5+5 inverse 10-CIPs which has described in Eq. (2.28) & (2.31) as well as in Eqs. (10.7) & (10.8) irrespective of scales. In Eq. (10.7) there would be an approximate magnitude for

$$\frac{k}{\psi} = 5.99599 \times 10^{-130} \ cm^{10} \qquad (10.9)$$

and that would be ultimate smallest possible value for every 10-Dimensional quantum-reality in quantized spatial length for entire unified quantum-real-state-of-energies UQRSE in Eqs. (8.1) & (8.2). That can also assume as minimum quantum-real *Length* and below of which there seems to be a territory of all 'continuum' of Non-quantized-energy as Non-quantized-virtual-state-of-energy of Non-quantum-virtuality with all zero & infinity magnitudes for 5+5 inverse 10-CIPs. That is, a quantum-real whirl in its most fundamental level or for one particle of micro-most scale emerges through subjectivity of Non-quantized-virtual-state-of-energy of Non-quantum-virtuality approximately to have a quantized value for its $\Delta r = 5.99599 \times 10^{-130} \ cm$.

Consequently, any quantum-real signal, if trying to exchange beyond that quantum-real Length, cannot exchange with the Non-quantized-energy as VVSE in physical nature. Because, the physical nature as if imposes that quantum-real limitation for us as quantum-real-observer in its UQRSE as objective cognizable fold. In other way, since there cannot be possible to exist any quantum-real Length or space beyond that limit as in above Eq. (10.9), therefore gravitation is also limited to have any effects beyond that spatial Length for any quantum-real space.

Then, if there, for $\Delta r = 5.99599 \times 10^{-130} cm$ as smallest quantum-real Length as any quantum-real radius for any quantum-real-whirl or say for the micro-most scale of particle in cognizable fold of physical nature opposite of the macro-most scale as the BBCOU, then from the Eq. (2.17) there would have inverse quantized magnitude for its anti-radius or de Broglie wavelength with approximate magnitude

$$\Delta \lambda = \frac{k_3}{\Delta r} = 1.167875 \times 10^{104} \; cm \; . \tag{10.10}$$

That would be assumed as mirror-imaged scale-specific smallest quantum-real value of anti-radius that could create from or destroy into the VVSE for any micro-most scale of particle or whirl.

However, any such smallest quantum-real value for $\Delta r = 5.99599 \times 10^{-130} cm$ (for its inverse antiradius $\Delta \lambda = 1.167875 \times 10^{104} \; cm$) would be also the conceptual maximum extent of diminished or collapsed quantum-real radius for BBCOU after Bigcrunch but before the moment of next Bigbang. Because, one such smallest quantum-real value for any smallest scale of particle could exist after that immense collapse opposite to macro-most scale of one BBCOU.

Conversely, the same could achieve by macro-most scale of BBCOU before the start of its moment of Bigcrunch. When the inverse quantized magnitude for the anti-radius will be inversely reduced to $\Delta \lambda = 5.99599 \times 10^{-130} \; cm$ with inverse radius of $\Delta r = 1.167875 \times 10^{104} \; cm$ as maximum expansion radius of space.

Consequently, the whole BBCOU being the macro-most scale of particles or systems-of-particles must oscillate in-between spatial radii from its maximum smallest magnitude $\Delta r = 5.99599 \times 10^{-130} \; cm$ in one end to maximum biggest magnitude in other $\Delta r = 1.167875 \times 10^{104} \; cm$. Therefore, the smallest radius $\Delta r = 5.99599 \times 10^{-130}$ cm would be the radius for universal quantum-real whirl at the moment of its creation or destruction within in VVSE for breaking of that cyclic causal sequence.

In other senses, the same $\Delta r = 5.99599 \times 10^{-130}$ cm for smallest possible radius of the universe as BBCOU would have magnitude $\Delta \lambda = 1.167875 \times 10^{104}$ cm during its moment of Bigbang when its maximum possible biggest inverse anti-radius would have magnitude $\Delta \lambda = 1.167875 \times 10^{104}$ cm.

Similarly, from the smallest quantized magnitude for anti-radius or de Broglie wavelength in Eq. (10.10), there would be also a smallest possible mass-energy limit for any scale of quantum-real particle or system-of-particles from Eq. (2.1) as

$$\Delta m = \frac{k_1}{\Delta \lambda} = 1.79728 \times 10^{-127} gm. \quad (10.11)$$

From Eq. (10.11), there would also be the highest quantized inertial-motion for such smallest quantized mass-energy for the possible micro-most scale of particle or system-of-particles from Eq. (2.14)

$$\Delta v = \frac{k_2}{\Delta m} = 6.2618 \times 10^{120} cm.sec^{-1}. \quad (10.12)$$

All those smallest and highest limits of CIPs with approximate values in Eqs. (10.9)-(10.12) depend on the corresponding approximate magnitudes of k, k_3, k_1 & k_2 in Chapter-2.

7. Ten Spatial Dimensional Duality of Physical Nature

Finally, the whole physical nature enfolds to us in two parts or folds: one is 'unified quantum-real-objective' as UQRSE also as our cognizable part with all non-zero & non-infinity discrete values for 5+5 inverse quantum-real 10-CIPs and other is Non-quantum-real virtual continuum as VVSE as the non-cognizable part with zero & infinity values for same 5+5 inverse Non-quantum-virtual 10-CIPs.

For Non-quantum-virtuality fold of physical nature there can also appear from Eq. (10.8) as well as from the Eq. (9.1) as opposite to Eq. (10.8) as

$$\begin{bmatrix} (\Delta r = 0 \; or \; \infty)^5 \cdot (\Delta \lambda = \infty \; or \; 0)^5 = k/\left(\tfrac{3}{4}\pi^2\right)^2 = k/\psi \\ (\Delta \lambda = \infty \; or \; 0)^5 \cdot (\Delta r = 0 \; or \; \infty)^5 = k/\left(\tfrac{3}{4}\pi^2\right)^2 = k/\psi \end{bmatrix} = {}_\infty^0\Gamma$$

(10.13)

That subjective Non-quantized-virtual-state-of-energy fold with non-cognizablity in physical nature has shown in higher energy state beyond the effects or exchange of universal gravitation & antigravitation compare to any gauge & antigauge forces. Because, gravitation & antigravitation need to have any non-zero & non-infinity quantum-real values of spacetime.

The quantum-real observers like us can able to exchange any quantum-real signals to any similar type observance to appear as objective. Because, that is our limit of quantum-real cognizablity minus the infinity range of subjective fold with all zero & infinity magnitudes for same 5+5 inverse 10-Spatial-CIPs in Eq. (10.8).

But the quantum-real limits, whether are exchanged with the quantum-real signals by quantum-real observers should remain be objective.

As a result, the whole physical nature is sometimes objective up to its quantum-real limits, up to which we can exchange through quantum-real effects of gravitation & antigravitation, and rest out of it is subjective.

Then whole physical nature can be imagined as a web of total 10-spatial sheets instead of 5+5 inverse 10-CIPs from Eqs. (10.8) & (10.13) in a matrix say $\mathbb{N}$ for elements $\Delta\Gamma$ & ${}_\infty^0\Gamma$

$$\mathbb{N} = \begin{Bmatrix} \begin{bmatrix} (0 < \Delta r < \infty)^5 \cdot (\infty > \Delta\lambda > 0)^5 = k/\left(\frac{3}{4}\pi^2\right)^2 = k/\psi \\ (\infty > \Delta\lambda > 0)^5 \cdot (0 < \Delta r < \infty)^5 = k/\left(\frac{3}{4}\pi^2\right)^2 = k/\psi \end{bmatrix} = \Delta\Gamma \\ \begin{bmatrix} (\Delta r = 0 \text{ or } \infty)^5 \cdot (\Delta\lambda = \infty \text{ or } 0)^5 = k/\left(\frac{3}{4}\pi^2\right)^2 = k/\psi \\ (\Delta\lambda = \infty \text{ or } 0)^5 \cdot (\Delta r = 0 \text{ or } \infty)^5 = k/\left(\frac{3}{4}\pi^2\right)^2 = k/\psi \end{bmatrix} = {}^0_\infty\Gamma \end{Bmatrix}.$$

(10.14)

The same physical nature, in Eq. (10.14), ultimately would be a continuum of 5+5 inverse 10-Spatial-Dimensions. Where the same in lower energy states as unified quantum-real-state-of-energies would appear as objective with causal sequences with duality of quantum-reality. But alternately where the same would be in highest energy-state like Non-quantized-virtual-state-of-energy or VVSE could be subjective with all non-causality or broken causal sequences.

Summary

The Non-quantum-virtuality, that has deduced from the inverse relationships of unified quantum-reality in previous Chapter-9, has assumed equivalent to Non-quantum-energy (of todays physics). Both are involving all zero & infinity continuous magnitudes of same 5+5 inverse 10-common-internal-parameters and are assumed to belonged in higher state of energy compare to one in quantum-reality of mass-energies. The quantum-reality, in its most fundamental micro-most scales are with all non-zero & finite discrete magnitudes of 5+5 inverse 10-CIPs. In current Quantum Mechanical understandings, it assumes that all quantum-real particles or systems-of-particles can create from & destroy into the Subjectivity of Non-quantized-virtual-state-of-energy. Those quantum-real particles or systems-of-particles are with all non-zero & non-infinity discrete magnitudes of 5+5 inverse 10-CIPs while in contrary to the Non-quantized-energy possessed all zero & infinity continuous magnitudes for same 5+5 inverse 10-CIPs. Then, how a zero & infinity continuous magnitude (for Non-quantized-energy in higher state of energy) could transmute into non-zero & finite discrete magnitude (for quantum-reality in lower state of energy) or vice versa?

In this Chapter-10, finally the quantum-reality, in its micro-most form of scale as discrete particles, has described as subjective 'whirls' of Non-quantum-virtuality within the range of our conventional geometric configurations. As a result, that quantum-reality as one such whirl with all non-zero & non-infinity magnitudes of spatial extensions correspond to all its 5+5 inverse 10-dimensions of only space & anti-space out of the zero & infinite continuity of Non-quantum-virtuality i.e. Non-quantized-energy as Non-quantized-virtual-state-of-energy. There transmutation of that 'continuity' into the 'quantization' could be better explained through such convention of whirl. Consequently, that 'quantization' in unified quantum-reality is a result of subjective 'continuity' in Non-quantum-virtuality, but that 'quantization' would not be a 'continuous'.

In such a way, the approximate smallest quantum-real limits of observations for the observers like us roughly would have the smallest values for $\Delta r, \Delta \lambda, \Delta m$ & Δv in corresponding Eqs. (10.9), (10.10), (10.11) & (10.12) for exchange of any quantum-real signal. That is up to a limit of the gravitational force, involving non-zero & non-infinity spacetime and mass-energy, can influence or exchange in physical nature. A consequent non-zero & non-infinity, 5+5 inverse 10-spatial dimensional space & anti-space, and a unified definition for the entire unified quantum-reality has deduced in Eq. (10.7).

Present Chapter has ultimately unified the entire physical nature with its two folds starting from the Chapters 1 & 2 stepwise through a series of other eight Chapters. Those initial Chapters 1 & 2, once begun with some simple assumptions, based on common observational comprehensions in todays observational physics, by assuming **(i)** total ten common-internal-parameters among diversity of scales and **(ii)** their respective seven inverse relationships and constants. However, all these assumptions were basically made on the basis of understandings those are involved in the realms of visible matters in BBCOU that estimates now merely 5% where rest 95% invisibles, beyond the exchange range of 'light' but within effects of gravitation, as dark matters & dark energies. Next, in the Chapter-6 it has also appeared that in that totality of BBCOU, as unified quantum-reality (unified quantum-real-state-of-energies), possesses its simultaneous 'duality' of LE-C & RE-C where observers like us are integrated along the LE-C only, and who never could see the RE-C through exchange of any similar quantum-real LE-C types of signals. Then, Chapter-9 has finally deduced an entire realm of Non-quantum-virtuality beyond that unified quantum-real duality. Then finally, the whole physical nature that has ultimately appeared to us with two folds – "unified quantum-real duality" and "Non-quantized-virtual".

Therefore, such physical nature never could be exchanged directly by the observers like us who is an integrated part of the unified quantum-real-state-of-energies as unified-quantum-real duality. Moreover, within the same unified quantum-real duality, the observers like us cannot also exchange directly with anything onward RE-C, since those observers like us are integrated parts along the LE-C direction. Even, within LE-C part of the unified quantum-real duality, we can observe directly only 5% of its visible matters through exchange of 'light'.

So, our scopes of direct quantum-real-observations in the totality of entire physical nature also seems very much restricted and microscopic through such non-zero & non-infinity limited signal-exchange capacities, and also such restriction does not mean that such an observer like us do not having their sufficient level of current technological achievements. Instead, such restriction actually seems imposed by the physical nature for us which never could be over ridden. But in other side, such restriction would not make any sense in way of observations of any so-called Non-quantized-virtual observers in same physical nature with all zero & infinity capacity of any signal-exchange.

CONCLUSION

> "The laws of nature are but the mathematical thoughts of God."   - Euclid

Finally, the entire issues in monograph has unfolded in total ten chapters. Although, each of those Chapters are apparently independent but are closely interlinked. The Chapter-1 has considered few universal phenomena, based on some present observational comprehensions in current physics mostly have emerged in last one hundred years or so, as the new basis for any potential theoretical model. That can unify the present physics by linking Theories of Relativity and Quantum Mechanics. The Chapter-2 is the core postulates part of the model that found seven universal invariant new inverse-relations with seven new invariant inverse constants for everything having discrete magnitudes as well as non-zero & non-infinite values. The rest of the Chapters from 3 to 10 are actually all eight sets of inferences. Those inferences have either resolved some existing inconsistencies in current physics or made lot of new predictions about the physical nature. Most importantly, those inferences have never challenged any of the predictions made by Special Relativity Theory, General Relativity Theory as well as Quantum Mechanics. But some of those new inferences have added more extended predictions with those theories.

The monograph has considered some earlier presumptions differently depending on present understandings without changing the same. The earlier ideas of 'constancy in magnitude of inertial speed of light' has assumed instead as the 'quantized magnitude of inertial speed of a photon-particle' in Special Relativity Theory. Earlier ideas of 'material-bodies' in General

Relativity Theory as precise 'systems-of-particles'. These small alterations never have altered anything in the respective foundations of those two Theories but outcomes of same are huge in Chapters 3 & 4. Subsequently, the General Relativity Theory has equated to the Standard Model of Particle Physics of Quantum Mechanics in Chapter-5.

Any event has assumed as inversely linked total 5+5 common-internal-parameters as 10-dimensions compare to merely 3+1 space & time 4-dimensional.

The Chapters 6 to 8 have defined all those particles or systems-of-particles have unified in previous Chapter as quantum-reality or unified quantum-reality. Each of those quantum-reality would have simultaneous left-handed & right-handed duality, always as causal events and independent of observations or objective in type. There also appeared to have all quantum-real limits of the observers like us, signals up to gravitational range of effects and all similar exchangeable observances as integrated parts of such unified quantum-reality. Chapter-9 has inferred one non-causal and observation-dependent or subjective virtual state of non-quantized continuum beyond that unified quantum-reality. That is with no space, no time, no matter but infinity motion in one infinite anti-space and infinite anti-time in physical nature. Consequently, the physical nature has ultimately appeared with two folds in Chapter-10: one is quantum-real and other as virtual., and the quantum-reality is nothing but the virtual whirls.

Another interesting outcome of this unified model is that it has defined cognition or intelligent-mind of observers like us as quantum-real in type. That is objective in type rather than anything subjective. The quantum-real description of cognition in observers like us has defined as if an unfolding of quantum-real algorithms in Chapter-8. Because it cannot observe or think anything or event that has any zero and/or infinity values for same 5+5 inverse 10-Dimensions. Therefore, any quantum-real observer like us must not have any subjectivity. As a consequence, if there any such quantum-real subjective observer in physical nature that could not be the integrated part of the unified quantum-reality in physical nature. The subjectivity is the part of Non-quantum-virtuality beyond exchange limit of such quantum-real cognition of observers like us. Observer who is integrated with Non-quantum-virtuality could possess a subjective mind with *will* of all broken sequences or non-causal logics. Therefore, one such observer, with subjective mind or 'will' and all zero & infinity magnitudes for its same 5+5 inverse 10-CIPs, cannot be the integrated and exchangeable parts of the unified quantum-reality described in Eq. (6.15). Alternately, one observer, with objective cognition and all non-zero & non-infinity magnitudes of same 5+5 inverse 10-CIPs, cannot be the subjective for its integration with unified quantum-reality defined in Eq. (6.15).

The Physical laws are Causal Laws particularly through Chapters 8 to 10. The unified quantum-reality fold of physical nature has all discrete values of non-zero & non-infinite 5+5 inverse 10-dimensions. The same is deterministic and objective due to 7 inverse invariant UCs, and are followed by all cause & effect sequence of Causal Laws.

The Non-quantum-virtuality fold has all non-quantized values of zero & infinite same 5+5 inverse 10-dimensions. Non-quantum-virtuality seems to have higher energy state compare to quantum-reality, and beyond the communication of any quantum-real-signals, as indeterministic and subjective and followed by the Non-causal Wills. But spreads within puzzling zero volume of space, no tick of time and no matter but infinity of anti-space, anti-time and motion.

The notion about the physical nature, that has emerged ultimately from the present observational comprehensions of physics in Chapters 8-10, also sounds in broader sense similar to the notions of *Brahman* in philosophical texts of *Vedanta* over the millennia before. Where the term Brahman has described as a state of *Chaitanya* (like a pure state of consciousness or something that has Will) and also to have two folds: the *Nirguna Brahman* and *Saguna Brahman*. The Nirguna Brahman (like a state of Non-quantum-virtuality) is subjective, intangible and inexplicable. That never could be physically measured in terms of any so-called finite measuring units of spatial magnitudes, finite count of ticks in clocks or time, and finite form of matter. Although has assumed to 'spread' up to the infinity with an infinity of 'motion'. The Saguna Brahman (alike entire unified quantum-reality i.e. ΔM_6) is objective, tangible and explicable through all similar non-zero & non-infinity values of space, time, matter & motion.

Are such resemblances in notions of Brahman as well as todays ideas of physical nature because of our identical process of quantum-real (algorithmic) thinking in spite of distances in thousands of years of time? Also, because, are both of those as integrated parts of the same unified quantum-real fold of physical nature with similar algorithm of cognition as mentioned in Chapters 8 to 10? Or, is such similarity in two notions merely a coincidence?

Another important thing that can realize within entire unified quantum-reality of Chapters 6 to 8 is about our maximum range of direct scope of observations within it like an observer. Because one quantum-real-observer like us, obviously never could directly exchange signals to observe the realm of Non-quantum-virtuality in same physical nature. That is, one entire fold of physical nature never can be directly observed. Next can be within the own realm of unified quantum-reality. The half of it is one right-handed-entropy part or RE-C part which never can be directly observed from left-handed-entropy part or LE-C part where we are integrated except some equations. Even, within that LE-C part, we can directly observe only the range of visible-matters, those are comprising all those scales of particles or systems-of-particles, which have estimated so far as 5% of total LE-C part of quantum-reality. Therefore, all the outcome equations, those have construed from (1.1) to (10.14) in all preceding Chapters 1 to 10, are actually the consequences of a fraction of that tiny 5% for the visible-matters through direct observations in entire physical nature. Therefore, we have very limited options to observe entire physical nature directly beyond that 5% of visible-matters except perceiving it only through some logics behind its causal laws alike Eqs. (6.15), (8.1) & (10.7).

Even such range of our logical basis for causal-logic with objectivity also seems to has 'no use' in the realm of non-causal subjectivity of Non-quantum-virtuality in Eq. (9.1) beyond that unified quantum-reality. But if there would have any form of physics within that subjectivity of Non-quantum-virtuality remain be incomprehensive to us due to our quantum-real exchange restrictions to communicate with it for any limited objective-cognition of us.

Finally, any physics that might be within that subjective Non-quantum-virtual continuum seems not to have our like Causal Laws but instead all Non-causal Wills. But that Wills of 'what' or 'whom'? Is it the Will of any physical-entity within mere a higher-energy-state continuum? Or, is it the Wills of God? And, God's Physics beyond the edge of quantum-reality?

REFERENCES

[1]. Bhunia, Dipak Kumar. "A Common Definition For All Particles in Nature". *Galilean Electrodynamics*, vol.24, no. 04, 2014, pp.41-53.

[2]. Bhunia, Dipak Kumar. "Quantized Curvature of Spacetime in all Scales of Gravitating Bodies". *Galilean Electrodynamics*, vol.28, no. 05, 2017, pp.49-63.

[3]. Bhunia, Dipak Kumar. "A Unified Definition for all Basic Forces in Nature". *Galilean Electrodynamics*, vol.29, no. 01, 2018, pp.3-1.

[4]. Linda T. Elkins-Tanton, et all, *"Planetesimals – Early Differentiation and Consequences for Planets". Cambridge University Press.* 2017. ISBN 9781107118485.

[5]. Nimtz, Dr. Gunter and Stahlofen, Dr. Alfons. "Photons flout the speed limit". New Scientist, 2617 (17 August 2007).

[6]. Barrow, J.D. and Silk, J. Prologue. "The Left Hand of Creation: The Origin and Evolution of the Expanding Universe.pub. by William Heinemann Ltd, London, UK; 1983 (1st edn.), SBN 434047600, Prologue.

[7]. Narlikar, Jayant V. (1960). *"Elements of Cosmology"*. Universities Press. ISBN 81-7371-043-0.

[8]. *"The Equivalence Principle"*, at Mathpages. www.mathpages.com

[9]. Pössel, Markus, AEI, an Article by. *"The elevator, the rocket, and gravity: the equivalence principle"*. Einstein online, https://www.einstein-online.info

[10]. Einstein, Albert. *"The Foundation of the General Theory of Relativity"*. Annalen der Physik. **354** (7): 769. 1916.

[11]. Escape Velocity. Wikipedia. http://en.wikipedia.org/wiki/Escape_velocity

[12]. Kutner, Marc. *"Astronomy: A Physical Perspective"*. Cambridge

University Press, pp. 148. 2003.nISBN 9780521529273.
[13]. "The matter-antimatter asymmetry problem", *CERN*. Retrieved April 3, 2018.
[14]. Einstein, A.; Podoslsky, B.; Rosen, N.. *"Can Quantum-Mechanical Description of Physical Reality Be Considered Complete?" Physical Review*. American Physical Society.
[15]. Rafelski, Johann and Muller, Berndt. *"The Structured Virtual – thinking about nothing"*. (free pdf copy). 1985. ISBN 3-87144-889-3.
[16]. Chester, Marvin. *"Primer of Quantum Mechanics"*. John Willey. 1987. ISBN 0-486-42878-8.
[17]. Battersby, S. (2008). *"It's Confirmed: Matter is Merely Virtual Fluctuations"*. newscientist.com. *New Scientist*. Archived from the original on 27 May 2017.
[18]. Cheng, T.-P.; Li, L.-F. (1983). *"Gauge Theory of Elementary Particle Physics"*. Oxford University Press. ISBN 0-19-851961-3.

GLOSSARY

Absolute Rest: In Newtonian Mechanics it was assumed that there would be a State of Inertia for one Inertial Frame of Reference respect which all other Inertial Frames of Reference would have relative motions, i.e. universal aether.

Absolute Motion: The inertial-motion which is independent to the motions of all other relatively moving material-bodies or inertial frames of reference assumed basic foundation of Special Relativity Theory, i.e. inertial speed of light c.

Absolute Value: The value that would not be a function of any other value.

Acceleration, Quantized: As the infinitesimal discrete rate of scale-specific changes in quantized inertial-motions of one accelerating particle or system-of-particles onward direction of applying forces.

Acceleration, Quantized Gravitational: The Quantized Acceleration that occurs under effects of gravitational forces on same accelerating particles or systems-of-particles.

Acceleration, Quantized Non-Gravitational: The Quantized Acceleration that occurs under effects of non-gravitational forces like electromagnetic and nuclear forces on same accelerating particles or systems-of-particles.

Algorithm, Quantum-Real: Every living or non-living quantum-real entity as integrated part of the entire quantum-real-fold of the physical nature, that includes the whole universe as macro-most scale, would have cyclic sequence of causality. That causal sequence seems like an objective algorithm pre-fixed to unfold from its beginning to end. Then, after end, through permutation and combination of such algorithm, fresh beginning to another end occur to complete the cycle.

Algorithm, Unfolding of Quantum-Real: Progresses in 'Life' sequence of any quantum-reality would be the unfolding of any quantum-real algorithm.

Astronomical Objects: Anything as quantum-real particles or systems-of particles as integrated parts of the universe.

Astronomical Objects, List of: The different micro to macro scales of particles or systems-of-particles those are observed / perceived to exist in astronomical space of universe.

Axial-Rotation, Left-Handed: Axial rotation of every particles or systems-of-particles in intrinsic left-handed direction.

Axial-Rotation, Right-Handed: A conceptual simultaneous axial rotation of all same particles or systems-of-particles in opposite intrinsic right-handed direction.

Basic Forces: Gravitation and three other non-gravitational forces in physical nature. The gravitation as curvature of spacetime of one particle or system-of particles and other three non-gravitation forces e.g. electromagnetic, strong and weak nuclear forces as material or gauge-fields of forces.

Broken Sequence of Causality: With no cause and subsequent effects for any event.

Causality: With all causes and subsequent effects for occurring events.

Causal Sequence: All causes and effects of any occurring event in cycle from beginning to end and end to beginning without breaking of the chain.

Causal Laws: The predictable quantum-real theorems follow ultimately any quantum-real algorithm.

Causality, Non-: The occurrence of an event with no sequence of cause and effect, i.e. with no predictability.

Causal Will, Non-: The physical laws with no predictability.

Centre-of-Mass: The point where parallelly moving two unescaped identical particles would converge in a gravitational **field of gravitating body.**

Cognition, Quantum-Real: The quantum-real origin of the intelligence for any entity that having integration with unified quantum-reality fold of physical nature.

Cognition of Quantum-Real Observers: The intelligence of observers like us who are integrated with unified quantum-realty fold of physical nature and followed specific quantum-real algorithm from its beginning to end.

Cognition, Psycho-Neural Basis of: The intelligence of any quantum-real observer like us on the basis of exchange of electro-chemical impulses among neurons.

Cognizable Mind, Quantum-Real: The thinking or analysing process or mind itself that has only quantum-real basis in terms any non-zero & non-infinity magnitudes limitation of 10-CIPs, but no practical involvements with any zero & infinity for same 10-CIPs.

Common-Internal-Parameters (CIPs): The ten-dimensional web of any event in physical nature in the form of either for matter-1, motion-1, space-3, antispaa-3, time-1 and anti-time or for spatial-5 and anti-spatial-5 co-ordinates.

Constants (SSUCs), Scale-Specific Universal: Each of the CIPs has scale-specific quantized magnitude that is universally invariant to all similar scales of particles or systems-of-particles in universe.

Constants (UCs), Universal: The seven numbers of inverse constants like k_1, k_2, k_3, k_4, k_5, k_6 & k are universally invariant irrespective of changes in scales of the particles or systems-of-particles in physical nature.

Continuum, Non-Quantized: Other fold of the physical nature beyond all communication limit of the quantum-real fold that has all zero & infinity magnitudes for all those same 10-CIPs or Spatial-5 and Antispatial-5 co-ordinates for an event, i.e. with no space, no time & no matter but infinity anti-space, infinity anti-time & infinity motion or vice versa.

Curved Spacetime, Scale-specific: Every gravitating body has scale-specific magnitudes as any scale-specific particle or system-of-particles. So, each of those scales would have also scale-specific spacetime for its scale-specific matter.

Cyclic Oscillation of Universe, Left-Handed: Cyclic Oscillation of Bigbang-Bigcrunch Universe parallel to its direction of increment-decrement in entropy.

Cyclic Oscillation of Universe, Right-Handed: Conceptually, a simultaneous Cyclic Oscillation of Bigcrunch-Bigbang Universe parallel to its direction of decrement-increment in anti-entropy.

Determinism: Predictability in the causes and effects in occurrence of any event.

Dimensions, Unfolded: Those dimensions in co-ordinate system are tangible or separately exchangeable. Apart from those 3+1 dimensions of space and time other six dimensions like 1 for matter, 3 for ant-space, 1 for motion and 1 for anti-time can be separately defined in any exchange.

Discrete Values: The intrinsic non-continuous values for all scale-specific particles or systems-of-particles.

Duality, Quantum-Real: The simultaneous left-handed and right-handed mirror-imaged existence as well as similar intrinsic values of every particles or systems-of-particles.

Electromagnetic Spectrum: That starts with shortest wavelength of a heaviest mass-energy gamma-ray photons and ends with longest wavelength of a lightest mass-energy radio-wave photons.

Effects of Forces: The non-inertial state where forces are acting on particles or systems-of-particles as a result the same have acceleration or deceleration.

Energy-State, Quantum-Real: The entire state of energies of unified quantum-real fold of physical nature definable by the non-zero & non-infinity quantized values.

Energy-State, Virtual: The assumption of a higher state of energy is comprising unified Non-quantized Continuum fold of physical nature with all zero & infinity values beyond exchange limits of observers like us.

Entropy, Left-Handed: That increases with the increments of expansion after Bigbang of universe.

Entropy, Right-Handed: That decreases with the increments of expansion after Bigbang of universe.

Escape Velocity, Scale-Specific: The scale-specific escape limits of quantized motions for the smallest-bound or unescaped particles within a gravitating-body.

Event, Quantum-Real: That occurs with non-zero & non-infinity quantized value, causality, and objectivity in association any scale of particles or systems-of-particles those are integrated with unified quantum-real fold of physical nature.

Forces, Quantum-Real: Those are involved with all non-zero & non-infinity quantized signals exchanges with similar observances and observers as integrated parts of the unified quantum-real fold of physical nature.;

Forces, Non-Gravitational: Material forces or Gauge-field of forces, e.g. electromagnetic, strong and weak nuclear forces.

Gravitational Field: That describes by Einstein Field Equations referring the gravitating bodies.

Gravitational Field Strength: The amount of curvature corresponds to total mass-energies.

Gravitation, Scale-Specific Quantized: Mass-energies of particles or systems-of-particles as any gravitating-body or gravitationally-shaped-body would have precise scale-specific magnitudes, so corresponding curved spacetime would have scale-specific values i.e. scale-specific values for gravitation.

Gravitation, Anti-: The mirror-imaged right-handed counter part of gravitation (as curved spacetime) for every scale of gravitating-body or gravitationally-shaped-body in anti-spacetime.

Gravitating-Body: Any scale of micro to macro particles or systems-of-particles with quantized mass-energies.

Gravitationally Shaped Body: Macro-scales of systems-of-particles, particularly from the scales of planetesimals and above, where the gravitating-body starts to shape itself under dominating gravitational force over all other quantum-real forces.

Gauge-Fields of Forces: All the quantum-fields of material forces other than gravitation as curved spacetime.

Gauge-Fields of Forces, Anti-: Mirror-imaged gauge-fields of forces.

Hydrostatic Equilibrium: Balancing state of outward forces from gauge-fields and inward from collapsing spacetime in every scale of gravitationally shaped body.

Inertial Frame of Reference: Conceptually isolated one co-ordinate system involving in particle or system-of-particles to measure any physical law or quantity.

Inertial Motion of Light: Intrinsic or observer-independent quantized inertial magnitude for a specific scale of photon in electromagnetic spectrum.

Inertial State: Conceptually state for any particle or system-of-particles in isolation where no force is affecting its state.

Inertial State, Non-: Any particle or systems-of-particles under effects of forces i.e. acceleration or decelerations and in other discrete rate of changes.

Infinity and Zero Values: The values beyond the limits of any quantum-real exchange of signals with non-zero & non-infinity quantized values.

Intrinsic Property: The observer-independent property.

Intrinsic Value: The observer-independent values of anything.

'Life', sense of: A sense of dynamism releases from the unfolding of the quantum-real algorithm within one quantum-real system.

Light, (Constancy in) inertial Speed of: Intrinsic quantized inertial magnitude for a specific scale of photon in electromagnetic spectrum.

Matter (or mass-energy): That has defined as scale-specifically quantized in magnitudes, associated with every quantum-real particles or systems-of-particles, inversely related with quantized motion, and finally revealed as a left-handed circumferential area of the 'whirl' in a Non-quantized virtual continuum and expressed in unit of cm i.e. unit of space.

Matter, Anti-: Mirror-imaged counter part of every matter in particles or systems-of-particles.

Material-Body: Non-scale-specific assumptions of matter in Classical Physics.

Mass: Something like condensed form of energy.

Mass-energies, Inertial: Mass in isolation from all effects of forces.

Mass-Energy, Equivalence: Equivalence of same quantity from mass to energy or vice versa through equation like $E = mc^2$ where c is inertial speed of light.

Mass, Quantized: the discrete values of mass.

Mass, Relativistic: Increments in mass under relativistic increments of non-luminous motions.

Motion, Inertial: Motion in isolation from effects of all forces.

Non-Zero and Non-Infinity Magnitudes: The values within the limits of all quantum-real exchange of signals with non-zero & non-infinity quantized values.

No Radius or Zero State of Radius: The Non-Quantized Continuum State of Virtuality where every quantum-real entity would have no real existence except all zero & infinity values including radius $\Delta r = 0$.

No Matter or Zero Matter: The Non-Quantized Continuum State of Virtuality where every quantum-real entity would have no real existence except all zero & infinity values including material parameter of any particles or systems-of-particles $\Delta m = 0$.

No Space or Zero Space: The Non-Quantized Continuum State of Virtuality where every quantum-real entity would have no real existence except all zero & infinity values including three spatial dimensions of space for any particles or systems-of-particles as $\Delta s\ (\Delta x \cdot \Delta y \cdot \Delta z) = 0$.

No Time or Zero Time: The Non-Quantized Continuum State of Virtuality where every quantum-real entity would have no real existence except all zero & infinity values including any 'tick' of time $\Delta t = 0$.

Objective Event: The occurrence of event that is independent to the exchange of any signals for observation or of observer.

Objectivity, Quantum-Real: Observer or observation independent occurrence of an event and its causal sequence of cause and effects.

Observances, Quantum-Real: The events under quantum-real range of limits in unified quantum-real fold of physical nature can exchange through quantum-real signals by one quantum-real observer.

Observers like us, Quantum-Real: The observers, those are followed by any quantum-real algorithm and quantum-real cognition.

Observational Limit (of us): Quantum-Real limit up to any non-zero & non-infinity quantized magnitudes of all 10-CIPs.

Particles, Anti-: Simultaneous mirror-images of particles.

Particles, Smallest-Bound (Mass-Energy): Those have intrinsic quantized inverse inertial-motion just lower the respective scale-specific escape-velocity to escape out from the gravitating-body.

Particles, Highest-Bound (Inescapable) Motion: Those have intrinsic quantized inverse inertial mass-energy just higher the respective scale-specific mass-energy for escape-velocity to escape out from the gravitating-body.

Particles, System-of-: Scale-specific composite form of micro scales of particles, where a Particle is an Intrinsic discrete form of any material-body with quantum-real values.

Physical Nature: The ultimate form of everything that could have two folds to comprise everything.

Physical Nature, Quantum-Real Fold of: That comprises entire unified quantum-real everything including whole universe.

Physical Nature, (Non-Quantized Virtual) Fold of: That seems like a Continuum of Wills or Subjectivity, beyond the exchange limits of quantum-real materialistic fold wherewith observers like us are integrated.

Quantum-Real (or Quantum-Reality): The intrinsic discrete and integer state of any materialistic existences in the form of any particles or systems-of-particles with non-zero & non-infinity values. Quantum means any integer & discrete value and Reality means up to the limit of all non-zero & non-infinity magnitudes of 5+5 inverse 10-CIPs of such a Quantum.

Quantized Magnitudes $'\Delta'$: The mathematical symbol has used to define any quantum.

Quantum-Real Limit: Up to the limit of any non-zero & non-infinity values for the $'\Delta'$.

Radius, Quantized: The spatial value for any radius having non-zero & non-infinity discrete magnitude.

Radius, Anti-: The mirror-imaged spatial inverse value for any quantized radius is anti-radius having non-zero & non-infinity discrete magnitude.

Relative Rest (or zero relative motions): The zero relative motions in-between two or more relatively moving Inertial Frames of Reference.

Relative Motion: Differences in motion in-between two or more relatively moving Inertial Frames of Reference.

Scale of Particles or Systems: The intrinsic group or class of identical particles or systems-of-particles scattered over universe having corresponding identical intrinsic quantized magnitudes for 5+5 inverse 10-CIPs.

Scale-Specificness, Intrinsic Property of: Since 'scale-specific' magnitude for particles or systems-of-particles is intrinsic property, that could be assumed as Intrinsic Property of Scale-Specificness in every scale of particles or systems-of-particles in universe.

Signals, Quantum-Real: Any Signal that one observer like us or as any quantum-real basic force carrying particles could exchange or interact with other similar kind quantum-real observances is quantum-real signal.

Space, Quantum-Real: Any quantum-real magnitude i.e. non-zero & non-infinity discrete values for three spatial dimensions of space involving with any scale of particles or systems-of-particles. Such space cannot exist without any particles or systems-of-particles.

Space, Anti-: Simultaneous, mirror-imaged inverse quantum-real discrete values of space involving with any scales of particles or systems-of-particles.

Subjectivity: The observer-dependent occurrence of any event. Without intervention of such observer's exchange or Will, the corresponding event never could occur. There would have all broken sequences in occurrence of that event.

Subjective-Observer: The non-causal or observer-dependent event, with broken sequence, that occurs through intervention of such observer.

Virtuality: A Non-quantized Continuum with all zero & infinity magnitudes of same 5+5 inverse 10-CIPs, with all broken causality or subjectivity in occurrence of any quantum-real events out of it in the form its 'whirls', and with no space, no time and no matter but infinite anti-space, infinite anti-time and infinite-motion.

Time, Quantized Sense of: Quantum-real magnitude i.e. non-zero & non-infinity discrete values for all scale-specific 'ticks' of time, and that time cannot exist without any particles or systems-of-particles.

Time, Anti-: Simultaneous, mirror-imaged inverse quantum-real discrete values of quantized time is anti-time that cannot be existed without any scales of particles or systems-of-particles.

Time, Spatial Unit for: Scale-Specific unit for time as unit circumferential rotation would be in spatial distance, i.e. in *cm*.

Universe, Scale of: Macro-most scale of system-of particles in unified quantum-real fold of physical nature.

Universe, Bigbang-Bigcrunch Cyclic Oscillating: From starting from Bigbang expansions and incrementing in entropy assumed to have cyclic oscillations from Bigbang to Bigcrunch then Bigcrunch to Bigbang in a causal cycle.

Unification of Physics: Includes all the forces as well as scales of particles or systems-of-particles within quantum-real limit have appeared to webbed by non-zero & non-infinity discrete magnitudes of 5+5 inverse 10-CIPs and linked through causal sequences. So, entire quantum-real fold of physical nature could be unified through one unified causal equation.

Whirls of Quantum-Reality: The basic forms of any quantum-reality have assumed with ten-spatial dimensions and also in 5+5 inverse 10-CIPs geometric structures. Subjectively occurs out of Non-Quantized Continuum of Virtuality beyond exchange limits of quantum-reality.

Wills, Non-Casual: The Subjectivity of Non-Quantized Continuum with all zero & infinity magnitudes of 5+5 inverse 10-CIPs having all broken causality for occurrence of any event within it. That seems like a 'Will' of same from the unified quantum-real fold of the physical nature.

BIBLIOGRAPHY

[1] Bambi, B. and Dolgov. A.D. (2015) *Introduction to Particle Cosmology: The Standard Model of Cosmology and its Open Problems.* New York: Springer.

[2] Barrow, J. D. and Silk, J. (1983) *The Left Handed Creation: The Origin and Evolution of the Expanding Universe.* London: William Heinemann Ltd.

[3] Bohm, D. (1996) *The Special Theory of Relativity.* London: Routledge.

[4] Bohr, N. and Planck, M. (2019) *Quantum Theory Great Works that Shape our World.* London: Flame Tree Publishing.

[5] Born, M. (2012) *Einstein's Theory of Relativity.* New York: Dover Publications, Inc.

[6] Broglie, Louis de (2018) *Matter and Light the New Physics.* Warszawa: Franklin Classic Trade Press.

[7] Cheng, T. -P. and Li, L. -F. (1983) *Gauge Theory of Elementary Particle Physics.* Oxford: Oxford University Press.

[8] Committee on the Physics of the Universe, National Research Council of the National Academies (2003) *Connecting Quarks with the Cosmos: Eleven Science Questions for the New Century.* USA: The National Academies Press.

[9] Cottingham, W. N. and Greenwood, D. A. (1998) *An Introduction to the Standard Model of Particle Physics.* Cambridge: Cambridge University Press.

[10] Duric, N. (2004) *Advanced Astrophysics* (page 94). Cambridge: Cambridge University Press.

[11] Einstein, A. (1922) *The Meaning of Relativity*. Cambridge (USA): Princeton University Press.

[12] Elkins-Tanton, L. T., et all (2017) *Planetesimals - Early Differentiation and Consequences for Planets*. Cambridge: Cambridge University Press.

[13] Feynman, R. (2008) *The Feynman Lectures on Physics, The Definitive Volume 1, 2/E* (page 52). Chennai: Pearson Education India.

[14] Fraser, G. (ed) *The Particle Century* (page 155). Florida: CRC Press.

[15] French, S. and Kamminga, H. (eds) *Correspondence, Invariance and Heuristics Essays of Heinz Post*. New York: Springer.

[16] Galilei, G. (1954) *Dialogues Concerning Two New Sciences, (Dover 1st Edition, 1954)*. New York: Dover Publications, Inc.

[17] Gamow, G. (1966) *Thirty Years That Shook Physics: The Story of Quantum Theory*. New York: Dover Publications, Inc.

[18] Gorbunov, D. S. and Rubakov, V. A. (2011) *Introduction to the Theory of the Early Universe: Hot Big Bang Theory*. Singapore: World Scientific.

[19] Gregory, R. D. (2006) *Classical Mechanics*. Cambridge: Cambridge University Press.

[20] Hawking, S. and Israel, W. (ed) *Three Hundred Years of Gravitation*. Cambridge: Cambridge University Press.

[21] Heilbron, J.L. (2003) *Ernest Rutherford: And the Explosion of Atoms*. Oxford: Oxford University Press.

[22] Heisenberg, W. (1971) *Physics and Beyond: Encounter and Conversations*. New York: Harper Tochbooks.

[23] Kolata, J. J. (2015) *Elementary Cosmology: From Aristotle's Universe to the Big Bang and Beyond*. San Rafael (USA): Morgan & Claypool Publishers.

[24] Krauss, L. M. (2012) *A Universe from Nothing: Why There is Something Rather Than Nothing*. New York: Simon & Schuster.

[25] Kunter, M. (2003) *Astronomy: A Physical Perspective*. Cambridge: Cambridge University Press.

[26] Maxwell, J. C. (2015) *Matter and Motion – Scholar's Choice Edition*. Sacramento, California: Creative Media Partners, LLC.

[27] Michalowiez, J. V., Nichols, J. M. and Bucholtz, F. (2013) *Handbook of Differential Entropy* (page 19). Florida: CRS Press.

[28] Muga, G., Ruschhaut, A. and del Campo, A. (eds) *Time in Quantum Mechanics, Vol. 2* (page 175). New York: Springer.

[29] Murdoch, D. (1989) *Niels Bohr's Philosophy of Physics* (page 11). Cambridge: Cambridge University Press.

[30] Narliker, J. V. (1960) *Elements of Cosmology*. Thumba, India: Universities Press.

[31] Nikhilananda, S. (2003) *The Principal Upanishads* (page 39). North Chelmsford, Massachusetts: Courier Corporation.

[32] Nimitz, G. and Haibel, A. (2008) *Zero Time Space: How Quantum Tunnelling Broke the Light Barrier*. Hoboken, New Jersey: Wiley.

[33] Penrose, R. (2010) *Cycles of Time an Extraordinary New View of the Universe*. London: The Boldly Head.

[34] Pross, A. (2012) *What is Life? : How Chemistry Becomes Biology*. Oxford: Oxford University Press.

[35] Quin, H. R. and Nir, Y. (2010) *The Mystery of the Missing Antimatter* (page 88). Princeton: Princeton University Press.

[36] Rafelski, J. and Berndt, M. (1985) *The Structured Vacuum: Thinking about Nothing*. Stuttgart: H. Deutsch.

[37] Schrijver, K. and Schrijver, Iris 2015) *Living with the Stars: How the Human Body is Connected to the Life Cycles of the Earth, the Planets, and the Stars*. Oxford: Oxford University Press.

[38] Schrödinger, E. (1992) *What is Life? : With Mind and Matter and Autobiographical Sketches.* Cambridge: Cambridge University Press.

[39] Smorodinskiĭ, Î. A. (1976) *Particles, Quanta, Waves* (page 107). Mosco: Mir Publishers.

[40] Thayer, S. H. (ed) *Newton's Philosophy of Nature: Selections from His Writings.* North Chelmsford, Massachusetts: Courier Corporation.

[41] Valtaoja, E. and Valtonen, M. (ed) *Variability of Blazers* (page 234). Cambridge: Cambridge University Press.

[42] Vedral, V. (2012) *Decoding Reality: The Universe as Quantum Information.* Oxford: Oxford University Press.

[43] Watson, J. D. and Berry, A. (2008) *DNA: The Secret of Life.* Ct. Paradise, CA, USA: Paw Prints.

[44] Weinberg, S. (1995) *The Quantum Theory of Fields, Volume 1.* Cambridge: Cambridge University Press.

[45] Weir, J. (2007) *Max Planck: Uncovering the World of Matter.* Huntington Beach, CA, USA: Teacher Created Materials Publishing.

[46] Xing, Z. and Zhou, S. (2011) *Neutrinos in Particle Physics, Astronomy, and Cosmology* (page 60). New York: Springer.

[47] Yamagishi, A., Kakegawa, T., Usui, T. (2019) *Astrobiology From the Origins of Life to the Search for Extraterristrial Intelligence.* New York: Springer.

[48] Yarus, M. (2010) *Life from an RNA World The Ancestor Within.* Cambridge, USA: Harvard University Press.

[49] Zimmerman, M. J. (2001) *The Nature of Intrinsic Value.* Marland, USA: Rowman & Littlefield.

INDEX

'future' to 'past', 105

1

10-quantum-real spatial dimensions, 352

5

5+5 inverse 10-Spatial-Dimensions, 360

A

a unified or common inertial equation, 111
absolute, 8, 12, 13, 15, 21, 22, 54, 56, 69, 70, 76, 113, 120, 122, 126, 127, 128, 129, 130, 133, 134, 138, 140, 143, 144, 149, 165, 200, 201, 202, 221, 332
absolute 'rest', 15
accelerated '*Lift*', 179
acceleration, 166, 167, 169, 172, 179, 180, 181, 184, 185, 191, 192, 206, 212, 227, 228, 246, 258, 266, 374, 375
accelerations as infinitesimally discrete rate changes, 178
aether, 15, 127, 128, 129, 143, 147, 371
Albert A. Michelson, 15, 128
Albert Einstein, 16, 90, 128, 131
algorithm, 301, 302, 303, 305, 306, 307, 308, 309, 367, 371
all complex neuro-somatic processes, 146
annihilated, 261, 264
anti-clockwise axial rotation, 73
anti-entropy, 26, 44, 46, 47, 92, 110
antigauge-fields, 252, 255, 283
antigravitation, 252, 253, 254, 256, 257, 282, 283, 359
anti-gravitational fields, 199, 212, 214
anti-matter, 27, 28, 268
anti-particles, 27, 262, 263, 264
anti-radius, 75, 80, 83, 101, 103, 108, 112, 113, 117, 229, 350, 357, 358, 378

anti-space, 24, 72, 74, 75, 83, 102, 113, 209, 214, 229, 257, 318, 328, 337, 341, 350, 355, 361, 365, 366, 373, 379

anti-time, 24, 80, 81, 83, 105, 113, 117, 209, 229, 257, 263, 264, 318, 328, 337, 341, 350, 351, 355, 365, 366, 373, 374, 379

Aristotle, 145

astronomical objects, 34, 40, 55, 81, 173, 231, 306

astronomical range of universe, 172

astronomical scales, 21, 64

asymmetry, 28, 264, 265, 268, 370

axial rotation, 23, 25, 45, 46, 52, 68, 70, 71, 73, 74, 75, 79, 80, 82, 100, 104, 205, 372

B

Basic Forces, 6, 13, 218, 239, 241, 369, 372

beyond exchangeable edge, 202

Bigbang, 11, 24, 25, 26, 27, 35, 42, 44, 91, 308, 343, 344, 357, 358, 373, 374, 379

Bigbang-Bigcrunch Cyclic Oscillating, 11, 25, 35, 91, 379

Bigbang-Bigcrunch Cyclic Oscillating Universe, 11, 25, 35, 91

Big-collapse, 42, 43, 44

Bigcrunch, 11, 24, 25, 26, 27, 35, 42, 43, 44, 91, 308, 343, 344, 357, 373, 379

biological evolutions, 146

black board, 301

Black Body Radiation experiments, 130

black hole, 177, 194, 195, 199, 207, 208, 231, 233, 238, 240

blackness, 199, 207, 208, 209

Bohr radius, 20

boson, 9, 59, 145, 154, 160, 201, 231, 234, 238, 240, 242

bosonic fields, 240

bosons, 17, 36, 53, 55, 56, 59, 63, 64, 65, 66, 67, 90, 96, 97, 99, 150, 233, 238, 307

Brahman, 366, 367

broken sequence of causality, 50, 303

C

Causal Laws, 13, 14, 28, 29, 31, 285, 312, 313, 316, 366, 368, 372

causal sequence, 50, 286, 287, 290, 294, 298, 300, 303, 320, 322, 343, 344, 354, 358, 371, 377

causality, 42, 50, 119, 277, 283, 286, 287, 288, 289, 292, 293, 294, 296, 309, 312, 315, 320, 321, 322, 324, 337, 344, 352, 360, 371, 374, 379, 380

center-of-mass, 175, 184, 185, 186, 187, 188, 211

Chaitanya, 366

Christiaan Huygens, 143

Classical Mechanics, 22, 24, 35, 40, 41, 51, 86, 88, 90, 97, 111, 168, 169, 178, 199, 215, 286

Classical Physics, 17, 34, 65, 97, 173, 376

clusters of galaxies, 21, 173

cognition, 30, 287, 289, 298, 301, 302, 332, 365, 367

cognizable quantum-real range, 260

cognizable-mind, 294, 299, 300, 301, 302, 305, 307, 308, 311, 316, 355

collapse-entropy, 26, 27
collapsing phase, 25, 75, 84, 93
common internal parameters, 17
common internal-parameters, 28, 29
Common Understandings, 32, 33, 88, 89
Common-Internal-Parameters, 11, 33, 37, 38, 51, 220, 373
Common-Properties-of-Inertia, 11
consciousness, 303, 305, 366
convention of infinite speed, 201
Copernicus, 125
corpuscular hypothesis about light, 16
corresponding *optimum homogeneity*, 189
crushing in the scales, 231
curved spacetime, 30, 166, 183, 191, 192, 193, 197, 198, 200, 201, 203, 207, 210, 214, 216, 217, 218, 219, 220, 221, 222, 225, 226, 230, 234, 235, 236, 237, 238, 239, 240, 241, 244, 246, 247, 248, 250, 251, 254, 255, 306, 344, 375
cyber software-hardware systems, 307
cyclic oscillation of universe, 42

D

dark energies, 17, 35, 66, 67, 97, 163, 241, 242, 243, 244, 245, 249, 362
dark matters, 17, 66, 67, 97, 163, 239, 241, 242, 243, 244, 245, 249, 362
de Broglie wavelength, 17, 52, 55, 103, 117, 274, 357, 358
Determinism, 281, 292, 373
dilation in relativistic time, 141
direct proportional relationship, 96, 101, 193, 275
discrete duration of time, 192, 206, 212, 229, 234, 266
discrete values, 29, 41, 42, 166, 189, 228, 248, 254, 255, 256, 257, 266, 271, 274, 275, 288, 291, 314, 317, 340, 358, 366, 376, 378, 379
DNA, 300, 301, 302, 307
duality, 30, 45, 46, 47, 48, 91, 246, 247, 248, 259, 260, 268, 269, 270, 273, 283, 284, 285, 289, 291, 312, 317, 346, 347, 353, 354, 360, 362, 365
dynamism, 302, 303, 304, 305, 306, 308, 311, 376

E

Edward W. Morley, 15, 128
effect of forces, 167
Einstein Field Equation, 29, 30, 168, 169, 193, 196, 197, 198, 199, 201, 204, 207, 216, 217, 222, 223, 235
Einsteinian Invariance, 127, 134
Electromagnetic, 11, 15, 128, 131, 143, 164, 230, 232, 241, 374
electromagnetic fields, 143, 307
electromagnetic force, 35
electromagnetic spectrum, 19, 38, 51, 121, 234, 238, 285, 376
electromagnetic wave, 19, 58, 128, 143, 147

electron and positron, 57, 58, 59
electroweak star, 238
elements, 114, 115, 142, 359
encoding of 'life', 300
energy of a vacuum, 8
entropy, 25, 26, 27, 43, 44, 46, 47, 52, 53, 92, 100, 110, 367, 373, 379
envelope, 70, 191
E-P-R paradox, 28, 266
equatorial distance, 77, 78, 80
Equivalence Principal, 178
escape velocity, 30, 148, 169, 189, 207, 235
escape velocity of earth, 148
Event Horizon, 177, 184, 186, 188, 189, 194, 208, 209, 211, 239
exotic boson star, 238, 240
expansion phase, 24, 25, 65, 66, 74, 79, 100, 111
expansion-entropy, 25, 26, 27
extromission, 145

F

fermion, 9, 64, 66, 67, 96, 97, 152, 154, 161, 163, 201, 231, 233, 237
flat fabric of spacetime, 202
fourth square, 192, 196, 211

G

galaxies, 18, 21, 36, 55, 173, 224, 231
Galen, 146
Galilean Electrodynamics, 6, 9, 369
Galilean invariance, 126
Galilean Theory of Relativity, 124, 125, 126
Galileo Galilei, 15, 125
gamma-ray-photons, 57, 58, 59, 60, 98, 233
gauge fields of forces, 164, 235, 243, 248, 307
General Relativity Theory, 11, 12, 14, 29, 118, 119, 166, 168, 169, 217, 218, 219, 245, 246, 364
Geo-centric concepts for universe, 125
geodesic, 102, 142, 250, 256
geodesic of entire antispacetime, 256
gluon-star, 237, 238
Grand Unification of Everything, 12
gravitating-body, 30, 166, 167, 169, 217, 218, 219, 220, 223, 227, 245, 374, 375, 377
gravitational acceleration, 175, 178, 179, 181, 184, 191, 220, 234
gravitational collapses, 220, 230, 231, 232, 242
gravitational crushing, 232, 233, 237, 238
gravitational fields, 167, 197, 198, 199, 212, 214, 229, 306
gravitational mass-energies, 197, 198

gravitationally-shaped-body, 166, 169, 219, 245, 375

H

Higgs Boson fields, 233, 237
highest limit of convergence, 175
highest-bound *inescapable particle motions*, 176
homogeneous highest-bound discrete motions, 166
homogeneous smallest-bound discrete mass-energies, 166, 203
hydrogen molecule, 153
hydrostatic-equilibrium, 187

I

icy-observer, 331, 347
imagine, 24, 25, 26, 53, 68, 77, 174, 176, 232, 241, 308, 331, 335, 344, 345, 346, 347, 348, 349
indeterminism, 271, 276, 279, 283, 288, 311, 317, 323
inert-gases, 153
inertial acceleration, 168, 169, 170, 171, 172, 178, 179, 191, 192
inertial acceleration ultimately appears as quantized, 172
inertial frames of reference, 8, 15, 41, 111, 113, 121, 163, 199, 200, 279, 371
inertial mass-energies', 9
inertial motions, 9, 19, 37, 98, 121, 122, 167, 172, 211, 352
inertial speed of light, 121, 122
inertial state, 22, 38, 53, 91, 94, 107, 110, 117, 118, 123, 134, 149, 192, 199, 205, 213, 216, 217, 220, 229, 250, 253, 256, 258, 265, 268, 270, 272, 275, 281, 283, 298, 317, 374
inertial-light, 16
inertial-motion for a neutrino, 152
inertial-motion of light, 129, 200
inertial-motions, 8, 9, 13, 16, 17, 19, 37, 38, 39, 41, 55, 56, 57, 58, 59, 61, 62, 63, 64, 65, 67, 75, 96, 97, 100, 102, 121, 122, 124, 126, 129, 130, 131, 132, 133, 137, 138, 144, 145, 146, 149, 150, 151, 152, 153, 154, 155, 156, 158, 159, 161, 162, 164, 165, 166, 167, 168, 172, 174, 175, 176, 177, 178, 181, 182, 187, 188, 189, 190, 191, 192, 194, 201, 203, 205, 206, 207, 208, 210, 215, 227, 251, 252, 275, 371
inertial-time, 22, 75, 76, 141, 256
infinite, 15, 28, 31, 40, 49, 87, 127, 129, 130, 137, 142, 199, 201, 202, 257, 260, 318, 322, 331, 355, 361, 364, 365, 366, 379
infinitesimal quantum *change in motion*, 170
integer sum, 76, 135, 159, 176, 182, 183, 197, 198, 202, 203, 219, 224, 226, 230, 245
intrinsic direction of entropy, 43
intrinsic left-handedness, 23, 44, 46, 71, 209
intrinsic values, 8, 16, 17, 20, 22, 38, 67, 374
intromission, 145
invariant, 15, 17, 19, 28, 29, 38, 39, 91, 94, 95, 96, 97, 103, 106, 107, 117, 121, 122, 123, 124, 125, 126, 127, 128, 129, 130, 131, 132, 133, 134, 137, 138, 139, 141, 144, 147, 151, 156,

157, 158, 159, 160, 162, 163, 166, 198, 199, 200, 202, 203, 204, 205, 206, 207, 208, 209, 229, 230, 269, 271, 272, 273, 288, 292, 293, 313, 364, 366, 373
Inverse Invariance, 132, 133
inverse relationship, 9, 17, 19, 51, 74, 94, 95, 96, 97, 100, 106, 124, 131, 132, 133, 145, 147, 148, 163, 170, 271, 274, 278, 280
inverse relationships, 9, 29, 72, 94, 107, 108, 109, 131, 205, 278, 290, 336, 339, 360, 362
inward gravitational pressures, 241
Isaac Newton, 16, 127
isolated, 123, 190, 191, 192, 375

J

James Clerk Maxwell, 15, 128
Jupiter's moon *Io*, 142

L

left-handed directions in axial rotations, 23
Lefthanded Entropy–Cycle, 27
Left-Handed-Entropy of Cyclic-Rotation, 11
Left-Handedness-in-Rotation, 11
limitations, 14, 22, 23, 51, 86, 177, 178, 207, 208, 209, 256, 257, 260, 287, 296, 297, 348
limited objective-cognition, 368
limits of quantum-reality, 312, 329, 380
List of Astronomical Objects, 18
living and non-living boundary, 303
Lorentz Transformations, 112, 130
Louis De Broglie, 131

M

macro-most, 13, 18, 24, 25, 26, 30, 35, 36, 40, 42, 43, 44, 45, 48, 50, 51, 66, 67, 68, 85, 88, 91, 92, 94, 97, 99, 108, 110, 117, 173, 174, 202, 217, 234, 236, 242, 244, 245, 248, 249, 254, 255, 260, 267, 268, 271, 308, 343, 349, 357, 371
material body, 170, 172
mathematics, 301, 302
matrix, 243, 244, 359
Max Planck, 16, 51, 120, 130
metric tensor, 193
Michelson and Morley, 15, 129
micro-most, 35, 36, 48, 66, 67, 68, 88, 91, 92, 99, 239, 249, 254, 255, 260, 340, 343, 347, 348, 350, 354, 356, 357, 358, 360, 361
microwave photons, 61
mirror-imaged fields, 211
mirror-imaged observers, 214, 259
mutual mirror-images, 24, 72, 118, 199, 214, 259, 266, 267, 269

N

neutron star, 231, 236
Newtonian Invariance, 126
Newtonian Laws of Motions, 127
Newtonian Mechanics, 15, 371
Newtonian Theory of Relativity, 127, 129
Nirguna Brahman, 367
no start no end, 320
Non-causal 'Wills', 13
non-causal creation of causal event, 326
Non-causal Wills, 14, 28, 29, 366, 368
non-causality, 50, 322, 324
non-conventional units, 79
non-gravitational forces, 173, 178, 191, 198, 203, 223, 230, 232, 245, 371, 372
non-inertial state, 199, 271, 275, 281
non-inertial unifications, 260
non-quantized continuum, 13, 257, 258, 336, 337, 365
Non-quantized-virtual, 31, 50, 112, 314, 320, 321, 322, 323, 324, 325, 327, 328, 329, 330, 332, 335, 341, 342, 344, 347, 348, 349, 350, 351, 352, 353, 355, 356, 359, 360, 361, 362, 363
Non-quantum-energy, 316, 342
non-zero and non-infinity magnitudes, 13, 70
null result, 15, 128

O

objectivity, 14, 285, 286, 289, 311, 315, 316, 328, 332, 338, 342, 352, 368, 374
observances, 48, 49, 87, 112, 248, 256, 262, 266, 267, 276, 294, 298, 323, 325, 345, 365, 374, 378
observational background, 20, 35, 130
observational limit, 50
observer like us, 22, 25, 49, 82, 93, 113, 207, 208, 256, 259, 264, 266, 276, 284, 293, 294, 295, 311, 316, 320, 321, 325, 326, 327, 337, 341, 342, 349, 363, 365, 367, 373, 378
observer-independence, 15, 73, 144
observer-independent constants, 149
Ole Rømer, 15
opposite observers, 84

P

Particle Physics, 16, 19, 34, 44, 54, 64, 90, 109, 132, 198, 220, 233, 370
particle-antiparticle pair, 269
Photoelectric Effects, 16, 144
photon-particles, 121, 164
physical nature, 8, 9, 13, 14, 17, 18, 20, 21, 22, 23, 24, 28, 29, 30, 31, 33, 34, 35, 37, 38, 40, 42, 44, 45, 46, 48, 49, 50, 51, 86, 87, 88, 91, 99, 108, 122, 125, 130, 131, 132, 142, 144,

164, 166, 168, 170, 177, 200, 207, 208, 210, 219, 220, 246, 247, 248, 256, 257, 258, 260, 261, 262, 266, 268, 269, 270, 272, 277, 279, 280, 283, 284, 285, 287, 289, 293, 299, 309, 312, 314, 315, 316, 317, 319, 322, 325, 328, 330, 332, 333, 336, 337, 338, 339, 340, 341, 347, 356, 357, 358, 359, 360, 361, 362, 364, 365, 366, 367, 371, 372, 373, 374, 377, 379, 380

physically tangible or 'unfolded', 112

Planetesimals, 18, 369

Plank's Constant, 273, 274, 275

property, 9, 16, 19, 34, 86, 123, 142, 143, 144, 169, 170, 297, 300, 319, 376, 378

proportionality constant, 190, 196, 210, 253, 282

psycho-somatic observers, 305

Ptolemy, 125

Q

quantized acceleration, 180, 227, 228, 229, 230, 246, 256

quantized inertial-motions, 56, 121, 122, 161, 166, 176, 215

quantized magnitudes, 9, 18, 19, 20, 23, 34, 37, 38, 41, 51, 52, 53, 54, 55, 56, 57, 61, 64, 65, 66, 67, 68, 69, 71, 73, 74, 75, 76, 79, 82, 83, 85, 87, 93, 95, 96, 97, 98, 100, 101, 102, 103, 104, 106, 108, 109, 112, 113, 114, 115, 121, 122, 131, 132, 133, 134, 138, 140, 141, 148, 149, 150, 151, 152, 153, 154, 155, 156, 157, 159, 161, 162, 166, 169, 171, 172, 178, 181, 184, 186, 189, 190, 191, 194, 201, 205, 206, 207, 211, 220, 225, 227, 230, 258, 266, 267, 271, 273, 274, 275, 276, 277, 279, 281, 284, 291, 292, 293, 297, 309, 310, 317, 333, 341, 356, 377, 378

quantized mass, 9, 12, 16, 18, 38, 56, 62, 65, 135, 137, 145, 146, 149, 153, 162, 164, 167, 176, 183, 186, 188, 189, 196, 197, 203, 206, 210, 219, 224, 226, 227, 233, 241, 273, 274, 280, 358, 375

quantized radius, 40

quantized sense of time, 23

quantized-spaces or volumes, 21

Quantum Electrodynamics, 11

Quantum Invariance, 130

Quantum Mechanics, 11, 14, 15, 16, 17, 28, 29, 33, 53, 88, 95, 100, 124, 131, 132, 133, 167, 219, 246, 271, 276, 278, 286, 288, 311, 364, 365, 370

quantum-real, 13, 14, 21, 22, 23, 24, 29, 30, 31, 40, 42, 43, 44, 48, 49, 50, 51, 69, 86, 87, 89, 91, 94, 112, 113, 115, 117, 118, 123, 256, 257, 258, 260, 261, 262, 263, 264, 265, 266, 268, 269, 270, 271, 272, 276, 277, 278, 280, 281, 282, 283, 284, 285, 287, 288, 289, 290, 291, 292, 293, 295, 296, 297, 298, 299, 300, 301, 302, 303, 304, 305, 306, 307, 308, 309, 311, 312, 314, 315, 316, 317, 318, 319, 320, 321, 322, 323, 324, 325, 326, 327, 328, 329, 330, 331, 332, 333, 334, 335, 336, 337, 339, 340, 341, 342, 343, 344, 345, 346, 347, 348, 349, 350, 351, 352, 353, 354, 355, 356, 357, 358, 359, 360, 361, 362, 365, 366, 367, 368, 371, 372, 373, 374, 375, 376, 377, 378, 379, 380

quantum-real algorithm, 300, 301, 303, 304, 305, 307, 311, 371, 372, 376, 377

quantum-real cognition, 13, 366, 377

quantum-real events, 272, 284, 287, 320, 322, 379

quantum-real fold, 14, 40, 48, 49, 50, 266, 268, 367, 373, 374, 377, 379, 380

quantum-real fold of physical nature, 367

quantum-real forces, 293, 375
quantum-real limitations, 257
quantum-real objectivity, 289
quantum-real Observers like us, 299, 300, 305, 307
quantum-real signals, 21, 22, 23, 24, 30, 48, 49, 50, 87, 112, 284, 287, 288, 289, 297, 298, 320, 322, 323, 325, 339, 345, 354, 359, 377
quantum-real space, 319, 320, 321, 349, 350, 357
Quantum-Real-Causalities, 310
quantum-real-event, 280, 281, 283, 284, 291, 292, 293, 296, 297, 311, 314, 316, 319, 321, 323, 324, 325, 326, 327, 330, 331, 335, 342, 343, 347, 348
quasi-living micro-organism, 303
quintessence, 35, 36

R

radio-wave photon, 19, 151, 238, 241, 242
realm of visible-matter-scales, 36
relative motions, 104, 125, 126, 134, 147, 371, 378
relativistic contraction, 139
relativistic contraction of space, 139
relativistic mass, 135, 138, 150, 156
René Descartes, 143
Ricci curvature tensor, 193
Righthanded Entropy–Cycle, 27
right-handed mirror-image, 24, 71
Right-Handed-Entropy of Cyclic-Rotation, 11
Right-Handedness-in-Rotation, 11
RNA, 300, 301, 302, 307

S

Saguna Brahman, 367
scale of universe, 173
scales of photons, 59, 60, 61, 62, 96, 98, 99, 149, 150, 233
scale-specific 'quantized curvatures of spacetime', 216
scale-specific convergence of spacetime, 183, 186
scale-specific crushing, 233
scale-specific escape-velocity, 175, 230, 377
scale-specific Event Horizons, 185
scale-specific magnitudes of time, 22
scale-specific maximum limits of convergence, 175
scale-specific maximum limits of homogeneity, 176
scale-specific quantization, 35, 79, 80, 114, 176, 205, 224
scale-specific quantized magnitudes, 19, 34, 67, 74, 76, 104, 151, 227, 230
scale-specific spatial structures, 68
scale-specificness, 18, 34, 35, 37
scale-specific-universal-constancies, 18

Scale-Specific-Universal-Constants, **11**
Schwarzschild radius, **208**
sense of 'life', **289**
sense of living, **301, 302, 303, 304, 305, 307, 308**
simultaneous flow of anti-time, **81**
simultaneous opposite directional flow, **81**
smallest-bound particles, **189, 216**
smallest-bound-gauge-fields, **237, 241**
Solar-centric universe, **125**
spatial unit for time, **77**
Special Relativity Theory, **9, 11, 12, 15, 16, 17, 19, 100, 119, 121, 122, 124, 125, 129, 163, 166, 364, 371**
speed of light, **15, 16, 59, 121, 122, 128, 129, 134, 142, 143, 147, 148, 204**
Standard Model of Particle Physics, **8, 11, 12, 29, 30, 95, 118, 167, 183, 198, 219, 221, 233, 235, 236, 241, 244, 245, 248, 365**
states of energies, **329, 330, 337, 345**
Stress-energy-momentum Tensor, **193**
strong nuclear force, **36**
Subjective Event, **296, 309**
subjective mind or 'will', **366**
Subjective observers, **296**
Subjective Wills of virtuality, **119**
Subjectivity, **13, 28, 29, 287, 299, 305, 309, 316, 321, 328, 335, 360, 378, 379, 380**
superluminal speeds, **19, 208**
Supersymmetric quantized fields, **183**
systems-of-particles, **9, 13, 16, 17, 18, 19, 20, 21, 22, 23, 24, 25, 26, 28, 29, 30, 33, 34, 35, 36, 37, 38, 39, 40, 41, 43, 44, 45, 46, 47, 48, 49, 51, 52, 64, 88, 89, 90, 91, 92, 96, 117, 118, 121, 130, 131, 132, 133, 140, 141, 145, 147, 154, 156, 163, 166, 167, 168, 172, 174, 182, 186, 187, 188, 189, 191, 203, 205, 206, 212, 213, 215, 216, 218, 220, 221, 246, 247, 249, 262, 263, 264, 265, 266, 268, 269, 270, 271, 272, 273, 275, 279, 280, 281, 283, 284, 288, 291, 293, 295, 296, 298, 299, 302, 303, 306, 308, 309, 317, 335, 339, 343, 347, 349, 350, 352, 353, 357, 360, 365, 367, 371, 372, 373, 374, 375, 376, 377, 378, 379**

T

tasks for living, **304**
thermonuclear reactions, **21**

U

Uncertainty Principles, **271, 276, 277, 278, 311**
unfolded 10-dimesions, **14**
unfolded dimensions, **13, 29, 92, 113**
unicellular virus, **303**
unification of physics, **125, 217, 234**
unified mechanical theory, **8, 29, 34**

unified quantum-reality, 13, 30, 31, 246, 248, 260, 272, 275, 277, 278, 281, 283, 285, 287, 288, 289, 291, 292, 293, 296, 297, 298, 299, 300, 301, 302, 303, 305, 306, 308, 311, 313, 315, 316, 317, 319, 321, 322, 324, 325, 326, 327, 330, 332, 333, 334, 336, 337, 339, 341, 342, 360, 361, 362, 365, 366, 367, 368, 372
Universal Constant, 19, 272
Universal Invariances, 159
universal omni-present time, 23, 127
Universal-Constants, 11
unknown basic forces, 260

V

vacuum-virtuality, 258
Vedanta, 366
virtual continuum, 319, 321, 325, 335, 341, 343, 344, 345, 347, 348, 349, 358, 368, 376
Virtual Wills, 316
Virtuality, 14, 28, 277, 316, 317, 333, 335, 340, 353, 376, 377, 379, 380
visible-matters, 90, 99, 244, 249, 332, 367
void, 21, 22, 69, 70, 71, 73, 76, 104, 105, 111, 112, 140, 153, 201, 202, 221

W

Wave Equation, 132
Wave Theory of Light, 143
wave-corpuscular duality, 9, 91, 117, 142, 145, 146, 148, 151
wave-corpuscular-phenomena, 52, 145
weak nuclear force, 36
whirl, 31, 347, 348, 349, 350, 351, 352, 353, 354, 355, 356, 357, 358, 361, 376
whirl of lower energy, 348
whirls of quantum-reality, 379
Will, 13, 367, 368, 379, 380
Wills of God, 368

Z

zero and infinity magnitudes, 13, 112, 295, 296, 297, 298, 319, 323, 335, 336
zero relative motions, 134, 378
zero-matter, 318
zero-space, 318
zero-time, 318

Made in United States
Orlando, FL
17 January 2023